34th International Symposium on Superconductivity (ISS 2021)

Journal of Physics: Conference Series Volume 2323

Online
30 November – 2 December 2021

Editors:

Atsutaka Maeda
Hiraku Ogino
Yoshiyuki Yoshida
Yasuhiro Iijima
Yasunori Mawatari
Mutsuo Hidaka
Tsunehiro Hato
Naoyuki Amemiya
Tsuyoshi Wakuda
Hiroshi Miyazaki

ISBN: 978-1-7138-6043-3
ISSN: 1742-6588

Printed from e-media with permission by:

Curran Associates, Inc.
57 Morehouse Lane
Red Hook, NY 12571

CURRAN ASSOCIATES INC. proceedings.com

Some format issues inherent in the e-media version may also appear in this print version.

Institute of Physics
Dirac House, Temple Back
Bristol BS1 6BE UK

Phone: 44 1 17 929 7481
Fax: 44 1 17 920 0979

techtracking@iop.org

Additional copies of this publication are available from:

Curran Associates, Inc.
57 Morehouse Lane
Red Hook, NY 12571 USA
Phone: 845-758-0400
Fax: 845-758-2633
Email: curran@proceedings.com
Web: www.proceedings.com

TABLE OF CONTENTS

1.4 VORTEX PHYSICS

2. WIRES AND BULK

2.1 HTS WIRES AND JOINT

2.2 BULK MATERIALS

3.4 DIGITAL

3.5 NOVEL DEVICES

4. LARGE SCALE SYSTEM APPLICATIONS

4.1 ELECTRIC POWER APPLICATIONS

4.2 MAGNETIC LEVITATION AND BEARING

Journal of Physics: Conference Series **2323** (2022) 011001 doi:10.1088/1742-6596/2323/1/011001

34th International Symposium on Superconductivity

I·S·S 2021

WEB-BASED SYMPOSIUM
AIST and VUW
November 30 – December 2, 2021

34th International Symposium on Superconductivity (ISS2021)

Preface

It is our pleasure to report that the Proceedings of the 34th International Symposium on Superconductivity (ISS2021) held on November 30–December 2, 2021, are now published in *Journal of Physics: Conference Series* (JPCS). ISS2021 was co-organized by the National Institute of Advanced Industrial Science and Technology (AIST) and the Victoria University of Wellington (VUW). At first, ISS2021 was planned to be held on site in Wellington, New Zeeland. However, owing to the continued travel restrictions by COVID-19 pandemic, the symposium was held online with using the "Cisco Webex Meetings" application. Although most of presentations were smoothly made, several presentations were cancelled probably due to limited speed of a communication line, some of which were re-scheduled and presented at later dates.

Total of 342 scientists, engineers, students, corporate executives and so on (107 from abroad) made advanced registrations for ISS2021, and fruitful discussions were made to promote superconductor sciences and technologies. Total of 215 papers were presented, which included 5 plenary lectures (45 min), 76 invited talks (25–30 min) and 134 contributed oral presentations (15–20 min). Among the papers presented in ISS2021, 41 manuscripts were submitted for ISS2021 Proceedings, and 39 papers were accepted for publication. The papers published in JPCS were categorized into relevant topics in the following four fields; (a) Physics and Chemistry, (b) Wires and Bulk, (c) Electronic Devices, (d) Large Scale System Applications.

To ensure the high publication standard mandated by JPCS, every paper was peer reviewed by a reviewer with expertise before it was accepted for publication. As editors of the

Proceedings, we would like to express our sincere appreciation to all the reviewers involved in the evaluation of the papers for their invaluable contribution.

The Editors

Physics and Chemistry: Atsutaka Maeda (The University of Tokyo), Hiraku Ogino (AIST)

Wires and Bulk: Yoshiyuki Yoshida (AIST), Yasuhiro Iijima (Fujikura Ltd.), Yasunori Mawatari (AIST)

Electronic Devices: Mutsuo Hidaka (AIST), Tsunehiro Hato (SUSTEC)

Large Scale System Applications: Naoyuki Amemiya (Kyoto University), Tsuyoshi Wakuda (Hitachi Ltd.), Hiroshi Miyazaki (Kyushu University)

Journal of Physics: Conference Series **2323** (2022) 011001 doi:10.1088/1742-6596/2323/1/011001

Organization of ISS2021

General Chair:
Dr. Haruhiko Obara
Executive Officer, Director-General, National Institute of Advanced Industrial Science and Technology (AIST)

Co-Chair:
Dr. Nicholas Long
Associate Professor, Robinson Research Institute, Victoria University of Wellington (VUW)

Special Advisor
Dr. Toshihiko Kanayama
Special Emeritus Advisor, AIST

Secretary General
Dr. Michiya Okada
Deputy Director, TIA Central Office, AIST
michiya.okada@aist.go.jp

Local Organizing Committee:
Members:
Michiya Okada (AIST), Nicholas Long (VUW), Zhenan Jiang (VUW), Jeannie Redman (VUW), Mutsuo Hidaka (AIST), Yoshiyuki Yoshida (AIST), Mitsuho Furuse (AIST), Nao Takeshita (AIST), Shigeyuki Ishida (AIST), Susumu Ogawa (AIST), Akio Yoshizawa (AIST), Tsuyoshi Yanagibayashi (AIST)

Supported by:
New Energy and Industrial Technology Development Organization (NEDO)
Japan Science and Technology Agency (JST)
Cryogenics and Superconductivity Society of Japan (CSSJ)
The Institute of Electrical Engineers of Japan (IEEJ)
The Japan Society of Applied Physics (JSAP)
Applied Superconductivity Constellations of Tsukuba (ASCOT)
TIA—a platform for open innovation

Journal of Physics: Conference Series **2323** (2022) 011001 doi:10.1088/1742-6596/2323/1/011001

Committees of ISS2021

Program committee

Physics and Chemistry

Chair: Atsutaka Maeda (The Univ. of Tokyo), Deputy-Chair: Hiraku Ogino (AIST)

Members: Tetsuo Hanaguri (RIKEN), Minoru Nohara (Okayama Univ.), Tsutomu Nojima (Tohoku Univ.), Tsuyoshi Tamegai (The Univ. of Tokyo), Hiroyuki Yamase (National Inst. for Materials Science), Teppei Yoshida (Kyoto University)

Advisors: Akira Iyo (AIST), Hiroshi Eisaki (AIST), Shen Chong (VUW), James Storey (VUW), Grant William (VUW)

Wires and Bulk

Chair: Yoshiyuki Yoshida (AIST), Deputy-Chair: Yasuhiro Iijima (Fujikura Ltd.), Deputy -Chair: Yasunori Mawatari (AIST)

Members: Satoshi Awaji (Tohoku Univ.), Hiroshi Ikuta (Nagoya Univ.), Takeshi Kato (Sumitomo Electric Industries, Ltd.), Takanobu Kiss (Kyushu Univ.), Akiyoshi Matsumoto (National Inst. for Materials Science), Kaname Matsumoto (Kyushu Inst. of Technology), Hisaki Sakamoto (Furukawa Electric Co., Ltd.), Hideki Tanaka (Hitachi Ltd.), Masaru Tomita (Railway Technical Research Inst.)

Advisors: Nicholas Long (VUW), Nick Strickland (VUW), Stuart Wimbush (VUW)

Electronic Devices

Chair: Mutsuo Hidaka (AIST), Deputy-Chair: Tsunehiro Hato (SUSTEC)

Members: Yoshimi Hatsukade (Kindai Univ.), Masashi Ohno (The Univ. of Tokyo), Naoto Sekiya (Univ. of Yamanashi), Yutaka Tabuchi (RIKEN), Masamitsu Tanaka (Nagoya Univ.), Hirotake Yamamori (AIST)

Advisors: Masataka Ohkubo (AIST), Eva Anton (VUW), Guy Dubuis (VUW), Simon Granville (VUW)

Large Scale System Application

Chair: Naoyuki Amemiya (Kyoto Univ.), Deputy-Chair: Hiroshi Miyazaki (Kyushu Univ.), Deputy-Chair: Tsuyoshi Wakuda (Hitachi Ltd.)

Members: Kohei Higashikawa (Kyushu Univ.), Kazuhiro Kajikawa (Kyushu Univ.), Kei Koyanagi (Toshiba Corporation), Shinji Matsumoto (National Inst. for Materials Science), Naoko Nakamura (MAYEKAWA MFG. Co., Ltd.), Taketsune Nakamura (Kyoto Univ.), Tetsuo Oka (Shibaura Inst. of Technology), Shun Tonooka (Mitsubishi Electric Co.), Makoto Tsuda (Tohoku Univ.), Tomonori Watanabe (Chubu Electric Power Co., Inc.)

Advisors: Mitsuho Furuse (AIST), Rod Badcock (VUW), Chris Bumby (VUW), Zhenan Jiang (VUW)

International Advisory Committee

Physics and Chemistry

Jun Akimitsu (Okayama Univ.), Hideo Aoki (The Univ. of Tokyo/AIST), Xianhui Chen (Univ. of Science and Technology of China), Shen Chong (VUW), Atsushi Fujimori (The Univ. of Tokyo), Zenji Hiroi (The Univ. of Tokyo), Takekazu Ishida (Osaka Prefecture Univ.), Maarit Karppinen (Aalto Univ.), Wai-Kwong Kwok (Argonne National Laboratory), Victor Moshchalkov (KU Leuven), Satoshi Okuma (Tokyo Inst. of Technology), Hermann Suderow (Autonomous Univ. of Madrid), Setsuko Tajima (Osaka Univ.), Jeffrey L. Tallon (VUW), Takami Tohyama (Tokyo Univ. of Science), Dirk van der Marel (Univ. of Geneva), Hai-Hu Wen (Nanjing Univ.), Grant William (VUW), Eli Zeldov (Weizmann Inst. of Science)

Wires and Bulk

Amalia Ballarino (CERN), David Cardwell (Univ. of Cambridge), Leonardo Civale (Los Alamos National Laboratory), Michael Eisterer (Vienna Univ. of Technology), René-Louis Flükiger (Univ. of Geneva/CERN), Timothy J. Haugan (Air Force Research Laboratory), Bernhard Holzapfel (Karlsruhe Inst. of Technology), Teruo Izumi (AIST), Hiroaki Kumakura (National Inst. for Materials Science), David Larbalestier (Florida State Univ.), Qiang Li (Brookhaven National Laboratory), Nicholas Long (VUW), Xavier Obradors (Institut Ciencia de Materials de Barcelona), Marina Putti (Univ. of Genoa), Venkat Selvamanickam (Univ. of Houston), Jun-ichi Shimoyama (Aoyama Gakuin Univ.), Nick Strickland (VUW), Michael Sumption (The Ohio State Univ.), Yutaka Yoshida (Nagoya Univ.)

Electronic Devices

Eva Anton (VUW), Guy Dubuis (VUW), Pascal Febvre (Univ. of Savoie Mont Blanc), Simon Granville (VUW), Masataka Ohkubo (AIST), Shigetoshi Ohshima (Yamagata Univ.), Horst Rogalla (Univ. of Colorado at Boulder), Keiichi Tanabe (SUSTEC), Zhen Wang (Chinese Academy of Sciences)

Large Scale System Application

R. G. Buckley (VUW), Chris Bumby (VUW), Toru Fukushima (SuperPower Inc.), Jianzhao Geng (VUW), Zhenan Jiang (VUW), Mathias Noe (Karlsruhe Inst. of Technology), Hiroyuki Ohsaki (The Univ. of Tokyo), Minwon Park (Changwon National Univ.), Christopher M. Rey (Energy-to-Power Solutions), Pascal Tixador (Grenoble-INP), Hubertus Weijers (VUW), Liye Xiao (Chinese Academy of Sciences)

Journal of Physics: Conference Series **2323** (2022) 011002 doi:10.1088/1742-6596/2323/1/011002

Peer Review Statement

All papers published in this volume have been reviewed through processes administered by the Editors. Reviews were conducted by expert referees to the professional and scientific standards expected of a proceedings journal published by IOP Publishing Publishing.

- **Type of peer review:** Single Anonymous
- **Conference submission management system:** ISS2021 web site
- **Number of submissions received:** 41
- **Number of submissions sent for review:** 41
- **Number of submissions accepted:** 39
- **Acceptance Rate (Submissions Accepted / Submissions Received × 100):** 95.1
- **Average number of reviews per paper:** 1.9
- **Total number of reviewers involved:** 41
- **Contact person for queries:**
  Name: Hirofumi Yamasaki
  Email: hirofumi.yamasaki@iss-secretariat.org
  Affiliation: Secretariat of International Symposium on Superconductivity

Journal of Physics: Conference Series **2323** (2022) 012001 doi:10.1088/1742-6596/2323/1/012001

Unconventional superconductivity after the BCS paradigm and empirical rules for the exploration of high temperature superconductors

Hai-Hu Wen

Center for Superconducting Physics and Materials, National Laboratory of Solid State Microstructures and Department of Physics, Nanjing University, Nanjing 210093, China

E-mail: hhwen@nju.edu.cn

Abstract. Superconducting state is achieved through quantum condensation of Cooper pairs which are new types of charge carriers other than single electrons in normal metals. The theory established by Bardeen-Cooper-Schrieffer (BCS) in 1957 can successfully explain the phenomenon of superconductivity in many single-element and alloy superconductors. Within the BCS scheme, the Cooper pairs are formed by exchanging the virtual vibrations of lattice (phonons) between two electrons with opposite momentum near the Fermi surface. The BCS theory has dominated the field of superconductivity over 64 years. Many superconductors discovered in past four decades, such as the heavy Fermion superconductors, cuprates, iron pnictide/chalcogenide and nickelates seem, however, to strongly violate the BCS picture. The most important issue is that, perhaps the BCS picture based on electron-phonon coupling are the special case for superconductivity, there are a lot of other reasons or routes for the Cooper pairing and superconductivity. In this short overview paper, we will summarize part of these progresses and try to guide readers to some new possible schemes of superconductivity after the BCS paradigm. We also propose several empirical rules for the exploration of high-temperature unconventional superconductors.

1. Introduction

Superconductivity is a very interesting quantum state which exhibits zero resistivity and expulsion of magnetic field (Meissner effect). It has been well understood that this fascinating phenomenon is induced by the condensation of electron pairs (termed as Cooper pairs) at a finite temperature. Up to date, the superconducting state can only be achieved with these electronic Cooper pairs although some other novel pictures concerning other type of charged bosons have been proposed. Since its first discovery in April 1911 by Kamerling Onnes et al., the field of superconductivity has experienced glorious developments and achievements in past 111 years. Owing to the deep and interesting physics involved by the phenomenon of superconductivity, and the potential of large scale applications, the field has been pushed forward step by step, most of time by the discovery of new superconductors. We can expect more interesting phenomena and materials, which will eventually lead to the revolution, both on the fundamental understanding of condensed matter physics and bring in the dreamful industrial applications of superconductors.

The development of the field of superconductivity is however not straightforward, it is always mixed with some excitement and frustrations. In Fig.1, we present the time dependence of the

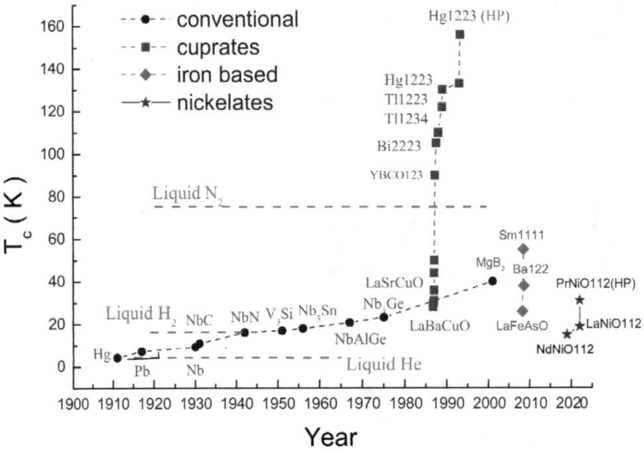

Figure 1. Correlation between the superconducting transition temperatures and the year for them to be found. The three horizontal red dashed lines represent the boiling temperatures of liquid helium, hydrogen, and nitrogen. The circles, squares, diamonds, stars represent the conventional, cuprate, iron-based and nickelate superconductors, respectively.

superconducting transition temperatures T_c of some representative superconducting systems when they were discovered. One can see that, the value of T_c ramps up with a small slope before October 1986. After the tremendous efforts in 75 years, the T_c finally reached about 23.2 K in Nb_3Ge system. Most superconductors discovered before this time were single element metals or alloys. The basic reason for superconductivity, called as the superconducting mechanism, was unraveled in 1957 by Bardeen, Cooper and Schrieffer and named as the BCS theory[1]. In this short review, we omit the description about the BCS paradigm. A short review can be found in our earlier review paper[2]. Since the field is rapidly moving forward, the understanding on the so-called unconventional superconductivity have not come to the mature status yet. Thus this paper serves only as a brief introduction and may be taken as a supplementary to other beautiful reviews in this fascinating field. Nowadays, it is known that either in conventional or unconventional superconductors, we have an energy gap that protects the superconducting condensate from breaking Cooper pairs or exciting so-called quasiparticles. In the BCS theory, the gap is given by

$$\Delta_s = 2\hbar\omega_D exp^{-1/N(0)V}. \quad (1)$$

Here $\hbar\omega_D = k_B\Theta_D$ is the Debye energy; N(0) is the density of states (DOS) near the Fermi energy; V is the interaction between two electrons involved in the pair-scattering process. In the weak coupling case, it is predicted that $\Delta_s = 1.75k_BT_c$. Therefore the superconducting state can be achieved at a finite temperature. One can see from the above description that, one of the basic conditions in the establishment of the BCS scheme is that $\Delta_s << \hbar\omega_D << E_F$. Once we have a system with Δ_s comparable with the Fermi energy E_F, this basic requirement is thus lost and we run into a new scheme of superconductivity. For superconductors with conventional superconducting pairing mechanism, the one with the highest transition temperature at ambient pressure is MgB_2 which was discovered in 2001[3]. In some pressurized hydrides[4], it was reported that the superconducting transition temperature rises to about 200K or beyond, but the pairing mechanism can still be described by the phonon mediated BCS picture.

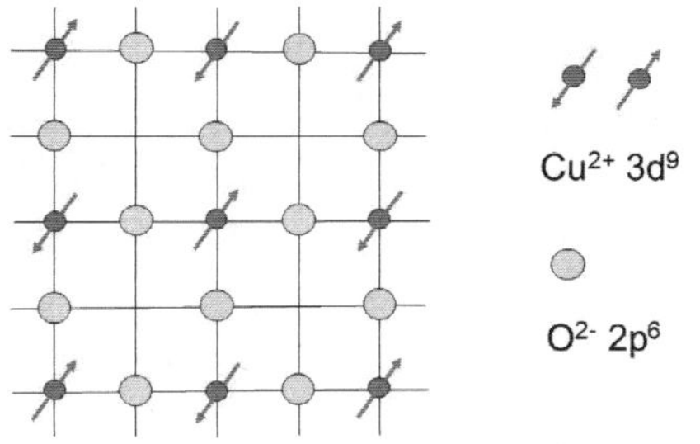

Figure 2. Sketch of the CuO_2 planes of the cuprate superconductors. The small filled circles with arrows represent the Cu ions with the $3d^9$ as the outermost orbital of electrons. The large circles represent the oxygen ions with the $2p^6$ outermost orbitals.

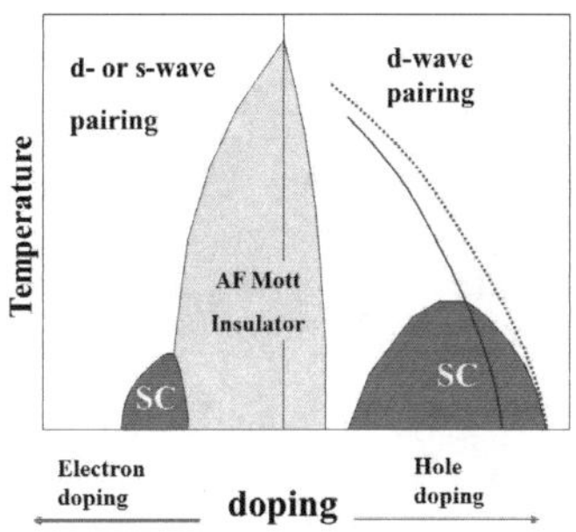

Figure 3. A general electronic phase diagram of the cuprate superconductors. The central yellow region shows the parent phase with a long range ordered antiferromagnetic insulating phase. The two dome like highlighted by red color show the regions for superconductivity. In the electron doped side, the AF order may enter the superconducting dome, showing a coexistence of superconductivity and AF order. The pairing symmetry has been proved to be d-wave in the hole doped region, it may be non-monotonic d-wave or s-wave pairing in the electron doped region. The solid and dashed lines represent the pseudogap temperatures according to different definitions or experimental measurements.

2. Unconventional superconductivity in cuprates and iron pnictides/chacogenides

In past four decades, many new superconductors were discovered. These systems exhibit some features which are clearly at odds with the BCS theory. These superconductors include the heavy Fermion[5], organic[6], cuprates, iron based superconductors and nickelates. Among them, the cuprate is a typical one which was firstly discovered by K. A. Müller and J. G. Bednorz[7]. They found that the superconductivity occurred at about 35 K in a typical copper oxide $La_{2-x}Ba_xCuO_4$. This discovery ignited an explosion in the field of superconductivity. In later 7-8 years, many other new superconducting systems were discovered, for example the $YBa_2Cu_3O_{7-\delta}$ ($T_c \approx 90K$)[8, 9], the bismuth family $Bi_2Sr_2CaCu_2O_8$ ($T_c \approx 90K$) and $Bi_2Sr_2Ca_2Cu_3O_{10}$ ($T_c \approx 125K$)[10], the thallium family[11] and the mercury family [12] etc. So far the highest superconducting transition temperature in cuprates is $T_c = 164$ K, which was discovered in the $HgBa_2Ca_2Cu_3O_{8+\delta}$ under a high pressure of 45 GPa.[13].

No sooner after the discovery by Müller and Bednorz, it was found that the major players for superconductivity in the cuprates are the CuO_2 planes in the middle of the pyramid constructed by the octahedra of Cu and oxygen atoms. This plane is sketched in Fig.2. If we have a look at the formulae of some parent phase of cuprate systems, such as La_2CuO_4, a natural counting on the ionic state would lead to Cu^{2+} with $3d^9$, and O^{2-} with $2p^6$ as the outermost shell of electron orbitals. Taking the structural parameters of the system into consideration, a simple density-functional-theory (DFT) calculation will find out that the material should be a metal with $Cu-3d$ orbitals as the conduction band, forming a large electron like Fermi surface surrounding the Γ point, while later experiments show that it is an insulator due to Mottness, thus it is called as a Mott insulator[14]. Many transport properties indicate that the resistivity close to the undoped case cannot be described by a band insulating picture, rather it is described by a law $\rho(T) \propto log(1/T)$ which manifests itself some exotic reasons. Almost simultaneously, researchers found that the undoped phase, here called as the parent phase of the cuprate superconductors, has a long range anti-ferromagnetic order with an ordering temperature at around 300 K and the super-exchange energy $J \approx 150meV$. Clearly the insulating behavior of the parent phase is due to the strong correlation effect. A Hubbard model based picture, namely the Zhang-Rice singlet[15] model, was proposed to interpret the electric conduction in the doped cuprates. In the cuprate system, due to its crystalline field effect, the $3d_{x^2-y^2}$ orbital is lifted to the highest energy and is half filled. However, because of the strong Hubbard U (6-8 eV), this half-filled band will be split further into lower and upper Hubbard bands with a large U between them. In this case, the oxygen $2p$ band will hybridize with the lower Hubbard band, thus the doped holes would accommodate in this Cu-3d and O-$2p$ hybridized band. The Zhang-Rice singlet model postulates that each doped hole on the oxygen site will couple to the Cu^{2+} cation and form a spin singlet state, it is this composed singlet that moves, leading to electric conduction. Meanwhile there are many other models to interpret this interplay of charge and spin in the doped cuprate, for example the t-J model incorporating a SU(2) symmetry[16], and the phase string model[17], etc. All these different models can also interpret the finite coherent DOS at the Fermi energy and a superconducting state with a diluted superfluid density arising from a strongly correlated background. A typical electronic phase diagram is shown in Fig.3. More interestingly, this insulating feature will leave the way to a strange metallic behavior if we dope it with holes (hole doping) or electrons (electron doping). Above a certain threshold of doping ($p \approx 0.05$ for example in $La_{2-x}Sr_xCuO_4$) the superconductivity emerges at a finite temperature. The T_c versus doping normally exhibits a dome like shape with the optimal T_c around a doping level of about $p = 0.16$ in the hole doped side. In the overdoped side the T_c drops down again and superconductivity vanishes at about $p \approx 0.25 - 0.30$. Phase separation picture was proposed to explain the dropping down of the superfluid density and the superconducting transition temperature in the overdoped region[18]. Recently, mutual inductance measurements on the systematically doped thin films illustrate also a rough linear correlation between the superfluid density and T_c in the overdoped region, manifesting also an non-BCS mechanism[19].

The normal state of the cuprate superconductors is also drastically deviating from the very base for building up the BCS paradigm, the Landau-Fermi liquid picture. One of the exotic features here is the appearance of the so-called pseudogap far above T_c, as shown in Fig.3 by the solid and dashed lines. According to the different definitions, the pseudogap temperature T^* is also quite different. For example, the Knight shift in the NMR measurements shows a decrease at a quite high temperature. This has been regarded as the opening of a spin gap[21]. The pseudogap can also be detected in the temperature dependence of resistivity in many cuprate systems. Below the pseudogap temperature, the resistivity shows a linear behavior with a kink at the pseudogap temperature. Similar behavior occurs in the specific heat data which exhibits also a decrease of the electronic specific heat coefficient below the pseudogap temperature[22]. While the Nernst effect measurements reveal a strong signal which may correspond to the existence of the simultaneous vortex-antivortex excitations[23]. The more direct evidence for the pseudogap is coming from the angle resolved photoemission spectroscopy (ARPES)[24]. One appealing picture for the normal state is the gapped feature near the anti-nodal area due to the correlation effect, while the electrons near the nodal point have still finite life-time recovering a certain weight of quasiparticles. This leads to the so-called truncated "Fermi surface" or Fermi arc. Although to reckon on the number of the conduction electrons through the well known Luttinger theorem is very questionable, while the superconducting condensation in the cuprates seem to be achieved by building up a gap on the so-called Fermi arc near the nodal point. When temperature is increased above T_c, the gap on the Fermi arc vanishes and the "Fermi arc metal" is recovered as the normal state[16]. Connected with this pseudogap feature, there are many other "intertwined orders/phases" existing in this region, such as the charge density way (CDW) order, pair-density wave (PDW) order, spin-glass, etc. A recent overview gives a nice summary about this complexity[25]. Thus we see many very exotic features of the normal state of the cuprate superconductors. These abnormal features in the normal state of cuprate superconductors are certainly shaking the very base for the BCS picture. It is still under a hot debate how these facts can be put together to reach a reconciled picture. Although some people suggest that the BCS picture may be still applicable to interpret the superconductivity in the cuprate system, while significant modifications have to be undertaken, especially the focus is about whether the retarded weak coupling picture, as the case for phonon mediated pairing, is still applicable here.

The discovery of iron based superconductors[26] seems to help unraveling the puzzle of unconventional superconductivity. Up to date, many FeAs or FeSe based superconductors have been synthesized, which enriches the families of the iron based superconducting systems[27]. A common knowledge now about the iron based superconductors is that the major electric conduction and the superconductivity are fulfilled by the FeAs or FeSe based planes. Theoretical calculations indicate that the Fe has a cationic state of Fe^{2+} with 6 electrons in the five 3d orbitals, namely d_{xy}, d_{xz}, d_{yz}, $d_{x^2-y^2}$ and $d_{3z^2-r^2}$. One can estimate the crystalline field splitting of these five 3d orbitals in the system, which gives a quite weak effect. Therefore in principle, all the five 3d orbitals cross the Fermi energy leading to several sets of Fermi surfaces, two hole-like cylindrical Fermi pockets and a small 3D like electron pocket in the Brillouin zone center (Γ point), and two more electron Fermi pockets in the zone corner (M point). For most FeAs based superconductors, the Fermi surface contains a set of hole pockets centering at Γ and a set of electron pockets near M point, as sketched in Fig.4(b). Early but swift theoretical calculations indicate that the phonon spectrum and the electron-phonon interaction does not support to give superconductivity above 1 K[28]. Interestingly the parent phase, for example, RFeAsO (R= rare earth elements) and $BaFe_2As_2$ all show an antiferromagnetic order at temperatures between 100-200 K. The wave vector of these AF phase measures exactly the momentum displacement between the electron pockets located at $M(\pm\pi, 0)$ or $(0, \pm\pi)$ to $\Gamma(0,0)$ in most families of iron based superconductors. There are some preliminary evidence to show that the iron based superconductors also have the pseudogap effect[29, 30], although it is quite rare compared with that in the cuprate superconductors. Meanwhile the normal state, especially those in phosphorous doped systems (LaFePO, or $BaFe_2As_{2-x}P_x$) indeed indicates a super-linear temperature dependence of resistivity near the possible quantum critical point where the AF vanishes[31]. Therefore we have the strong reason to believe that the superconductivity in the iron based superconductors should be closely related to the antiferromagnetic phase[32], and may have the same origin as the cuprates.

Recently, superconductivity up to 15 K was found in hole doped $Nd_{1-x}Sr_xNiO_2$[33]. Interests arose immediately after this discovery since this infinite layer system may also have a $3d^9$ orbital as the dominant one for the electric conduction. The discovery of superconductivity has also been extended to other compounds, for example $Pr_{1-x}Sr_xNiO_2$ and $La_{1-x}Ca_xNiO_2$. The calculations based on DFT reveal that for the parent phase $Nd(La)NiO_2$, the major bands crossing Fermi energy are $Ni-3d_{x^2-y^2}$, $Nd-5d_{3z^2-r^2}$, and $Nd-5d_{xy}$ with some

Journal of Physics: Conference Series **2323** (2022) 012001 doi:10.1088/1742-6596/2323/1/012001

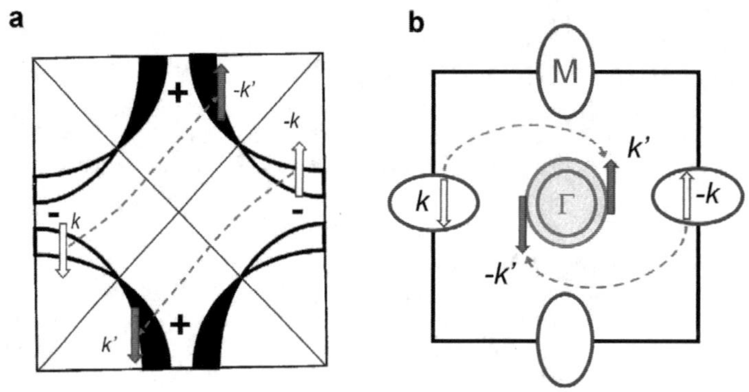

Figure 4. Cartoon picture for the pair-scattering process based on a retarded weak coupling in (a) cuprates and (b) iron based superconductors. The open arrows represent two electrons with opposite momenta and spins before the pair-scattering; the filled arrows show the final states after the pair-scattering. The dashed lines show the pairing channel in the pair-scattering process through a retarded interaction. If the interaction V is positive, this kind of pairing will induce sign change of the superconducting gaps in the momentum space, for example d-wave in cuprates and $s^{\pm}$-wave in iron-based superconductors.

hybridizations. Thus it is clear that the major contribution of superfluid density comes from the orbital of $Ni - 3d_{x^2-y^2}$, but the $Nd - 5d$ bands also contribute partially to the DOS at the Fermi level[34, 35]. A common understanding is that the $Ni - 3d_{x^2-y^2}$ band constructs a large hole pocket around $\Gamma(0,0,0)$ at the $k_z = 0$, which changes into an electron pocket at the $k_z = \pi$ due to the dispersion along k_z direction. The hole pocket at $k_z = 0$ looks like that of underdoped cuprate; while on the cutting plane at $k_z = \pi$, the electron pocket shrinks and becomes similar to the Fermi pocket in overdoped cuprate. There are two more 3D electron pockets, one is surrounding Γ which is contributed by the hybridization of the $Nd - 5d_{3z^2-r^2}$ and $Ni - 3d_{3z^2-r^2}$, another one surrounding $A(\pi,\pi,\pi)$ is constructed by the hybridization of $Nd - 5d_{xy}$ and $Ni - 3d_{3z^2-r^2}$. A spin excitation spectrum has been measured in the undoped system $NdNiO_2$, which indicates a wide band spin excitations up to an energy of 200 meV[36]. Based on the scheme of repulsive interactions[35], *d*-wave pairing as the dominant one has been predicted on the central hole pocket. By measuring scanning tunneling microscopy, two types of gaps are observed[37], one has a *d*-wave gap function with gap maximum about 3.9 meV, another one exhibits a full gap about 2.35 meV. Some time a mixture of the two types of gap features appears on one single spectrum, which indicates that the $Nd_{1-x}Sr_xNiO_2$ thin film is a multigap system and probably with a *d*-wave pairing tendency as the dominant one. Due to the difficulties of synthesizing good samples, more results are desired to unravel the mystery of superconductivity in the nickelate systems.

3. Preliminary understanding of the unconventional superconductivity

If there exist strong Coulomb interactions U among electrons, or the ratio between the repulsive coulomb interaction and the band width t, i.e., U/t is large, the free electron gas or the so-called Landau-Fermi liquid picture immediately becomes questionable. In this case one cannot assume a negligible potential energy in the electronic Hamiltonian, the latter has been adopted in building up the BCS picture. With the presence of strong correlations, the Hamiltonian has been considered based on the $t - U$, or $t - J$ or $t - J - U$ models. Since tremendous number of electrons need to be considered in the system, this multibody problem is not solvable. However, this problem may be at least inspected by assuming a low energy approach. In this case, a microscopic Hubbard-type Hamiltonian with a four fermion interaction can still be written as[38]

$$H = \sum_{k,\alpha} \varepsilon_k \psi^+_{k\alpha} \psi_{k,\alpha} + \sum_{k_i,\alpha_i} U^{\alpha_1,\alpha_2,\alpha_3,\alpha_4}_{k_1,k_2,k_3,k_4} \psi^+_{k_1,\alpha_1} \psi^+_{k_2,\alpha_2} \psi_{k_3,\alpha_3} \psi_{k_4,\alpha_4}. \quad (2)$$

Here $U^{\alpha_1,\alpha_2,\alpha_3,\alpha_4}_{k_1,k_2,k_3,k_4}$ represents the four electron interactions, $\psi^+_{k_i,\alpha_i}$ (ψ_{k_i,α_i}) is the creation (annihilation) operator for electrons with the spin α_i and momentum k_i (i=1,2,3,4). For a single band Hubbard model with local Coulomb interactions, we have

$$U^{\alpha_1,\alpha_2,\alpha_3,\alpha_4}_{k_1,k_2,k_3,k_4} = U\delta_{k_1+k_2-k_3-k_4}(\delta_{\alpha_1\alpha_4}\delta_{\alpha_2\alpha_3} - \delta_{\alpha_1\alpha_3}\delta_{\alpha_2\alpha_4}). \quad (3)$$

If only the interactions in the low energy area are considered, the situation can be treated by using the so-called random phase approximation (RPA) method[39]. For a system with very good Fermi surface nesting, the electronic system will be instable to some long range ordered state, such as the charge-density-wave (CDW) or the spin-density-wave (SDW) state. In the iron based systems, the perfect nesting between the hole and electron pockets may just satisfy this situation leading to the formation of a long range SDW order in the parent phase in the FeAs based families. There are some arguments that the AF order is induced by the local super-exchange effect. By chemical doping, or applying a pressure, the perfect nesting situation deteriorates, and the long range ordered SDW state will be replaced by the state with strong antiferromagnetic spin fluctuations (AF-SF). The elementary excitations of this state are the so-called paramagnons. When an itinerant electron is passing through a position, it will polarize the spin cloud around it, this polarized spin cloud will be sensed by the second itinerant electron passing by. This process will build up a pairing interaction between the two electrons with a pairing interaction as

$$\Gamma_s(k,k') = \frac{3}{2}U^2 \frac{\chi_0(q)}{1 - U\chi_0(q)}, \quad (4)$$

where U is the repulsive energy, χ_0 is the bare spin susceptibility in absence of interactions, the subscript s here represents the singlet pairing channel. One can see that Γ_s is also peaked at the wave vector q but is positive. If we use this pairing interaction and follow the BCS approach, the gap function can be written as

$$\Delta_k = -\sum_{k'} \Gamma_s(k,k') \frac{\Delta'_k}{2E'_k} tanh \frac{E'_k}{2T} \quad (5)$$

Here $E_{k'} = \pm\sqrt{\varepsilon^2_{k'} + \Delta^2_{k'}}$. In iron based superconductors, this interesting pairing picture was first proposed by Mazin et al.[40], and later further formalized by several other groups using different theoretical approaches[41, 42, 43, 44]. More details about the pairing order parameter can be found in a recent review[45]. The basic idea of weak coupling picture is illustrated in Fig.4(a) and (b) for cuprates and iron based superconductors, respectively. It still remains unclear whether this weak coupling based picture applies to the cuprates, since the pairing is very strong and Fermi surface is incomplete (truncated). In iron based superconductors, especially in FeAs based superconductors, the two electrons on the electron (hole) FSs (marked with the thick arrows indicating opposite directions of spins) are scattered to the hole (electron) FSs by exchanging the antiferromagnetic spin fluctuations (AF-SFs). If the pairing is established through exchanging the AF-SFs, according to above equations, the so-called $S^{\pm}$ pairing model is expected, i.e., the DOS on individual FS are fully gapped (although the gap magnitude has some momentum dependent variation), but the signs of the superconducting gaps on the electron and hole pockets should be opposite. This kind of s-wave gap has been supported by the scanning tunneling experiment already[46]. The recent experiment in observing the in-gap quasi-particle states by non-magnetic Cu impurities in $NaFe_{0.97-x}Co_{0.03}Cu_xAs$ may provide a more

Journal of Physics: Conference Series **2323** (2022) 012001 doi:10.1088/1742-6596/2323/1/012001

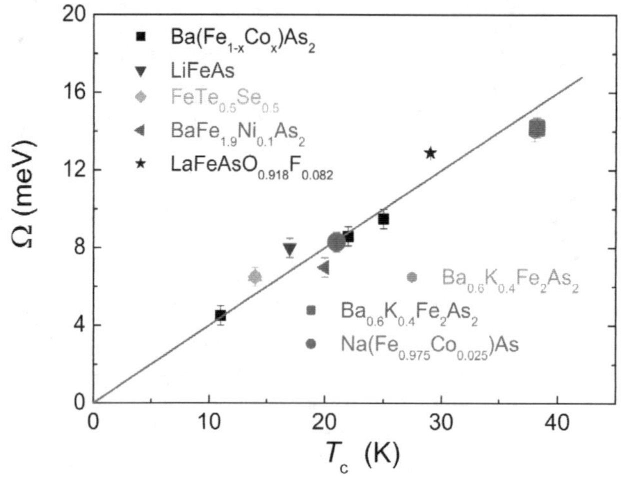

Figure 5. A roughly linear relation between the neutron resonance or bosonic mode energy and the superconducting transition temperature T_c in many iron based superconductors. The small sized filled symbols are data collected in the inelastic neutron scattering measurements, the two red bigger square and circle symbols are determined from the bosonic mode on the tunneling spectrum. This figure is made by quoting to the data in Ref.[52]

decisive evidence for the $S^{\pm}$ pairing[47]. Furthermore, this sign reversal of gaps on the electron and hole pockets in $FeTe_{1-x}Se_x$ has been well checked by the phase referenced quasiparticle interference technique[48]. This RPA based picture may also be extended to the case of cuprates for understanding the well known *d*-wave pairing symmetry. As shown in Fig.4(a), since there is only one set of Fermi surface in the cuprate, the pair-scattering would lead to a sign change on the different segments of the Fermi surface, this naturally leads to a *d*-wave gap function.

One of the consequences of the electronic system with the pairing gap with opposite signs is the appearance of a resonance peak in the imaginary part of the spin susceptibility at the wave vector of the SDW. This resonance peak with a typical energy $\Omega \leq 2\Delta_s$ can be detected with inelastic neutron scattering experiment and has been observed in cuprates[49], heavy Fermion[50] and recently in the iron pnictide superconductors[51]. This resonance peak may be understood as due to the spin flipping scattering when exciting a quasiparticle from the k to $k+q$ with the pairing gap with the opposite signs, since in this case the BCS coherence factor is given by[39, 45]

$$I_{res} \propto \frac{1}{2}[1 - \frac{\Delta(k)\Delta(k+q)}{E(k)E(k+q)}]. \tag{6}$$

Here I_{res} gives the intensity of the resonance peak, $E(k)$ is the dispersion of the Bogoliubov quasiparticles in the superconducting state $E(k) = \pm\sqrt{\varepsilon_k^2 + \Delta^2(k)}$. One can see that, if the superconducting gap has a sign reversal at momenta connected by vector q, that will produce a peak at this particular vector. The RPA calculations tell that this peak should locate at a energy of $\Omega \leq 2\Delta_s$ depending on the correlation strength. Although the resonance peak is regarded as the consequence of the sign-reversal gaps, sometimes it is called as the "egg" of the superconducting pairing state, not the pairing glue for superconductivity, we must emphasize that this "egg" can be naturally traced back to the scenario of the magnetic pairing, namely the origin of pairing. Therefore, although the resonance peak may be due to the spin-1 exciton mode

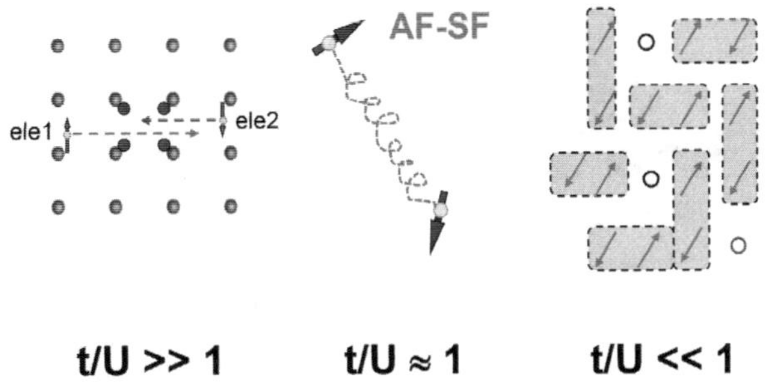

Figure 6. Evolution of the pairing from weak coupling to strong coupling, depending on the ratio U/t. The left hand picture shows the case for $U/t << 1$, the BCS scheme works well. Here the four blue circles represent the distortion of the atoms due to the electron-phonon interaction. This distortion will be "seen" by the second electron passing by in a retarded duration. The right hand one represents the possible situation of strong correlation effect where the local super-exchange may dominate the pairing. In the middle, the $U/t \approx 1$ which may be applicable for the case of AF-SF mediated pairing.

of the sign reversal gaps, we can conclude that it is only possible that the pairing is induced through the electronic origin, either through exchanging the AF-SFs, or the superexchange itself, but certainly not through some non-magnetic origins, like exchanging phonons. In Fig.5, we present a collection of many resonant energies versus the superconducting transition temperature T_c in different iron based superconductors. Here the filled symbols are data collected in the inelastic neutron scattering measurements, the two open symbols are determined from the bosonic mode on the tunneling spectrum. In some iron based superconductors, the scanning tunneling spectroscopy measurements reveal extra peaks (humps) outside the superconducting coherence peaks, which is explained as the coupling of the quasiparticles with the spin resonance mode. In Fig.5, a simple ratio $\Omega \approx (4.3 \pm 0.5)k_B T_c$ is shown[52]. This strongly suggests that the neutron resonance is indeed closely related to superconductivity. Similar linear relation between Ω and T_c was also found in cuprate superconductors[53]. A study on the superconducting condensation energy measured through specific heat and the intensity of the resonance peak in the $Pr_{0.88}LaCe_{0.12}CuO_{4-\delta}$ finds that both quantities decay with the applied magnetic field in the same way[53], indicating a close relationship between the resonance mode energy and superconductivity.

In above discussion, we have shown that the low energy approach, or called as the picture of weak coupling limit can give a reasonably good understanding of the superconductivity in many unconventional superconducting systems. In the systems with moderate Coulomb repulsion energy U, this seems to work quite well. In some systems with extremely large ratio U/t, it is questioned whether this picture based on retarded scattering can still be valid. In this case, some people argue that the local super-exchange J dominates the pairing process, leading to the simultaneous pairing[54, 55] with a very strong pairing strength. One of the extreme cases is the so-called resonance-valence-bond (RVB) spin liquid state for the $s = 1/2$ systems, for example in the cuprates. In this case, the spin singlet pairs are naturally formed yielding a highly fluctuating quantum RVB state.[56] This is certainly a very interesting picture which may get an indirect support from the strong gap, very strong superconducting fluctuation, or residual Copper pairs in the normal state far above T_c. The convincing evidence for this RVB state is still lacking. In Fig.6 we present a schematic show for the superconducting pairing for the cases ranging from the weak coupling (e.g., phonon mediated) pairing to the extreme case of the simultaneous pairing through the local super-exchange. The situation evolves with the ratio of t/U: (1) $t/U \gg 1$, weak coupling induced pairing; (2) $t/U \approx 1$, intermediate correlation induced

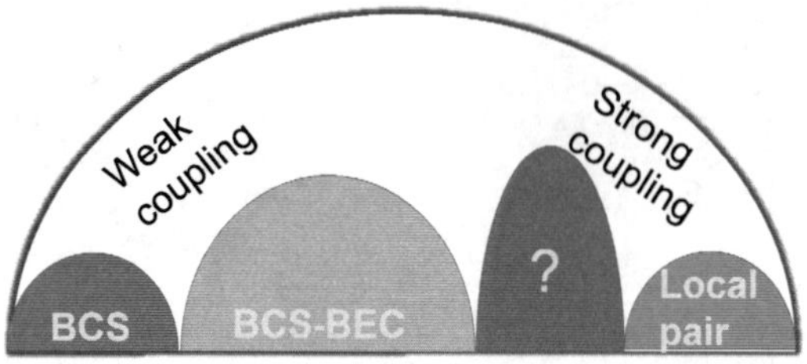

Figure 7. Schematic camp for superconductivity. The BCS theory with weak coupling occupies a small region in the left hand side. We need a general or global camp to cover all possibilities for superconductivity, from weak coupling to strong local pairing.

pairing; (3) $t/U \ll 1$ strong correlation and local pairing could occur. This picture follows a similar thought as that when considering the subtle balance between the correlation effect and the effective Drude weight for optimized superconductivity in a variety of compounds[57]. In the middle of Fig.6, we have a delicate balance between the two extreme cases: complete itinerant ($U/t << 1$) and strong correlation ($U/t >> 1$). The middle case may apply to the iron based superconductors, which still needs the retarded interaction as the pairing origin, but through exchanging the paramagnon or AF-SFs, not phonons. At the moment or in the near future, debate will go on about whether we need to have a more exotic picture which assumes the pairing completely coming from the local strong superexchange. Based on this idea, in Fig.7, we sketch a big camp for superconductivity. The BCS theory occupies only a small area in the weak coupling limit. It remains to know how the pairing can evolve smoothly from the weak coupling limit to the extremely strong coupling limit. It is expectable that the condensation process for superconductivity will also evolve from the BCS type to BEC type, the superconducting transition temperature is determined by the gap in the BCS case, but by the phase stiffness in the BEC case[58, 59].

4. Some empirical rules for the exploration of high temperature superconductors

It is always enthusiastic to explore new high temperature superconductors. However, beside those of pressurized hydrides, no theory can really predict how to get success, thus it has been a quite frustrated task. Even so, some empirical rules may be summarized here. Here we would like to propose several key points which may be the necessary ingredients for exploring high temperature superconductivity at ambient pressure.

4.1. Subtle balance between Mottness and itinerancy of d orbital electrons

As illustrated by Qazilbash et al.[57], the unconventional superconductors should have a delicate balance between Mottness and itineracy which governs the effective DOS at the Fermi energy. Strong Mottness may manifest a large superexchange energy J for the local pairing, but the ground state of that may be an insulator. Thus we need to effectively separate the dual roles of the d-orbital electrons. The $3d$ orbital electrons possess by them-self a possibility to balance between these two factors. Thus it may give more chances in the compounds with the $3d$ transition metal elements, like Cr, Mn, Fe, Co, Ni, Cu. Some $4d$ elements may also be considered, such as Ru, Pd, Rh etc.

4.2. Bad metal in vicinity of an AF order

Recalling both the iron based and cuprate superconductors, it is known that the parent phases are either Mott insulators or bad metals with or near an AF order. Most importantly, this AF order can be easily suppressed by chemical doping or applying a pressure. To satisfy this condition, it is important to have ions with lower magnetic moments, for example below $2\mu_B$. The spin density wave order due to the Fermi surface instability would help. To avoid the large magnetic moments, one may request the low filling or even number filling states of the $3d$ or $4d$ orbitals, such as the Cu-$3d^9$ or Fe-$3d^6$ cationic states. This is actually the case in iron based superconductors. In $BaFe_2As_2$, for example, the cationic state is Fe^{2+}, which yields a magnetic moment generally smaller than $2\mu_B$. In the cuprate, for example, each Cu ion in La_2CuO_4 has a magnetic moment of 0.5-1.0 μ_B. The strong quantum fluctuation of the gauge field near the quantum critical point (QCP) may also help. An odd filling number of 3, 5, 7 of the d orbital will induce a strong Hund's coupling leading to strong magnetic moments and localization of electrons.

4.3. Shallow and flat band near the Fermi energy

Some shallow and flat bands, or saddle points near the Fermi energy will result in the von Hove singularity of DOS, this is certainly helpful for the pairing instability. In iron based superconductors, it is known that some bands are very shallow, which yields a high DOS. In addition, a shallow band possesses a small Fermi energy E_F, which leads to the small ratio E_F/Δ_s. This quantity measures the Cooper pair numbers within a coherence length ξ or volume ξ^3. If this value is small, one may expect a BEC type superconducting transition. Therefore, to have both strong pairing and high superfluid density, we probably need a system with both shallow and wide bands and together with a strong interband scattering, the shallow band produces a strong local pairing, the wide band contributes a high superfluid density.

4.4. Anion mediated superexchange may help

In cuprates and iron based superconductors, there are anions to mediate the superexchange for the antiferromagnetic order or spin fluctuations. Thus in the dominant conductive planes, some atomic chains like T-A-T are necessary, where T represents a transition metal with d orbital electrons, A is an anion. For example, in cuprates the pairing through J_1 superexchange is established by the Cu-O-Cu bonds. While in iron based superconductors, the pairing may be established by the J_2 superexchange which is generated by the Fe-As-Fe bonds. However, this requirement may not be that strict for high T_c superconductivity. In many binary alloy compounds, unconventional superconductivity coexists with the magnetic excitations. Thus for unconventional superconductivity, the anion bridge may not be the necessary component, but may be helpful for stabilizing the structure.

Acknowledgments

We acknowledge the useful discussions with Qianghua Wang, Ilya Eremin, Peter Hirschfeld, Andrey Chubukov, Igor Mazin. This work was supported by NSF of China with Grant Nos. 11927809, NSFC-DFG1206113001(ER463/14-1), and the Ministry of Science and Technology of China.

References

[1] Bardeen J, Cooper L N and Schrieffer J R 1957 *Phys. Rev.* **108** 1175
[2] Wen, H.-H 2014 *Japn Phys. Soc. Conf. Proc.***1** 011005
[3] Nagaatsu J, Nakagawa N, Muranaka T *et al.* 2001 *Nature* **410** 63-65
[4] Drozdov A P, Eremets M I, Troyan I A, Ksenofontov V and Shylin S I 2015 *Nature* **525** 73C76

[5] Steglich F, Aarts J, Bredl C D, Lieke W, Meschede D, Franz W and H. Schäfer 1979 *Phys. Rev. Lett.* **43** 1892
[6] Jerome D, Mazaud A, Ribault M and Bechgaard K 1980 *J. Phys. Lett.* **41** L95. For a recent review on organic superconductors, see Ardavan A *et al* 2012 *J. Phys. Soc. Jpn.* **81** 011004
[7] Bednorz J G and Müller K A 1986 *Z. Phys.* **B64** 189
[8] Wu M K, Ashburn J R, Tirng C J, Meng R L, Gao L, Huang Z L, Wang Y Q and Chu C W 1987 *Phys. Rev. Lett.* **58** 908
[9] Zhao Z X *et al* 1987 *KeXue TongBao* **32** 412
[10] Maeda H, Tanaka Y, Fukutomi M and Asano T 1988 *Jpn. J. Appl. Phys.* **27** L209
[11] Sheng Z Z, Herman A M, Ali A, Almasan C, Estrada J, Datta T and Matsonet R J 1988 *Phys. Rev. Lett.* **60** 937
[12] Putilin S N, Antipov E V, Chmaissem O and Marezio M 1993 *Nature* **362** 226
[13] Gao L, Xue Y Y, Chen F, Xiong Q, Meng R L, Ramirez D, Chu C W, Eggert J H and Mao H K 1994 *Phys. Rev.* **B50** 4260(R)
[14] Anderson P W 1987 *Science* **235** 1196
[15] Zhang F C and Rice T M 1988 *Phys. Rev.* B **37** 3759
[16] Wen X G and Lee P A 1996 *Phys. Rev. Lett.* **76** 50; Lee P A and Wen X G 1997 *Phys. Rev. Lett.* **78** 4111
[17] Weng Z Y, Sheng D N, Chen Y C and Ting C S 1997 *Phys. Rev.* B **55** 3894; Sheng D N, Chen Y C and Weng Z Y 1996 *Phys. Rev. Lett.* **77** 5102
[18] Wen H H, Yang W L, Zhao Z X and Ni Y M 1999 *Phys. Rev. Lett.* **82** 410
[19] Bozovic I, He X, Wu J and Bollinger A T 2016 *Nature* **536** 309C311
[20] Newns D M and Tsuei C C 2007 *Nature Phys.* **3** 184
[21] Alloul H, Ohno T and Mendels D 1989 *Phys. Rev. Lett.* **63** 1700
[22] Loram J W, Mirza K A, Cooper J R and Liang W Y 1993 *Phys. Rev. Lett.* **71** 1740
[23] Xu Z A, Ong N P, Wang Y, Kakeshita T and Uchida S 2000 *Nature* **406** 486
[24] For a review of the ARPES study on the Fermi surface see Damascelli A, Hussain Z and Shen Z X 2003 *Rev. Mod. Phys.* **75** 473
[25] Keimer B, Kivelson S A, Norman M R, Uchida S and Zaanen J 2015 *Nature* **518** 179
[26] Kamihara Y, Watanabe T, Hirano M and Hosono H 2008 *J. Am. Chem. Soc.* **130** 3296
[27] Wen H H and Li S L 2011 *Annu. Rev. Condens. Matter Phys.* **2** 121
[28] Boeri L, Dolgov O V and Golubov A A 2008 *Phy. Rev. Lett.* **101** 026403
[29] Dai Y M, Xu B, Shen B, Wen H H, Hu J P, Qiu X G and Lobo R P S M 2012 *Phys. Rev.* B. **86** 100501(R)
[30] Shimojima T, Sonobe T, Malaeb W et al. *et al* 2014 *Phys. Rev.* B **89** 045101
[31] Hashimoto K *et al* 2012 *Science* **336** 1554
[32] Ning F L, Ahilan K, Imai T, Sefat A S, McGuire M A, Sales B C, Mandrus D, Cheng P, Shen B and Wen H H 2010 *Phys. Rev. Lett.* **104** 037001
[33] Li D F, Lee K, Wang B Y, Osada M, Crossley S, Lee H R, Cui Y, Hikita Y and Hwang H Y 2019 *Nature* **572** 624
[34] Botana A S and Norman M R 2020 *Phys. Rev.* **X 10** 011024
[35] Wu X X, Sante D D, Schwemmer T, Hanke W, Hwang H Y, Raghu S and Thomale R 2020 *Phys. Rev.* **B101** 060504(R)
[36] Lu H *et al* 2021 *Science* **373** 213
[37] Gu Q Q, Li Y Y, Wan S Y, Li H Z, Guo W, Yang H, Li Q, Zhu X Y, Pan X Q, Nie Y F and Wen H H 2020 *Nat. Commun.* **11** 6027
[38] Chubukov A V 2012 *Annu. Rev. Condens. Matter Phys.* **3** 57
[39] Scalapino D J 2012 *Rev. Mod. Phys.* **84** 1383
[40] Mazin I I, Singh D J, Johannes M D and Du M H 2008 *Phys. Rev. Lett.* **101** 057003
[41] Kuroki K, Onari S and Arita R 2008 *Phys. Rev. Lett.* **101** 087004
[42] Yao Z J, Li J X and Wang Z D 2009 *New J. Phys.* **11** 025009
[43] Maier T, Graser S, Hirschfeld P J and Scalapino D J 2009 *Phys. Rev.* **B79** 224510
[44] Wang F and Lee D H 2009 *Phys. Rev. Lett.* **102** 047005
[45] Hirschfeld P J, Korshunov M M and Mazin I I 2011 *Rep. Prog. Phys.* **74** 124508
[46] Hanaguri T, Niitaka S, Kuroki K and Takagi K 2010 *Science* **328** 474
[47] Yang H, Wang Z Y, Fang D L, Deng Q, Wang Q H, Xiang Y Y, Yang Y and Wen H H 2013 *Nat. Commun.* **4** 2749
[48] Chen M Y, Tang Q K, Chen X Y, Gu Q Q, Yang H, Du Z Y, Zhu X Y, Wang E Y, Wang Q H and Wen H H 2019 *Phys. Rev.* **B99** 014507
[49] Rossat-Mignod J, Regnault L P, Vittier C, Bourges P, Burlet P and Bossy J 1991 *Physica* **C185-189** 86
[50] Stock C, Broholm C, Hudis J, Kang H J and Petrovic C 2008 *Phys. Rev. Lett.* **100** 087001

[51] Christianson A D *et al* 2008 *Nature* **456** 930
[52] Wang Z Y, Yang H, Fang D L, Shen B, Wang Q H, Shan L, Zhang C L, Dai P C and Wen H H 2013 *Nature Phys.* **9** 42
[53] Wilson S D, Li S L, Zhao J, Mu G, Wen H H, Lynn J W, Freeman P G, Regnault L P, Habicht K and Dai P C 2007 *Proc. Natl. Acad. Sci. U.S.A.* **104** 15259
[54] Si Q M and Abrahams E 2008 *Phys. Rev. Lett.* **101** 076401
[55] Seo K, Bernevig B A and Hu J P 2008 *Phys. Rev. Lett.* **101** 206404
[56] Anderson P W, Lee P A, Randeria M, Rice T M, Trivedi N and Zhang F C 2004 *J. Phys. Cond. Matter* **16** R755
[57] Qazilbash M M, Hamlin J J, Baumbach R E, Zhang L, Singh D J, Maple M B and Basov D N 2009 *Nature Phys.* **5** 647
[58] Emery V J and Kivelson S A 1995 *Nature* **374** 434
[59] Randeria M and Taylor E 2014 *Annu. Rev. Condens. Matter Phys.* **5** 209

Impurity effects on Bogoliubov Fermi surfaces: A comparison with conventional BCS pairing

Shintaro Hoshino and Tatsuya Miki

Department of Physics, Saitama University, Shimo-Okubo, Saitama 338-8570, Japan

E-mail: hoshino@mail.saitama-u.ac.jp

Abstract. The superconductors can have a Fermi surface even below the transition temperature, which is known as the Bogoliubov Fermi surface. Focusing on the degrees of freedom at low energies, which are described by the spinless Bogoliubov particles for the case without time-reversal symmetry, we consider the self-energy from the impurity scattering on the Bogoliubov Fermi surfaces. The physical quantities such as the density of states are evaluated based on the single-particle Green's functions. We systematically compare these results with those of the conventional s-wave spin-singlet pairing state, to illuminate the characteristic features of the Bogoliubov particles located at the Fermi level.

1. Introduction

The superconductivity is induced by the Cooper pair condensation near the Fermi level, which results in the energy gap formation in their excitation spectrum. Usually, this leads to a disappearance of the Fermi surface below the transition temperature. On the other hand, there exists another possibility for unconventional superconductors: The Fermi surfaces remain even in the superconducting state. This is known as Bogoliubov Fermi surface, where the gapless excitation near the Fermi level is described by the Bogoliubov quasiparticles (bogolons) and is first proposed by Volovik [1]. Furthermore, it has recently been proposed that the superconductors with multiple orbital degrees of freedom and broken time-reversal symmetry can have Fermi surfaces even below the transition temperature, which are topologically stable in the presence of the inversion symmetry [2]. A variety of physical systems with Bogoliubov Fermi surfaces has been proposed [3, 4, 5, 6] including the multi-orbital electrons [2, 7, 8, 9, 10, 11, 12, 13, 14, 15, 16]. These are classified further based on the symmetries that exist in the normal state and in the superconducting state. The effects from the interaction and impurity scattering, which modify the ideal system to non-ideal one, have also been studied recently [17, 18, 19, 20, 21].

In the following, we consider the system where the spatial inversion and time-reversal symmetries are present in the normal phase, and the time-reversal symmetry is spontaneously broken in the superconducting state. Such a situation is possibly realized in the actual material Fe(Se,S) where the remaining density of states is observed even below the superconducting transition temperature [22, 23, 24]. Since the quantum oscillation is observed in this material [25], the disorder effect is expected to be weak. In our previous work, we have proposed that the characteristic feature of the Bogoliubov Fermi surface different from the usual Fermi surface can be seen in the self-energy effects originating from the disorders and/or interactions

[20]. The anomalous self-energy, which describes the pair potential among the bogolons, the frequency dependence in the low-energy region is *purely odd-function.* Hence, the systems with the Bogoliubov Fermi surface are a suitable playground for the study of the odd-frequency pairing [26, 27, 28, 29, 30, 31, 32, 33, 34], although the mechanism is specific to bogolons and is different from the previously proposed ones for electrons.

Here, we report on a characteristic impurity effect on the Bogoliubov Fermi surface by making a contrast against the conventional s-wave superconductor. In the next section, we review our previous work for the disordered spinless bogolons given in Ref. [20], and we compare the results systematically with the disordered conventional BCS superconductor in §3.

2. Impurity effect on Bogoliubov Fermi surfaces

2.1. The bogolon model

We consider the bogolons near the Bogoliubov Fermi surface in the inversion symmetric system with broken time-reversal symmetry as discussed in Refs. [2, 18]. The non-interacting Hamiltonian is given by

$$\mathscr{H}_0 = \sum_{\boldsymbol{k}} \varepsilon_{\boldsymbol{k}} \alpha^\dagger_{\boldsymbol{k}} \alpha_{\boldsymbol{k}}, \tag{1}$$

which is composed of a spinless fermion without internal degrees of freedom, whose annihilation operator is $\alpha_{\boldsymbol{k}}$ at the wavevector $\boldsymbol{k}$. The single-particle excitation energy is $\varepsilon_{\boldsymbol{k}}$ $(= \varepsilon_{-\boldsymbol{k}})$, and the Bogoliubov Fermi surface is given by the solution of the equation $\varepsilon_{\boldsymbol{k}} = 0$. Since we have assumed the inversion symmetry, the static pair potential $\alpha^\dagger_{\boldsymbol{k}}\alpha^\dagger_{-\boldsymbol{k}}$ cannot exist in Eq. (1). This is the important feature of the present system which generates the odd-frequency pairing among bogolons when the system is away from the ideal limit [20].

We add the impurity Hamiltonian

$$\mathscr{H}_{\rm imp} = \frac{1}{V}\sum_{\boldsymbol{kq}} \rho_{\boldsymbol{q}} u_1(\boldsymbol{k},\boldsymbol{q}) \alpha^\dagger_{\boldsymbol{k}+\boldsymbol{q}} \alpha_{\boldsymbol{k}} + \frac{1}{V}\sum_{\boldsymbol{kq}} \Big[\rho_{\boldsymbol{q}} u_2(\boldsymbol{k},\boldsymbol{q}) \alpha^\dagger_{\boldsymbol{k}+\boldsymbol{q}} \alpha^\dagger_{-\boldsymbol{k}} + \text{H.c.}\Big], \tag{2}$$

where $\rho_{\boldsymbol{q}} = \sum_i \mathrm{e}^{-\mathrm{i}\boldsymbol{q}\cdot\boldsymbol{R}_i}$, and V being a system volume. The summation with respect to i is taken over all the impurity position $\boldsymbol{R}_i$. The total Hamiltonian is then given by $\mathscr{H} = \mathscr{H}_0 + \mathscr{H}_{\rm imp}$ (we call it 'the bogolon model') which describes the low-energy properties of the system. Although the impurity scattering event is spatially local for electrons [20], for bogolons a non-local nature (i.e., wavevector dependence) enters through the unitary transformation from the original electrons. The second term with $\alpha^\dagger\alpha^\dagger$ in Eq. (2) is characteristic for the bogolon systems: the number of bogolons is not a conserved quantity. Note that the impurity potential in general breaks the inversion symmetry, but it is recovered after the averaging over the impurity configuration. The above Hamiltonian is given based on the symmetry of the system, and is regarded as a general expression.

2.2. Single-particle Green's functions

We define the normal and anomalous Green's functions of bogolons by

$$G_{\boldsymbol{k}}(\tau) = -\langle \mathcal{T} \alpha_{\boldsymbol{k}}(\tau) \alpha^\dagger_{\boldsymbol{k}} \rangle, \tag{3}$$

$$F_{\boldsymbol{k}}(\tau) = -\langle \mathcal{T} \alpha_{\boldsymbol{k}}(\tau) \alpha_{-\boldsymbol{k}} \rangle, \tag{4}$$

where $\mathcal{T}$ represents the imaginary-time ordering and we have introduced the Heisenberg picture $\alpha_{\boldsymbol{k}}(\tau) = \mathrm{e}^{\tau\mathscr{H}} \alpha_{\boldsymbol{k}} \mathrm{e}^{-\tau\mathscr{H}}$. We emphasize that not only the normal Green's function but anomalous

Green's function $F_{\boldsymbol{k}}(\tau)$ become finite because of the second term in Eq. (2). Because of the inversion symmetry, we can show the relation

$$F_{\boldsymbol{k}}(\tau) = -F_{\boldsymbol{k}}(-\tau). \tag{5}$$

Namely, the pair amplitude must be odd function in time (or frequency) for the present system. Thus the odd-frequency pairing is naturally appears for the bogolon model, and the impurity effect is the simplest way to generate this exotic pairing.

The Fourier transformation of the Green's function is given by

$$G_{\boldsymbol{k}}(\mathrm{i}\omega_n) = \int_0^{1/T} \mathrm{d}\tau G_{\boldsymbol{k}}(\tau)\mathrm{e}^{\mathrm{i}\omega_n\tau}, \tag{6}$$

and also for the anomalous Green's function $F_{\boldsymbol{k}}(\mathrm{i}\omega_n)$ where $\omega_n = (2n+1)\pi T$ is the fermionic Matsubara frequency. The normal (Σ) and anomalous (S) self-energies are introduced as

$$G_{\boldsymbol{k}}(\mathrm{i}\omega_n) = \frac{\mathrm{i}\omega_n - \Sigma_{\boldsymbol{k}}(\mathrm{i}\omega_n) + \varepsilon_{\boldsymbol{k}}}{[\mathrm{i}\omega_n - \Sigma_{\boldsymbol{k}}(\mathrm{i}\omega_n)]^2 - S_{\boldsymbol{k}}(\mathrm{i}\omega_n)S^*_{\boldsymbol{k}}(-\mathrm{i}\omega_n) - \varepsilon^2_{\boldsymbol{k}}}, \tag{7}$$

$$F_{\boldsymbol{k}}(\mathrm{i}\omega_n) = \frac{S_{\boldsymbol{k}}(\mathrm{i}\omega_n)}{[\mathrm{i}\omega_n - \Sigma_{\boldsymbol{k}}(\mathrm{i}\omega_n)]^2 - S_{\boldsymbol{k}}(\mathrm{i}\omega_n)S^*_{\boldsymbol{k}}(-\mathrm{i}\omega_n) - \varepsilon^2_{\boldsymbol{k}}}. \tag{8}$$

In order to see the lowest-order contribution of the impurity effect, we apply the Born approximation which is valid for weak disorders [20]. The normal and anomalous self-energies are explicitly given as follows:

$$\Sigma_{\boldsymbol{k}}(\mathrm{i}\omega_n) = n_{\mathrm{imp}} \int \frac{\mathrm{d}\boldsymbol{q}}{(2\pi)^3} \left[|u_1(\boldsymbol{q}, \boldsymbol{k}-\boldsymbol{q})|^2 G^0_{\boldsymbol{q}}(\mathrm{i}\omega_n) - 2|u_2(\boldsymbol{q}, \boldsymbol{k}-\boldsymbol{q})|^2 G^0_{\boldsymbol{q}}(-\mathrm{i}\omega_n)\right], \tag{9}$$

$$S_{\boldsymbol{k}}(\mathrm{i}\omega_n) = -2n_{\mathrm{imp}} \int \frac{\mathrm{d}\boldsymbol{q}}{(2\pi)^3} u^*_1(\boldsymbol{q}, \boldsymbol{k}-\boldsymbol{q}) u_2(\boldsymbol{q}, \boldsymbol{k}-\boldsymbol{q})[G^0_{\boldsymbol{q}}(-\mathrm{i}\omega_n) - G^0_{\boldsymbol{q}}(\mathrm{i}\omega_n)], \tag{10}$$

where $G^0_{\boldsymbol{q}}(\mathrm{i}\omega_n)$ is a zeroth-order Green's function for bogolon. The impurity average is performed as $\overline{\rho_{\boldsymbol{q}}\rho_{\boldsymbol{q}'}} = Vn_{\mathrm{imp}}\delta_{\boldsymbol{q},-\boldsymbol{q}'}$. In order to evaluate the wavevector integral, we replace the coupling constant by its averaged value with respect to $\boldsymbol{q}$. Then, we obtain

$$\Sigma_{\boldsymbol{k}}(\mathrm{i}\omega_n) = -\mathrm{i}\Gamma_{1\boldsymbol{k}}\mathrm{sgn}\,\omega_n, \tag{11}$$

$$S_{\boldsymbol{k}}(\mathrm{i}\omega_n) = -\mathrm{i}\Gamma_{2\boldsymbol{k}}\mathrm{sgn}\,\omega_n, \tag{12}$$

where

$$\Gamma_{1\boldsymbol{k}} = \pi n_{\mathrm{imp}} D_0 \big(\langle|u_1(\boldsymbol{q}, \boldsymbol{k}-\boldsymbol{q})|^2\rangle_{\boldsymbol{q}} + 4\langle|u_2(\boldsymbol{q}, \boldsymbol{k}-\boldsymbol{q})|^2\rangle_{\boldsymbol{q}}\big), \tag{13}$$

$$\Gamma_{2\boldsymbol{k}} = 4\pi n_{\mathrm{imp}} D_0 \langle u^*_1(\boldsymbol{q}, \boldsymbol{k}-\boldsymbol{q}) u_2(\boldsymbol{q}, \boldsymbol{k}-\boldsymbol{q})\rangle_{\boldsymbol{q}}, \tag{14}$$

with $\langle\cdots\rangle_{\boldsymbol{q}}$ being the average in the reciprocal space. The bare density of states D_0 has been introduced as $\int \frac{\mathrm{d}\boldsymbol{q}}{(2\pi)^3} = D_0 \int \mathrm{d}\varepsilon_{\boldsymbol{q}}$. One can confirm that there is the magnitude relation $\Gamma_{1\boldsymbol{k}} > |\Gamma_{2\boldsymbol{k}}|$ which indicates the necessity of the normal self-energy [20]. If this inequality is not satisfied, the negative spectral weight is obtained. Hence this relation is regarded as a fundamental relation for the thermodynamically stable state. In the following, we assume that the $\boldsymbol{k}$-dependence of the self-energies is weak and replace them as $\Gamma_{1\boldsymbol{k}} \to \Gamma_1$ and $\Gamma_{2\boldsymbol{k}} \to \Gamma_2$, meaning that the local s-wave pair is assumed. We also choose the phase of the pairing potential as real and positive. This setup is the simplest model for the odd-frequency pairing and can be realized in systems with Bogoliubov Fermi surfaces.

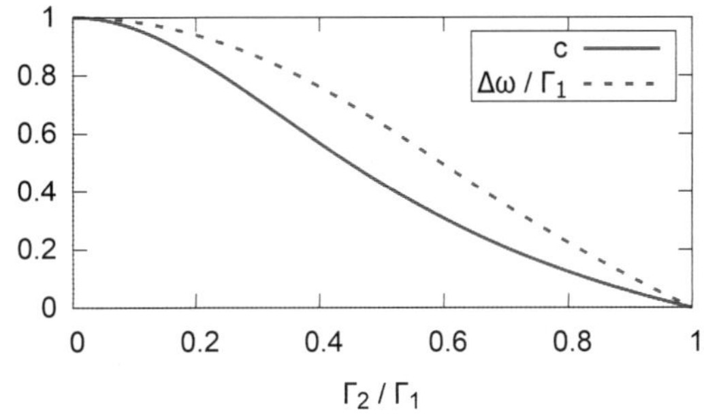

Figure 1. Coefficients for the dispersion relation at low energy and the spectral width defined in Eqs. (17) and (18), respectively.

2.3. Physical quantities

In this subsection, we summarize the physical quantities of the bogolon model. We first define the local Green's function $G_{\mathrm{loc}}(\tau) = \sum_{\boldsymbol{k}} G_{\boldsymbol{k}}(\tau)$. Using the spectral representation, the local Green's function is written as

$$G_{\mathrm{loc}}(\tau) = \int \mathrm{d}\omega D(\omega) \frac{\mathrm{e}^{-\omega\tau}}{\mathrm{e}^{\omega/T} + 1} \tag{15}$$

for $\tau < 0$. $D(\omega)$ is a density of state given by $D(\omega) = \sum_{\boldsymbol{k}} A_{\boldsymbol{k}}(\omega)$ where

$$A_{\boldsymbol{k}}(\omega) = -\frac{1}{\pi} \mathrm{Im}\, G_{\boldsymbol{k}}(\omega + \mathrm{i}0^+) \tag{16}$$

is the wavevector resolved single-particle spectrum which is relevant to experimental spectroscopies. The example of the spectrum will be shown later (in Figs. 3 and 4).

The single-particle spectrum Eq. (16) is characterized by two quantities; the position and the width of the peak as a function of ω at each $\boldsymbol{k}$ (see also the left panel of Fig. 3). The energy dispersion of quasiparticles at low-energy is determined by the peak position of the spectrum to give

$$\omega = \frac{\Gamma_1^2 - \Gamma_2^2}{\Gamma_1^2 + 3\Gamma_2^2} \varepsilon_{\boldsymbol{k}} \equiv c\varepsilon_{\boldsymbol{k}}. \tag{17}$$

On the other hand, the width, which implies the damping, is also defined by the energy at which the spectral height becomes half of the maximum value at $\varepsilon_{\boldsymbol{k}} = 0$. We can explicitly evaluate it as

$$\Delta\omega = \sqrt{\sqrt{(\Gamma_1^2 - \Gamma_2^2)^2 + 4\Gamma_2^4} - 2\Gamma_2^2}. \tag{18}$$

The parameter dependence of these characteristic values c, $\Delta\omega$ is plotted in Fig. 1. The decrease in c indicates the flatter dispersion relation, while the decrease in $\Delta\omega$ implies the sharper spectral width. Namely, with increasing the odd-frequency pair potential Γ_2 ($< \Gamma_1$), the quasi-particle nature is changed as they become heavier and damping-less. Whereas the normal part Γ_1 makes the damping stronger, it is interpreted as that Γ_2 counteracts Γ_1 to reduce the spectral broadening.

The internal energy can also be evaluated from the diagonal Green's function:

$$E = \langle \mathscr{H} \rangle = \mathrm{Re} \sum_{\boldsymbol{k}} \langle \alpha_{\boldsymbol{k}}^\dagger [\alpha_{\boldsymbol{k}}, \mathscr{H}] \rangle = -\partial_\tau G_{\mathrm{loc}}(\tau)|_{\tau=-0^+}. \tag{19}$$

This expression can be rewritten in the familiar form

$$E = \int \mathrm{d}\omega D(\omega)\omega f(\omega), \tag{20}$$

where $f(\omega) = 1/(\mathrm{e}^{\omega/T} + 1)$ denotes the Fermi distribution function. With the use of the concrete form for the density of states

$$D(\omega) = D_0 \mathrm{Re}\, \frac{|\omega| + \mathrm{i}\Gamma_1}{\sqrt{(|\omega| + \mathrm{i}\Gamma_1)^2 + \Gamma_2^2}}, \tag{21}$$

where D_0 is the bare density of states at Fermi level, we can calculate the specific heat as

$$C = \frac{\partial E}{\partial T} = \frac{\pi^2}{6} D(0) T = \frac{\pi^2 D_0 \Gamma_1}{3\sqrt{\Gamma_1^2 - \Gamma_2^2}} T, \tag{22}$$

which is evaluated at low temperature. This expression shows how the odd-frequency pair potential Γ_2 affects the specific-heat coefficient.

At the end of this section, we consider the characteristic length scale, which corresponds to the Cooper pair radius of bogolons. This is evaluated through the real-space representation of the anomalous Green's function [20]. The length scale is in general written as $\xi \simeq \hbar v_{\mathrm{F}}/E$ where E is a characteristic energy scale of the system. In the present case, this energy is determined as the pole of the anomalous Green's function as a function of $\varepsilon_{\boldsymbol{k}}$. The lowest energy determines the dominant behavior at long distance, so we obtain

$$\xi = \frac{\hbar v_{\mathrm{F}}}{\sqrt{\Gamma_1^2 - \Gamma_2^2}}, \tag{23}$$

where the real-space anomalous Green's function behaves as $F(r \to \infty) \sim \mathrm{e}^{-r/\xi}$. The Fermi velocity for bogolon is given by $v_{\mathrm{F}} = |\partial \varepsilon_{\boldsymbol{k}}/\partial \boldsymbol{k}|_{\varepsilon_{\boldsymbol{k}}=0}$. Here again, the longer correlation length at larger Γ_2 indicates that Γ_2 counteracts Γ_1 to reduce the effect of damping.

3. Comparison with conventional s-wave superconductor

Now we make a contrast against the conventional BCS pairing in the presence of the non-magnetic impurity scattering. We note that the pairing state considered in the previous section is regarded as *odd-frequency s-wave pairing* (spinless), while in the following we consider the usual *even-frequency s-wave spin-singlet pairing* for comparison. The total Hamiltonian is given by

$$\mathscr{H} = \sum_{\boldsymbol{k}\sigma} \varepsilon_{\boldsymbol{k}} c_{\boldsymbol{k}\sigma}^\dagger c_{\boldsymbol{k}\sigma} + \Delta \sum_{\boldsymbol{k}} (c_{\boldsymbol{k}\uparrow}^\dagger c_{-\boldsymbol{k},\downarrow}^\dagger + \mathrm{H.c.}) + \frac{1}{V} \sum_{\boldsymbol{k}\boldsymbol{q}\sigma} \rho_{\boldsymbol{q}} u_{\mathrm{imp}} c_{\boldsymbol{k}+\boldsymbol{q}\sigma}^\dagger c_{\boldsymbol{k}\sigma}, \tag{24}$$

where the local non-magnetic impurity is assumed. We call this Hamiltonian 'the BCS model'. The single-particle energy $\varepsilon_{\boldsymbol{k}}$ is measured from the chemical potential. This impurity Hamiltonian is contrasted against Eq. (2) for bogolons: the off-diagonal impurity term does not appear here. We note that the pair potential Δ is produced so as to decrease the total energy by the spontaneous symmetry breaking. In contrast, the pair potential Γ_2 for bogolons in the last section is just induced by the impurity scattering term without considering energetics.

The Green's functions for the BCS model are defined by

$$G_{\boldsymbol{k}}(\tau) = -\langle \mathcal{T} c_{\boldsymbol{k}\uparrow}(\tau) c_{\boldsymbol{k}\uparrow}^\dagger \rangle, \tag{25}$$

$$F_{\boldsymbol{k}}(\tau) = -\langle \mathcal{T} c_{\boldsymbol{k}\uparrow}(\tau) c_{-\boldsymbol{k}\downarrow} \rangle. \tag{26}$$

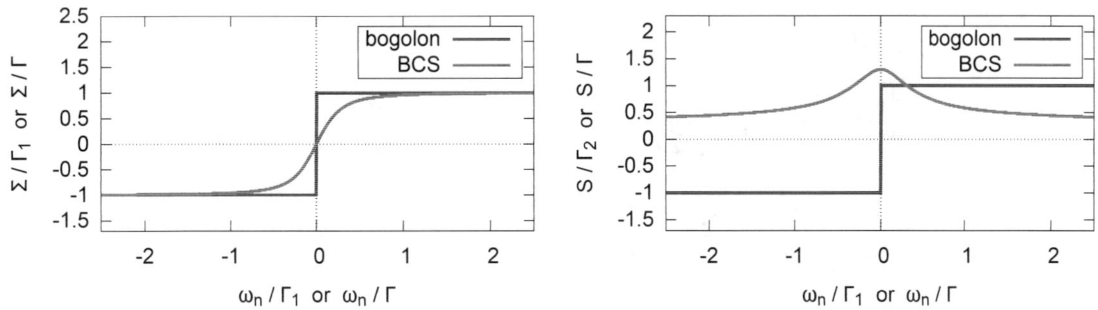

Figure 2. (Left) Normal and (Right) anomalous self-energies for the bogolon model and the BCS model in the presence of the disorder effects. The parameters are $\Delta/\Gamma = 0.3$ for the BCS model.

The inversion and spin-rotational symmetries are assumed. The self-energies are evaluated by the Born approximation as [35]

$$\Sigma(\mathrm{i}\omega_n) = -\Gamma \frac{\mathrm{i}\omega_n}{\sqrt{\omega_n^2 + \Delta^2}}, \tag{27}$$

$$S(\mathrm{i}\omega_n) = \Delta + \Gamma \frac{\Delta}{\sqrt{\omega_n^2 + \Delta^2}}, \tag{28}$$

where $\Gamma = 2\pi n_{\mathrm{imp}} D_0 u_{\mathrm{imp}}^2$ is the scattering rate. These are compared with Eqs. (11)-(14). Figure 2 shows the Matsubara frequency dependence of the normal and anomalous self-energies. For the BCS model, the frequency dependence enters through the impurity effect Γ, while for the bogolon model the self-energies are given by a simple sign function. The important difference is that the BCS model has the even functional form and the bogolon model has the odd functional form.

Once the self-energies are obtained in the frequency domain, the Green's functions are obtained by Eqs. (7) and (8). We can explicitly write down them in the following simple form:

$$G_{\boldsymbol{k}}(\mathrm{i}\omega_n) = \frac{1}{\eta(\mathrm{i}\omega_n)} \frac{\mathrm{i}\omega_n + \frac{\varepsilon_{\boldsymbol{k}}}{\eta(\mathrm{i}\omega_n)}}{(\mathrm{i}\omega_n)^2 - (\frac{\varepsilon_{\boldsymbol{k}}}{\eta(\mathrm{i}\omega_n)})^2 - \Delta^2}, \tag{29}$$

$$F_{\boldsymbol{k}}(\mathrm{i}\omega_n) = \frac{1}{\eta(\mathrm{i}\omega_n)} \frac{\Delta}{(\mathrm{i}\omega_n)^2 - (\frac{\varepsilon_{\boldsymbol{k}}}{\eta(\mathrm{i}\omega_n)})^2 - \Delta^2}, \tag{30}$$

where we have defined the dimensionless function by

$$\eta(\mathrm{i}\omega_n) = 1 + \frac{\Gamma}{\sqrt{\omega_n^2 + \Delta^2}} > 1. \tag{31}$$

From these expressions, we can draw the single-particle spectrum $A_{\boldsymbol{k}}(\omega)$ in Eq. (16). We show the results in Fig. 3. In the BCS model, the full-gap structure is robust against the presence of the disorder. On the other hand, the bogolon model shows the gapless structure and the dispersion relation becomes heavier in addition to the broadening of the spectrum. This behavior is consistent with the estimates in Fig. 1.

We also obtain the density of states as

$$D(\omega) = -\frac{1}{\pi} \sum_{\boldsymbol{k}} \mathrm{Im}\, G_{\boldsymbol{k}}(\omega + \mathrm{i}0^+) = D_0 \mathrm{Re} \frac{|\omega|}{\sqrt{\omega^2 - \Delta^2}}, \tag{32}$$

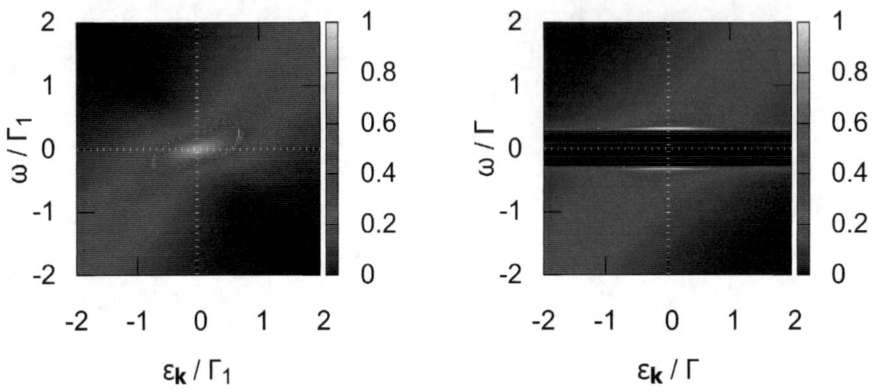

Figure 3. Wavevector resolved single-particle spectra $A_{\boldsymbol{k}}(\omega)$ for (Left) the bogolon model and (Right) the BCS model in the presence the impurity effects. The parameters are $\Gamma_2/\Gamma_1 = 0.8$ for the bogolon model and $\Delta/\Gamma = 0.3$ for the BCS model.

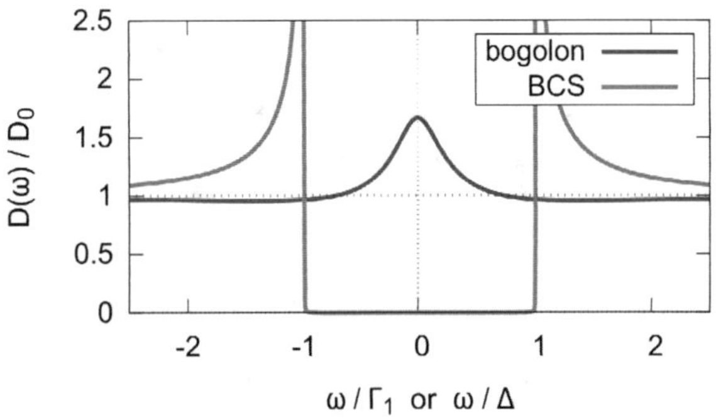

Figure 4. Densities of states for the bogolon model and the BCS model in the presence of the disorder effects. The parameter is chosen as $\Gamma_2/\Gamma_1 = 0.8$ for the bogolon model, while the BCS model does not depend on the impurity scattering strength.

where the effect of impurities does not enter for the BCS model. The functional form is plotted in Fig. 4. Whereas the number of states near the Fermi level is decreased in the BCS model, it is increased in the bogolon model as shown in Fig. 4 [see also Eq.(21)] which demonstrates the peculiarity of the odd-frequency pairing.

With the density of states, the specific heat is also evaluated in a manner similar to the bogolon's case as

$$C = \int \mathrm{d}\omega D(\omega)\omega \frac{\partial f(\omega)}{\partial T} = 2\sqrt{2\pi}\, D_0 \Delta \left(\frac{\Delta}{T}\right)^{3/2} \exp\left(-\frac{\Delta}{T}\right) \tag{33}$$

at low temperatures, which is in contrast with Eq. (22).

Finally, the Cooper pair radius is evaluated for the BCS model as

$$\xi = \frac{\hbar v_{\mathrm{F}}}{\Delta + \Gamma}, \tag{34}$$

which is to be compared with Eq. (23) for bogolons. Note that this length scale is different from the coherence length which determines spatial variation of the gap function with respect to the center-of-mass coordinate. For the BCS model, the length becomes shorter with increasing Δ, which corresponds to the larger characteristic energy scale. In contrast, the bogolon Cooper pair radius in Eq. (23) shows the nontrivial parameter dependence: the length scale becomes longer with increasing Γ_2 ($< \Gamma_1$). All those results for bogolons imply that the characteristic energy scale is $\sqrt{\Gamma_1^2 - \Gamma_2^2}$ instead of $\Gamma_{1,2}$ itself.

4. Summary

We have studied the disorder effect on the Bogoliubov Fermi surface and compared it with the conventional BCS pairing state. Based on the physical quantities such as angle-resolved single-particle spectrum, local density of states, specific heat, and correlation length, we have clarified the characteristic properties of the Bogoliubov quasiparticle with gapless excitation.

Acknowledgement

This work is supported by JSPS KAKENHI Grants No. JP18H01176, No. JP19H01842, and No. JP21K03459.

References

[1] Volovik G E 1989 *Phys. Lett.* A **142** 282
[2] Agterberg D F, Brydon P M R, and Timm C 2017 *Phys. Rev. Lett.* **118** 127001
[3] Volovik G E 1993 *JETP Lett.* **58** 469
[4] Liu W V and Wilczek F 2003 *Phys. Rev. Lett.* **90** 047002
[5] Gubankova E, Mishchenko E G and Wilczek F 2005 *Phys. Rev. Lett.* **94** 110402
[6] Autti S, Mäkinen J T, Rysti J, Volovik G E, Zavjalov V V and Eltsov V B 2020 *Phys. Rev. Res.* **2** 033013
[7] Brydon P M R Agterberg D F Menke H and Timm C 2018 *Phys. Rev.* B **98** 224509
[8] Yuan N F Q and Fu L 2018 *Phys. Rev.* B **97** 115139
[9] Sumita S, Nomoto T, Shiozaki K and Yanase Y 2019 *Phys. Rev.* B **99** 134513
[10] Menke H, Timm C and Brydon P M R 2019 *Phys. Rev.* B **100** 224505
[11] Faraei Z and Jafari S A 2019 *Phys. Rev.* B **100** 035447
[12] Link J M and Herbut I F 2020 *Phys. Rev. Lett.* **125** 237004
[13] Lapp C J, Börner G and Timm C, 2020 *Phys. Rev.* B **101** 024505
[14] Herbut I F and Link J M 2021 *Phys. Rev.* B **103** 144517
[15] Coleman P, Miranda E and Tsvelik A 1993 *Phys. Rev. Lett.* **70** 2960
[16] Hoshino S 2014 *Phys. Rev.* B **90** 115154
[17] Oh H and Moon E-G *Phys. Rev.* B 2020 **102** 020501(R)
[18] Tamura S-T, Iimura S and Hoshino S 2020 *Phys. Rev.* B **102** 024505
[19] Timm C, Brydon P M R and Agterberg D F 2021 *Phys. Rev.* B **103** 024521
[20] Miki T, Tamura S-T, Iimura S and Hoshino S 2021 *Phys. Rev.* B **104** 094518
[21] Oh H, Agterberg D F and Moon E G 2021 *Phys. Rev. Lett.* **127** 257002
[22] Sato Y, Kasahara S, Taniguchi T, Xing X, Kasahara Y, Tokiwa Y, Yamakawa Y, Kontani H, Shibauchi T and Matsuda Y 2018 *Proc. Nat. Acad. Sci.* **115** 1227
[23] Setty C, Bhattacharyya S, Cao Y, Kreisel A and Hirschfeld P J 2020 *Nat. Commun.* **11** 523
[24] For a review, see Shibauchi T, Hanaguri T and Matsuda Y 2020 *J. Phys. Soc. Jpn.* **89** 102002
[25] Coldea A I, Blake S F, Kasahara S, Haghighirad A A, Watson M D, Knafo W, Choi E S, McCollam A, Reiss P, Yamashita T, Bruma M, Speller S C, Matsuda Y, Wolf T, Shibauchi T and Schofield A J 2019 *npj Quant. Mater.* **4** 2
[26] Berezinskii V L 1974 *JETP Lett.* **20** 287
[27] Kirkpatrick T R and Belitz D 1991 *Phys. Rev. Lett.* **66** 1533
[28] Balatsky A and Abrahams E 1992 *Phys. Rev.* B **45** 13125
[29] Emery V J and Kivelson S 1992 *Phys. Rev.* B **46** 10812
[30] Schrieffer J R, Balatsky A V, Abrahams E and Scalapino D J 1994 *J. Supercond.* **7** 501
[31] For a review, Bergeret F S, Volkov A F and Efetov K B 2005 *Rev. Mod. Phys.* **77** 1321
[32] For a review, Tanaka Y, Sato M and Nagaosa N 2012 *J. Phys. Soc. Jpn.* **81** 011013
[33] For a review, Linder J and Balatsky A V 2019 *Rev. Mod. Phys.* **91** 045005
[34] For a review, Cayao J, Triola C and Black-Schaffer A M 2020 *Eur. Phys. J. Spec. Top.* **229** 545
[35] Kopnin N 2001 *Theory of Nonequilibrium Superconductivity* (Oxford: New York) chapter 4 pp 64-73

Nematic pinning of the superconducting state in the doped topological insulator $Sr_xBi_2Se_3$

M P Smylie[1], Z Islam[2], G D Gu[3], S Rosenkranz[4], J Z Dans[1], W-K Kwok[4], and U Welp[4]

[1]Department of Physics and Astronomy, Hofstra University, Hempstead, NY 11549, USA
[2]Advanced Photon Source, Argonne National Laboratory, Lemont, IL 60439, USA
[3]Condensed Matter Physics and Materials Science Department, Brookhaven National Laboratory, Upton, NY 11793, USA
[4]Materials Science Division, Argonne National Laboratory, Lemont, IL 60439, USA

Corresponding author: matthew.smylie@hofstra.edu

Abstract. The discovery of a rotational symmetry breaking, or nematic, state in the superconducting doped topological insulator $M_xBi_2Se_3$ (M = Cu, Sr, Nb) indicates a multicomponent superconducting order parameter which is topological. Many questions remain in this family of materials. In particular, the nematic axis of twofold symmetry is always pinned along one of three rotationally equivalent directions in the threefold symmetric basal plane in the rhombohedral crystal. Crystallographic strain has been proposed as the origin of the choice of pinning axis. Here, we discuss nematic pinning of the superconducting state, summarizing our magnetotransport, magnetization, penetration depth, and calorimetry measurements, which support the theoretical picture of topological superconductivity. We discuss the impact of controlled disorder via proton irradiation on the nematic superconducting state and discuss results of our multimodal technique for searching for crystallographic strain below 1 K via simultaneous diffraction and magnetotransport measurements.

1. Introduction

Superconductivity at comparatively high temperatures arises [1-3] upon doping the well-known topological insulator Bi_2Se_3 with either Cu (T_c = 3.5 K), Nb (3.5 K), or Sr (3.0 K). ARPES measurements confirm that the topological surface states are preserved after doping [4,5] while calorimetry [6-8] and μSR [9-11] measurements confirm that the superconductivity is a bulk effect. The dopants were originally thought to lie in the van der Waals gap between quintuple BiSe layers. However, recent results indicate that the microstructure may be more complicated [12-18]. Of the three members of the family, $Sr_xBi_2Se_3$ has recently received the most attention despite its lower T_c as air stable, mm-scale crystals approaching 100% SC volume fraction can be grown, unlike the other members. All three members demonstrate a striking twofold symmetry in bulk electronic properties below T_c seen by our group and many others via magnetotransport, magnetization, calorimetry, Knight shift, torque magnetometry, and other techniques [19]. This twofold symmetry appears in-plane despite the threefold symmetric crystallographic structure, represented below in Fig. 1(a). Conventional explanations such as vortex motion, anisotropic or spontaneous magnetization, anisotropic magnetoresistance or effective mass effects for the observed behavior have largely been ruled out [20-23].

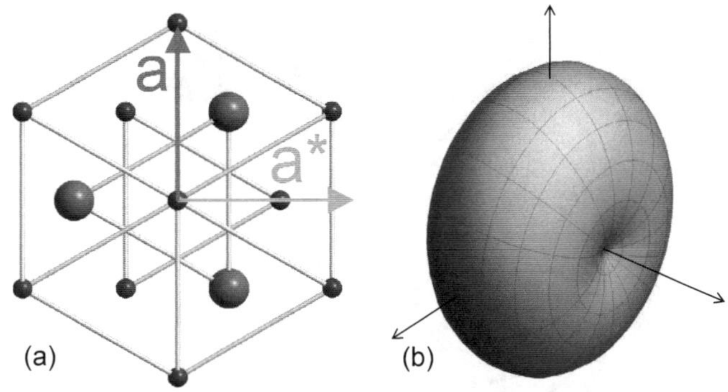

Figure 1. (a) The crystal structure of $M_xBi_2Se_3$, viewed along the c axis. There are three symmetric sets of a and a^* axes, separated by 120°. **(b)** Representation of the Δ_4 pairing state, which breaks rotational symmetry by having either a node along a or a deep minimum along a^*.

Symmetry arguments restrict possible superconducting order parameters to four possibilities [24]: the conventional fully gapped Δ_1, the odd-parity fully gapped Δ_2, the even parity nodal Δ_3, and the odd parity Δ_4 with a node or deep minimum in-plane. The Δ_2 state generated great interest as it was thought to satisfy the conventional criteria for a topological superconductor. Following the first observation of the twofold symmetry in Knight shift measurements [25] on $Cu_xBi_2Se_3$, it was realized that only the odd parity Δ_4 state, which has a two-component order parameter of the E_u representation, was consistent with the broken in-plane symmetry [26]. This introduced the $M_xBi_2Se_3$ family as a new class of unconventional superconductor, where both the superconducting phase and gap amplitude break rotational symmetry of the lattice, and potentially as the first true bulk topological superconductor.

While exciting, this presented a number of theoretical questions. It was first shown [26,27] that crystal anisotropy lifts the continuous degeneracy of the order parameter, forcing a discrete set of degenerate ground states with the nematic director either pinned parallel to (Δ_{4y}) or normal to (Δ_{4x}) the crystallographic mirror plane, resulting in nodes in the superconducting gap for Δ_{4x} and a strongly anisotropic gap with a deep minimum for Δ_{4y}. Furthermore, it was then shown that Ginzburg-Landau theory for a two-component order parameter predicts sixfold basal plane symmetry in a trigonal crystal, not two-fold [28]. Thus, a symmetry breaking pinning field that couples linearly to the order parameter and could then act to lift the sixfold degeneracy has been invoked. A likely deduction is that crystallographic strains are responsible. Thorough investigation of the superconducting properties of bulk $Sr_xBi_2Se_3$ and its homologs will hopefully lead to an answer, and confirmation of the multicomponent order parameter may cement the topological nature of the superconducting state.

2. Probing the superconducting state

The twofold symmetry in the superconducting state in the $M_xBi_2Se_3$ family of compounds has been observed by many groups via several techniques [19]; Fig. 2 demonstrates typical results from our group on samples of $Sr_xBi_2Se_3$ with T_c of ~3.0 K. Fig. 2(a) shows angular dependence of magnetoresistance in an applied field of 1 T slowly swept in-plane; different contours represent 0.1 K increments from 1.6 K (red) to 3.1 K (black); as T increases, isotropic behavior is recovered. Fig. 2(b) shows T_c as a function of applied field angle in-plane derived from $C(T)$ sweeps (blue) in 0.4 T and $M(T)$ sweeps in 0.6 T; T_c values are normalized to $T_{c,max}$ for each measurement type to fit both in the same panel. In all samples for which we have verified the crystallographic axis, the axis of high H_{c2} is along an a axis of the crystal, and no clear evidence for competing nematic domains is observed; other groups report variations batch to batch in the pinning direction of the nematic axis. All agree though that the nematic axis choice is preserved through repeated thermal cycling.

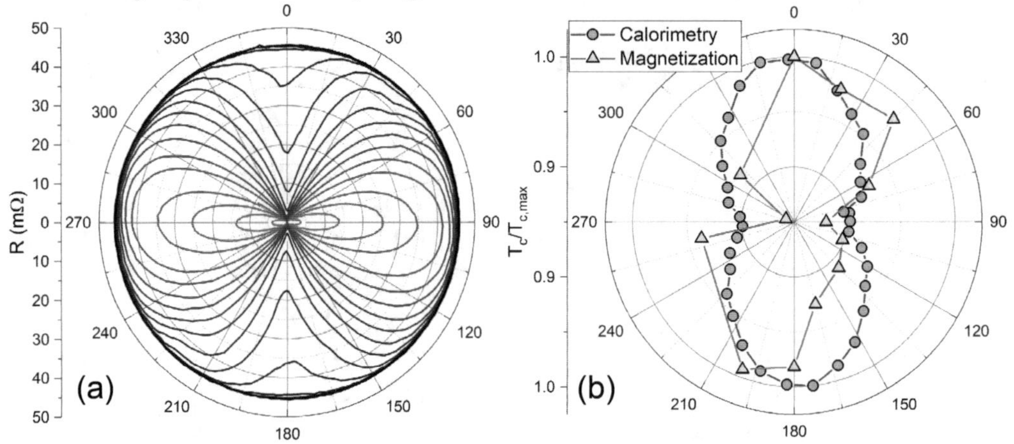

Figure 2. (a) $R(\theta)$ with $\mu_0 H = 1\,T$ swept in-plane at fixed temperatures; colors mark 0.1 K intervals starting at 1.6 K. **(b)** $T_c(\theta)$ from sets of $C(T)$ sweeps in 0.4 T (blue) and of $M(T)$ sweeps in 0.6 T (yellow). 0° marks the crystallographic a axis. In all data, the twofold axis of symmetry is clear.

The low-temperature variation of the penetration depth in the superconducting state is determined by both electron scattering and the distribution of the thermally excited quasiparticles on the Fermi surface. Conventional BCS theory predicts an exponential variation in $\lambda(T)$ at sufficiently low temperatures given by

$$\frac{\Delta\lambda(T)}{\lambda_0} = \left(\frac{\pi\Delta_0}{2T}\right)^{\frac{1}{2}} e^{-\frac{\Delta}{T}} \tag{1}$$

where Δ_0 is the zero-temperature value of the energy gap. Nodes in the gap, however, induce enhanced thermal excitation of quasiparticles, typically resulting [29] in a power law variation of λ: $\lambda(T) \sim T^n$, where the exponent depends on the type of the node; for line nodes, $n = 1$, for point nodes, $n = 2$. A power law fit with $n > 3$ is typically taken as indistinguishable from exponential and thus interpreted as fully gapped. The tunnel diode oscillator (TDO) technique [30] can be used to measure low-temperature variation of the London penetration depth $\Delta\lambda(T)$, as it is linearly proportional to the change in the TDO resonator frequency $\Delta f(T)$, with the rf field of the coil sufficiently low (<50 mG) to ensure the sample remains fully in the Meissner state during measurements. Fig. 3 shows normalized TDO frequency shifts plotted vs reduced temperature T/T_c for (a) a crystal of $Nb_xBi_2Se_3$ with T_c of 3.4 K and (b) a crystal of $Sr_xBi_2Se_3$ with T_c of 2.6 K. For both samples, a T^2 fit (red line) well describes the data, indicating the presence of point nodes [31] in the superconducting gap, consistent with the topological Δ_{4x} superconducting state. It is possible, however, that such data represents a very anisotropic full superconducting gap with a deep minimum well below the minimum temperature scale probed, which would be consistent with the topological Δ_{4y} state; TDO measurements alone are not necessarily sufficient to eliminate this second possibility. All $Nb_xBi_2Se_3$ samples examined show $T_c \sim 3.4$ K and T^2 behavior, whereas in $Sr_xBi_2Se_3$, the fit exponent varies between 1.2 and 2, depending on the bulk T_c. It has been shown that upon doping, the Fermi surface of $M_xBi_2Se_3$ evolves from closed to an open cylinder [32-34] along k_z; this is predicted to distort point nodes in the Δ_{4x} state into line nodes as the Fermi surface opens. The matter is more complicated in $Nb_xBi_2Se_3$, which may have multiple Fermi surfaces [35]. The evolution of the gap structure may provide further evidence for the existence of the topological

Δ_4 states in $M_xBi_2Se_3$, but more work at lower temperatures is required to ascertain the true systematics in $Sr_xBi_2Se_3$ due to its lower T_c.

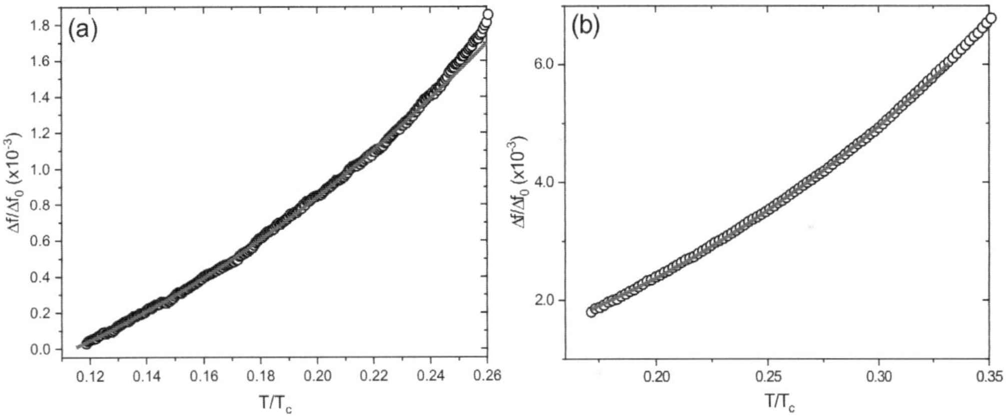

Figure 3. Normalized low-temperature frequency shift $\Delta f(T)$ in a single crystal of **(a)** $Nb_{0.25}Bi_2Se_3$ and **(b)** $Sr_{0.15}Bi_2Se_3$ plotted vs reduced temperature T/T_c. Both measurements show T^2 temperature dependence (red line), consistent with point nodes and thus the Δ_{4x} odd parity pairing state.

3. The effect of disorder

Disorder scattering is a useful tool to distinguish similar superconducting gap structures. In a nodeless anisotropic superconductor, increasing disorder scattering should systematically decrease anisotropy; in a nodal superconductor, increasing disorder should rapidly suppress the superconducting transition while minimally affecting the anisotropy [36-38]. As the nematic anisotropy observed in bulk probes is thought to be gap-driven, magnetotransport is an easy tool to test this. Repeated rounds of irradiation with 5 MeV protons were performed on single crystals of $Sr_xBi_2Se_3$ cut thin enough to ensure the particle beam fully traversed the sample. Fig. 4 shows the angular dependence of normalized magnetoresistance in a single crystal of Sr_xBi_2Se3 with $T_c(H = 0)$ of 3.0 K in an applied field of 1 T slowly swept in-plane on (a) a pristine sample and (b) on the same sample following irradiation-induced suppression of T_c/T_{c0} to 0.84. Colors mark identical temperatures in both panels in 0.1 K steps, from black (1.6 K) to purple (3.1 K). The superconducting anisotropy measured via multiple $R(T, H)$ curves along the maximum and minimum H_{c2} directions (not shown) remains essentially unchanged following repeated irradiations. The nematic axis is clearly unchanged. T_c was continuously, slowly suppressed with repeated irradiation doses, consistent with prior observations on $Cu_xBi_2Se_3$ samples by Ando's group [39] and $Nb_xBi_2Se_3$ samples by our group [40], indicating that $Sr_xBi_2Se_3$ is also anomalously robust against disorder scattering, as described by recent theory [41-45]. This supports the Δ_4 pairing in this compound, but some recent theoretical works [46,47] claim that only the even-parity Δ_1 and Δ_3 states should be robust against impurity scattering. This claim, however, is at odds with the observations of rotational symmetry breaking, which should only occur for the Δ_4 state, unless both states co-exist and compete.

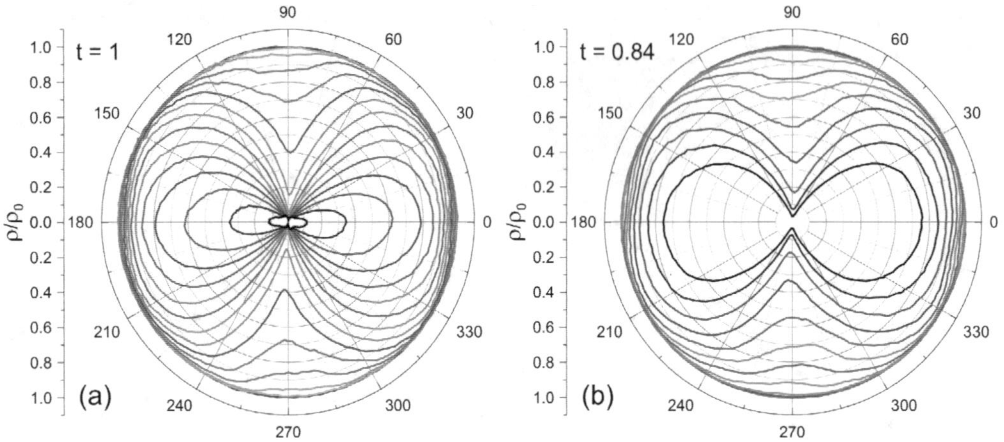

Figure 4. Normalized $R(\theta)$ measurements in a single crystal of $Sr_{0.1}Bi_2Se_3$ with $\mu_0 H = 1\ T$ swept in-plane at fixed temperatures **(a)** before and **(b)** after sufficient proton irradiation to suppress T_c to 84% of its pristine value. Colors represent 0.1 K increments from 1.6 K (black) to 3.1 K (purple). 0° marks the a^* axis of the crystal.

Examining the TDO frequency shift before and after proton irradiation can probe whether the gap is highly anisotropic with a deep minimum below the accessible minimum temperature, or if the gap is nodal. Disorder scattering should lower the anisotropy, increasing the minimum gap value with increasing scattering, making $\Delta\lambda(T)$ flatten with increasing dose. This is not seen in $Nb_xBi_2Se_3$ samples [40] as shown in Fig. 5; when the TDO frequency shift is plotted vs $(T/T_c)^2$, the data remains linear after several rounds of irradiation. This indicates that the gap indeed contains a point node, supporting the Δ_{4x} state. Similar measurements on $Sr_xBi_2Se_3$ will require currently under-development dilution fridge TDO techniques due to the lower pristine T_c.

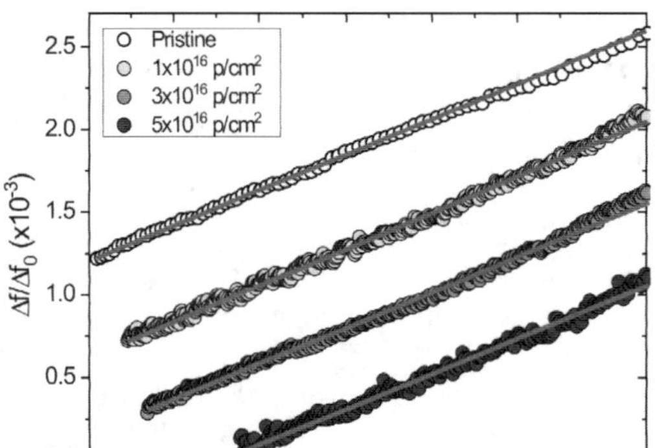

Figure 5. Normalized low-temperature frequency shift Δf plotted vs T^2 following repeated 5-MeV proton irradiation on a thin crystal of $Nb_{0.25}Bi_2Se_3$. The linearity of the plots indicates the T^2 behavior is not due to deep gap minima but rather to point nodes.

4. Looking for crystallographic strain via multimodal measurements

In a previous synchrotron diffraction study by our group [22], we examined high-L diffraction peaks at room temperature to look for violations of the $R\bar{3}m$ space group, finding none. We then considered low-temperature diffraction measurements of high-symmetry, in-plane (hk0) peaks, using a ^{3}He cold finger which allowed for application of magnetic field in-beam at the synchrotron. Due to the beam intensity, the unknown x-ray absorption rate, and the long thermalization pathway to the sample, reaching a base sample temperature below T_c in zero field (~3 K) was not a guarantee. As a result, we configured the cold finger to allow for simultaneous low-current (0.1 mA), 4-point resistivity measurements with the sample in-beam, with the beam spot restricted to between the voltage contacts (made of evaporated Au) on the sample. The crystallographic axes of the single crystal of $Sr_{0.1}Bi_2Se_3$ selected for the study was first verified by Laue measurements. The nematic axis was verified both before and after diffraction measurements via magnetotransport, which found the axis of high-H_{c2} along the crystallographic a axis with no secondary nematic domains.

Diffraction measurements of the (300) peak are shown below in Fig. 6. Simultaneous resistivity measurements showed that even in full beam of $\sim 10^{11}$ γ/sec, the superconducting transition onset was only suppressed by ~0.25 K. We take this to indicate that the base temperature reached on the ^{3}He thermometer, 0.6887 K, was within 0.25 K of the sample temperature, indicating that the base temperature reached was indeed deep in the superconducting state. Fig. 6(a) shows a high-resolution scan of the (300) peak both at base temperature of ~0.7 K and at a temperature, ~5.3 K, discernably above T_c. There is no splitting or shift of the peak that would indicate a monoclinic distortion from the rhombohedral $R\bar{3}m$ structure. Fig. 6(b) shows the same (300) peak at ~2.2 K in different magnetic fields applied along the crystallographic a axis, the axis of high H_{c2}. In zero applied field at this temperature, the sample still had zero resistance. Application of successive fields up to 4 T allowed for measurement of the peak through the transition into the fully resistive state. The same Pearson VII fit (black) essentially describes all the scans, with the fit parameters w and x_c shown in the insets. Again, there is no splitting or shift of the peak that would indicate a monoclinic distortion.

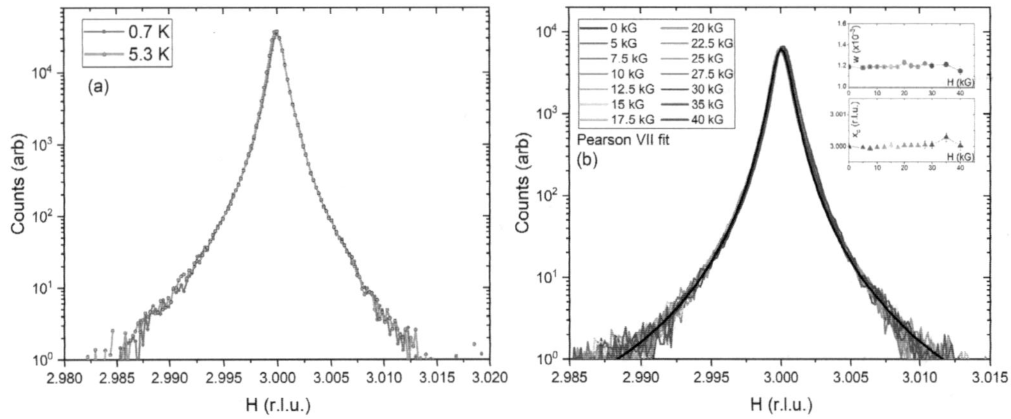

Figure 6. (a) The (300) peak in a single crystal of $Sr_{0.1}Bi_2Se_3$ below and above T_c. No change is observed. **(b)** The (300) peak in the same crystal at $T \approx 2.2\ K$ as a function of field applied along the a axis, the axis of high-H_{c2}. No change is observed as the sample exits the superconducting state; the insets show the Pearson VII fit parameters w and x_c not changing.

5. Conclusions

The lack of observable crystallographic distortion is consistent with recent neutron diffraction measurements on $Cu_xBi_2Se_3$ [48] which saw no distortion but is still a somewhat surprising result given recent dilatometry [49], magnetoresistance [18] and calorimetry [50] measurements which each show broken rotational symmetry at various temperatures just above T_c. The origin of this discrepancy is not yet clear, but it must be pointed out that the dilatometry result claims a $\Delta L/L_0$ of $\sim 10^{-7}$, which is below the resolution of our synchrotron diffraction measurements. It is surprising that such a small strain could cause the observed large superconducting anisotropy while being robust against thermal cycling to 200 °C (for attachment of silver contacts), repeated cycling from room temperature to liquid helium temperatures, and copious proton-induced disorder. This is also somewhat at odds with recent measurements showing [51] slight suppression of nematic domains which required large in-plain applied strain, as one would imagine it would be quite easy to switch nematic pinning axes with strain if the required distortion is so small.

There remain several open questions regarding nematic superconductivity in the $M_xBi_2Se_3$ family. Crystallographic strain was initially suggested as the symmetry breaking field responsible for the nematic axis pinning, but multiple groups have reported somewhat mutually exclusive data. A full structural refinement deep in the superconducting state would be valuable, especially under uniaxial strain, but it may not be possible to reach the zero-strain $\Delta a/a \sim 10^{-7}$ resolution claimed in dilatometry measurements. Regardless of the origin of the nematic pinning, all available evidence points to the existence of either the Δ_{4x} or Δ_{4y} pairing states, both of which are topological; some evidence points to both occurring, raising the curious possibility that the topology of the superconducting gap varies across an otherwise homogenous single crystal. Aside from the large interest in proposed topologically nontrivial qualities, the nematic state is fundamentally a new kind of unconventional superconductor with an additional broken symmetry, and as such deserves vigorous research going forward.

Acknowledgments

Tunnel diode oscillator and magnetotransport measurements at Argonne National Laboratory were supported by the U.S. Department of Energy, Office of Science, Basic Energy Sciences, Materials Sciences and Engineering Division. This research used resources of the APS, a US DOE Scientific User Facility, operated for the DOE Office of Science by Argonne National Laboratory under Contract DE-AC02-06CH11357.

References

[1] Hor Y S, Williams A J, Checkelsky J G, Roushan P, Seo J, Xu Q, Zandbergen H W, Yazdani A, Ong N P and Cava R J 2010 *Phys. Rev. Lett.* **104**, 057001
[2] Qui Y, Sanders K N, Dai J, Medvedeva J E, Wu W, Ghaemi P, Vojta T and Hor Y S 2015 *Preprint* arXiv/1512.03519
[3] Liu Z, Yao X, Shao J, Zuo M, Pi L, Tan S, Zhang C and Zhang Y 2015 *J. Am. Chem. Soc.* **137**, 10512
[4] Wray L A *et al.* 2010 *Nat. Phys.* **6**, 855
[5] Neupane M *et al.* 2016 *Sci. Rep.* **6**, 22557
[6] Yonezawa S, Tajiri K, Nakata S, Nagai Y, Wang Z, Segawa K, Ando Y and Maeno Y 2017 *Nat. Phys.* **13**, 123
[7] Willa K, Willa R, Song K W, Schneeloch S A, Zhong R, Koshelev A E, Kwok W-K and Welp U 2018 *Phys. Rev.* B **98**, 184509
[8] Sun Y, Kittaka S, Sakakibara T, Machida K, Wang J, Wen J, Xing X, Shi Z and Tamegai T 2019 *Phys. Rev. Lett.* **123**, 027002
[9] Leng H, Cherlan D, Huang Y K, Orain J-C, Amato A and de Visser A 2018 *Phys. Rev.* B **97**, 054503
[10] Neha P, Biswas P K, Das T and Patnaik S 2019 *Phys. Rev. Materials* **3** 074201
[11] Das D, Kobayashi K, Smylie M P, Mielke C, Takahashi T, Willa K, Yin J-X, Welp U, Hasan M Z, Amato A, Luetkens H and Guguchia Z 2020 *Phys Rev* B **102**, 134514
[12] Li *et al.* 2018 *Phys. Rev. Materials* **2**, 014201 2018
[13] Callaert C, Bercx M, Lamoen D and Hadermann J 2019 *Acta Cryst* B**75**, 717
[14] Yu S-H, Hung T L, Ou M-N, Chou M M C and Chen Y-Y 2019 *Phys. Rev.* B **100**, 174502

[15] Kuntsevich A Y, Rybal'chenko G V, Martovitskii V P, Bannikov M I, Selivanov Y G, Gavrilkin S Y, Tsvetkov A Y and Chizhevskii G 2020 *JETP Letters* **111**, 151
[16] Dalgaard K J, Kevy S M, Wollesen L, Ma Q, Wiedmann S, Sigfridsson Clauss K G V and Bremholm M 2021 *Phys. Rev.* B **103**, 184103
[17] Lin Y-R, Bagchi M, Soubatch S, Lee T-L, Brede J, Bocquet F C, Kumpf C, Ando Y and Tautz F S 2021 *Phys. Rev.* B **104**, 054506
[18] Kuntsevich A Y, Bryzgalov M A, Prudkoglyad V A, Martovitskii V P, Selivanov Y G and Chizhevskii E G 2018 *New J. Phys.* **20**, 103022
[19] Yonezawa S 2019 *Condens. Matter* **4**, 2
[20] Pan Y, Nikitin A M, Araizi G. K., Huang Y. K, Matsushita Y, Naka T and de Visser A 2016 *Sci. Rep.* **6**, 28632
[21] Du G, Li Y, Schneeloch J, Zhong R D, Gu G D, Lin H and Wen H-H 2017 *Sci. China Phys. Mech. Astron.* **60**, 037411
[22] Smylie M P *et al.* 2018 *Sci. Rep.* **8**, 7666
[23] Mi X, Jing Y, Yang K, Gan Y, Wang A, Chai Y and He M 2021 *Preprint* arXiv/2110.14447
[24] Fu L and Berg E 2010 *Phys. Rev. Lett.* **105**, 097001
[25] Matano K, Kriener M, Segawa K, Ando Y and Zheng G-Q 2016 *Nat. Phys.* **12**, 852
[26] Fu L 2014 *Phys. Rev.* B **90**, 100509(R)
[27] Venderbos J W F, Kozii V and Fu L 2016 *Phys. Rev.* B **94**, 180504(R)
[28] Venderbos J W F, Kozii V and Fu L 2016 *Phys. Rev.* B **94**, 094522
[29] Prozorov R and Giannetta R 2006 *Supercon. Sci. Technol.* **19**, R41
[30] Giannetta R, Carrington A and Prozorov R 2021 London penetration depth measurements using tunnel diode resonators *J. Low Temp. Phys.*
[31] Smylie M P, Claus H, Welp U, Kwok W-K, Qiu Y, Hor Y S and Snezhko A 2016 *Phys. Rev.* B **94** 180510(R)
[32] Lahoud E *et al.* 2013 *Phys. Rev.* B **88**, 195107
[33] Lawson B J, Li G, Yu F, Asaba T, Tinsman C, Gao T, Wang W, Hor Y S and Li, L 2014 *Phys. Rev.* B **90**, 195141
[34] Almoalem A *et al.* 2021 *Phys. Rev.* B **103**, 174518
[35] Lawson B J, Corbae P, Li G, Yu F, Asaba T, Tinsman C, Qiu Y, Medvedeva J E, Hor Y S and Li L 2016 *Phys. Rev.* B **94**, 041114(R)
[36] Gross F, Chandrasekhar B S, Einzel B, Andres K, Hirschfeld P J, Ott H R, Beuers J, Fisk Z and Smith J J 1986 *Z. Phys* B **64**, 175
[37] Borkowski L S and Hirschfeld P J 1994 *Phys. Rev.* B **49** 15404
[38] Fehrenbacher R and Norman M 1994 *Phys. Rev.* B 1994 **50** 3495
[39] Kriener M, Segawa K, Sasaki S and Ando Y 2012 *Phys. Rev.* B **86**, 180505(R)
[40] Smylie M P *et al.*, 2017 *Phys. Rev.* B **96**, 115145 (2017)
[41] Michaeli K and Fu L 2012 *Phys. Rev. Lett.* **109**, 187003
[42] Nagai Y 2015 *Phys. Rev.* B **91**, 060502(R)
[43] Anderson L, Ramires A, Wang Z, Lorenz T and Ando Y 2020 *Science Advances* **6** 9
[44] Cavanagh D C and Brydon P M R 2020 *Phys. Rev.* B **101**, 054509
[45] Cavanagh D C and Brydon P M R 2021 *Phys. Rev.* B **104**, 014503
[46] Huang H, Gu J, Tan M, Wang Q, Ji P and Hu X 2017 *Sci. Rep.* **7**, 45565
[47] Sato T and Asano Y 2020 *Phys. Rev.* B **102**, 024516
[48] Frölich T, Wang Z, Bagchi M, Stunault A, Ando Y and Braden M 2020 *Phys. Rev. Materials* **4**, 054802
[49] Cho C-W *et al.* 2020 *Nat. Commun.* **11** 3056
[50] Shi Z and Tamegai T 2019 *Phys. Rev. Lett.* **123**, 027002
[51] Kostylev I, Yonezawa S, Wang Z, Ando Y and Maneo Y 2020 *Nat. Commun.* **11**, 4152

Crystal structure and superconductivity of $NaAlSi_{1-x}Ge_x$ single crystals

Takahiro Yamada[1,*], Daigorou Hirai[2], Toshiya Ikenobe[2], Hisanori Yamane[1], and Zenji Hiroi[2]

[1]Institute of Multidisciplinary Research for Advanced Materials, Tohoku University, Katahira, 2-1-1 Aoba-ku, Sendai 980-8577, Japan
[2]Institute for Solid State Physics, University of Tokyo, Kashiwanoha 5-1-5, Kashiwa, Chiba 277-8581, Japan

e-mail address: takahiro.yamada.b4@tohoku.ac.jp

Abstract. NaAlSi and NaAlGe are nodal-line semimetals with an anti-PbFCl type structure. The former is a superconductor with a transition temperature T_c of 7 K, while the latter does not exhibit superconductivity. We prepared single crystals of solid solution $NaAlSi_{1-x}Ge_x$ by the Na–Ga flux method and investigated their crystal structures and superconductivity by X-ray structural analysis and magnetization measurements, respectively. The crystal structure parameters vary continuously with x, while the T_c initially decreases gradually from 6.9 K for $x = 0$ to 5.6 K for $x = 0.33$ and then decreases rapidly to below 1.8 K for $x = 0.45$. This sudden variation in T_c at $x \sim 0.4$ may not be explained by the decrease in phonon frequency due to the Ge substitution, but suggests an unusual change in the electronic structure, in spite that the calculated band structures of the end members are almost identical with nearly equal density of states.

1. Introduction

Topological nodal semimetals have attracted great interest because of their intriguing physical properties and technological application potential that could not be achieved with conventional materials [1–4]. The nodal-line semimetals have one-dimensional lines of band crossing points in the Brillouin zone and topologically-protected surface states. The surface states with a "drumhead" shape induce a large density of states (DOS), leading to the appearance of various macroscopic quantum phenomena including superconductivity [3–5]. Experimental studies on the superconductivity of the nodal-line semimetals are still limited except for $PbTaSe_2$ [6–8] and CaAgP [9–11], which have superconducting transition temperatures T_c of 3.72 and 1.7–1.8 K, respectively.

NaAlSi is an *sp* electrons-based superconductor (T_c = 7 K) with an anti-PbFCl type structure, as shown in figure 1 [12–14], and electronic state calculations have shown that it is a clean nodal-line semimetal [15–17]. On the other hand, an isostructural and isoelectronic NaAlGe does not exhibit superconductivity above 1.8 K [13, 14, 17, 18]. It was suggested that the superconductivity of NaAlSi is mediated by high-frequency phonons of Si. Moreover, the absence of

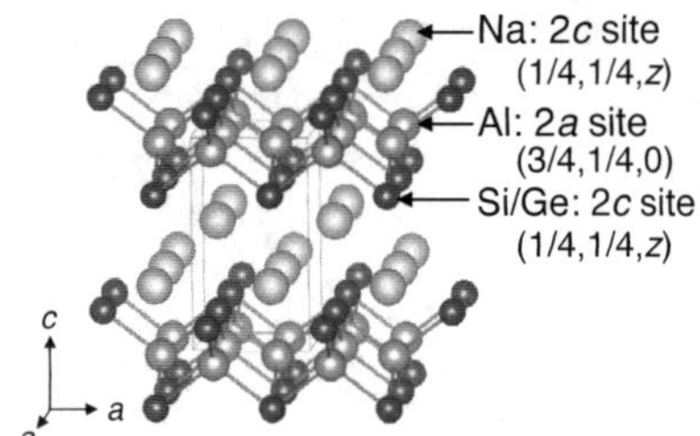

Figure 1. Crystal structure of $NaAlSi_{1-x}Ge_x$.

superconductivity in NaAlGe has been ascribed to the softening of the phonons by the substitution of Ge with a heavier mass [13, 14, 19].

We recently synthesized single crystals of NaAlSi using a Na-Ga flux method and suggested the existence of an anisotropic superconducting gap by measuring heat capacity [20]. In the present study, a series of single crystals of $NaAlSi_{1-x}Ge_x$ solid solution were prepared, and the effects of substituting Ge for Si on the crystal structure and superconductivity were investigated.

2. Experimental procedure

The metals Na, Al, Si, Ge, and Ga were weighed so that the composition ratio was Na:Al:Si:Ge:Ga = $3:1:1-p:p:0.5$ ($p = 0$–1.0); then, they were placed in a BN crucible under an Ar atmosphere and sealed in a stainless-steel reaction container. The container was heated to 1073 K and slowly cooled to 823 K for 80–100 h in an electric furnace. The sample was washed with ethanol to remove excess Na, and the single crystals were taken out from the crucible. The obtained single crystals were subjected to X-ray diffraction (XRD) measurements to refine the crystal structure and composition. The magnetic susceptibility of the single crystals was measured using a magnetic property measurement system (MPMS), and their superconductivity was evaluated. Electronic structure calculation was carried out for NaAlSi and NaAlGe using the "Advance/PHASE" software package based on density functional theory (Advance Soft Corp., Japan) with the lattice parameters and internal atomic coordinates experimentally determined in the present study. The details of the synthesis and measurements are described in a previous paper [20].

3. Results and discussion

3.1. Crystal structure

Plate-like crystals with well-developed {001} planes, primarily 1–4 mm squares with thicknesses of 0.1–0.5 mm, were formed from the raw materials of each composition. The structures of the obtained single crystals were refined by X-ray structure analysis as $NaAlSi_{1-x}Ge_x$ with an anti-PbFCl type structure. Table 1 shows the composition x of each crystal determined from Si/Ge site occupancy. The x values of crystals obtained from starting materials with $p = 0.125$–0.75 were about 10–30% smaller than the p values.

Figure 2 shows the x-dependence of the lattice parameters of the $NaAlSi_{1-x}Ge_x$ crystal and z coordinates of z(Na) and z(Si/Ge). The lattice constants and volumes of the end materials are in close agreement with the recent values reported by Schoop et al. for NaAlSi [21], but there are relatively large deviations from the previous values by Westerhaus and Schuster for NaAlSi and NaAlGe [12]. The lattice parameter and volume generally increase monotonically with increasing the content of larger Ge atoms, indicating that a complete solid solution with a continuously varying Si/Ge ratio was obtained. The z(Na) and z(Si) values for NaAlSi are in close agreement with those of Schoop et al. [21] [z(Na): 0.6346(3), z(Si): 0.20759(15)], but they are 2% larger and 7% smaller, respectively, than the previous values by Westerhaus and Schuster [12] [z(Na):0.622(6), z(Si):0.223(5)]. On the other hand, the values of z(Na) and z(Ge) for NaAlGe differ by 2–3% from those reported by Westerhaus and Schuster [12] [z(Na): 0.618(5), z(Ge): 0.217(5)]. The value of z(Na) in the solid solution varies from 0.63461(8) to 0.63552(11), which is almost constant regardless of the value of x, while the value of z(Si/Ge) gradually increases with increasing x from z(Si) = 0.20764(4) to z(Ge) = 0.21252(2).

In previous band calculations by Rhee et al. [14], it was reported that there are small differences in the Fermi surfaces of the end members using z(Si/Ge) given by Westerhaus and Schuster [12]. Although the shape and size of the main electron and hole surfaces are almost the same, a small electron pocket appears near the M point only in NaAlSi. The presence of this pocket has been suggested as a possible factor to the occurrence of superconductivity [14]. However, subsequent band calculations using the recent z(Si, Ge) values show that the hole pocket does not exist in either case [14–21]. Therefore, the end members have a common crystal structure and nearly identical Fermi surfaces.

Table 1. Crystal structure and superconducting properties of the $NaAlSi_{1-x}Ge_x$ single crystals prepared from the starting materials with a composition of Na:Al:Si:Ge:Ga = 3:1:1 − *p*:*p*:0.5.

Nominal composition *p*	0	0.125	0.25	0.375	0.50	0.75	0.875	1.0
Actual composition *x*	0	0.094(5)	0.181(2)	0.332(2)	0.446(2)	0.625(6)	0.874(8)	1.0
a-axis length/ Å	4.12170(10)	4.1226(4)	4.1329(4)	4.1367(3)	4.14010(10)	4.1382(2)	4.1479(2)	4.1634(2)
c-axis length/ Å	7.3629(2)	7.3758(8)	7.3913(6)	7.3894(2)	7.3868(2)	7.4045(5)	7.4300(4)	7.4146(4)
V / Å^3	125.084(7)	125.99(3)	126.48(2)	126.657(7)	126.558(7)	126.800(15)	127.834(14)	128.524(14)
z(Na)	0.63461(8)	0.6348(3)	0.63487(11)	0.63510(9)	0.63512(7)	0.6354(2)	0.6352(2)	0.63552(11)
z(Si/Ge)	0.20764(4)	0.20898(12)	0.20985(4)	0.21099(3)	0.21140(2)	0.21211(5)	0.21270(4)	0.21252(2)
T_c / K	6.9	6.8	6.0	5.6	< 1.8	3.0	3.6	< 1.8
VF* / %	120	123	123	110	0	3	19	0

* Shielding volume fraction

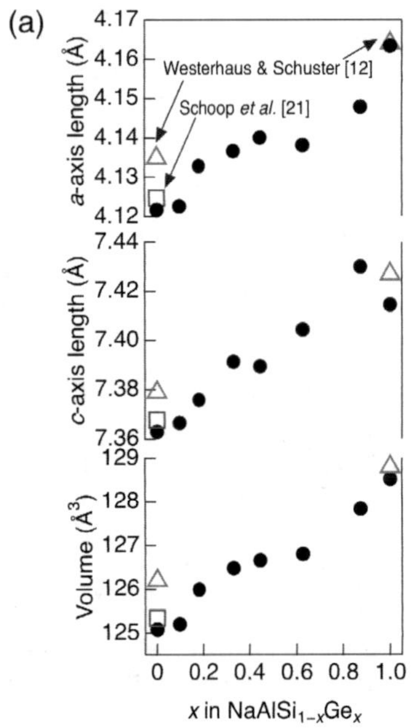

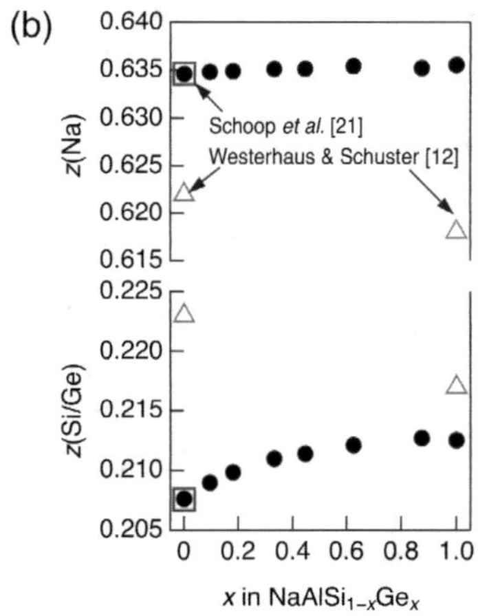

Figure 2. Lattice parameters (a) and the z coordinates of Na and Si/Ge sites (b) for $NaAlSi_{1-x}Ge_x$. The reported values in references 12 (x = 0 and 1.0) and 21 (x = 0) are shown by open triangles and squares, respectively.

3.2. Superconductivity

Figure 3 shows the temperature dependence of the magnetization of $NaAlSi_{1-x}Ge_x$ single crystals under an applied magnetic field of 1 mT. Single crystals with x = 0, 0.09, 0.18, and 0.33 show large diamagnetic responses with the T_c gradually decreasing. Their shielding volume fraction (VF) at 1.8 K is over 100% and the Meissner volume fraction is 5–9%, which demonstrates that bulk superconductivity occurs in these samples. On the other hand, single crystals of x = 0.45 and 1.0 (NaAlGe) do not exhibit diamagnetism above 1.8 K. NaAlGe was reported to be a non-superconductor in a previous report [13] (no measurement data were given), and it is certainly confirmed that NaAlGe does not exhibit superconductivity at 1.8 K or above. Weak diamagnetism is observed below 3.0 and 3.6 K, respectively, in the crystals with interim x = 0.63 and 0.88.

Figure 4 shows the x-dependence of the T_c and VF of $NaAlSi_{1-x}Ge_x$ deduced from magnetization measurements. T_c decreases gradually with increasing x from 6.9 K for x = 0 to 5.6 K for x = 0.33, and then decreases sharply to below 1.8 K for x = 0.45. In addition, superconductivity seems to be revived in the Ge-rich compositions (x = 0.63 and 0.88), but the shielding volume fraction is small, which makes it hard to call it bulk superconductivity. The cause of the appearance of partial superconductivity in the Ge-rich compositions is unknown. It is possible that there are regions of Si-rich composition that demonstrate microscopic superconductivity in the crystals, or that they are caused by fluxes or impurities incorporated in the crystals.

The disappearance of superconductivity in the $x = 0.33$–0.45 region is not easy to understand. It seems that there is a critical composition x_c at ~0.4. The refined crystal structure parameters continuously change at x_c. Furthermore, as shown in figure 5, the NaAlSi and NaAlGe band structures calculated using the crystal structure parameters obtained in this study are almost the same with each other, as in previous reports [14–21]; particularly, the DOSs are nearly equal. Thus, it is unlikely that any essential change in the electronic state occurs at x_c. On the other hand, if the superconductivity in NaAlSi is due to the phonon mechanism, the behavior of the gradual decrease of T_c with x may be because the phonons responsible for the superconductivity include the contributions of Si and Ge atoms, and the phonon frequency is decreased by the substitution of higher atomic mass Ge [13, 14]. However, it is difficult to explain the abrupt decrease in T_c at x_c by continuous changes in phonon frequency. Therefore, different factors other than phonons may play a major role in the superconductivity of $NaAlSi_{1-x}Ge_x$.

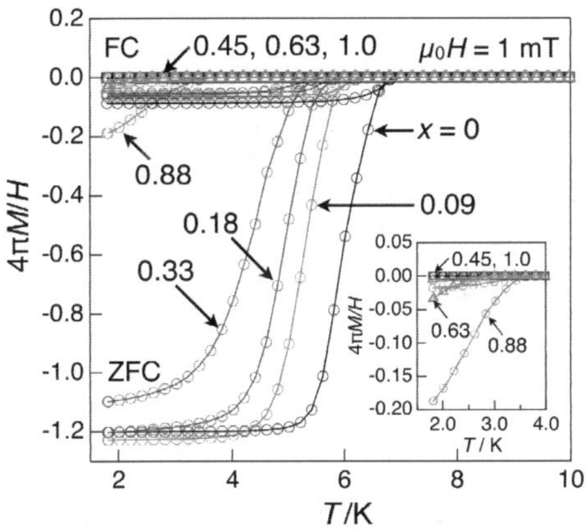

Figure 3. Magnetic susceptibility of the $NaAlSi_{1-x}Ge_x$ single crystals measured at 1 mT. The inset shows an enlarged view around the onset of superconducting transition.

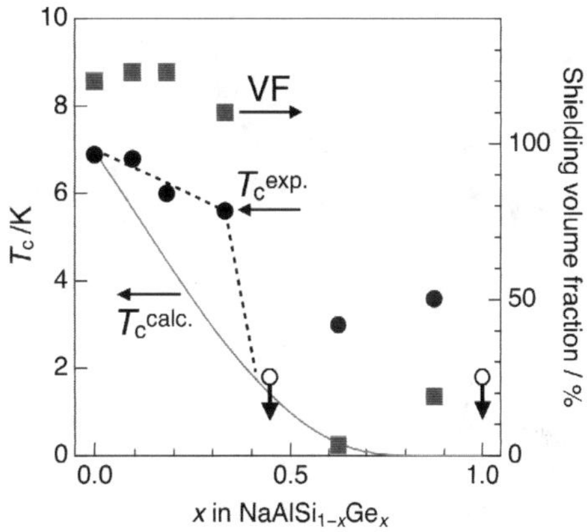

Figure 4. T_c defined from the onset of the M–H curves (solid and open circles), T_c calculated using the McMillan formula (red line), and the superconducting shielding volume fraction (solid squares) for the single crystals of $NaAlSi_{1-x}Ge_x$.

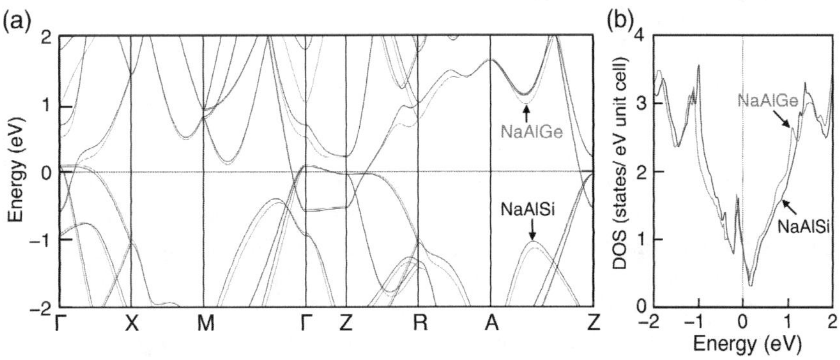

Figure 5. Band structures (a) and the DOS (b) of NaAlSi (black lines) and NaAlGe (red lines) calculated on basis of the structural parameters determined in this study.

The T_c in weakly-coupled BCS superconductivity depends on the phonon frequency and DOS at the Fermi level [22]. Since the phonon frequency should gradually decrease with the Ge substitution, the observed drop in T_c around x_c may be due to a decrease in DOS. This is the most plausible hypothesis because T_c is exponentially dependent on DOS. However, according to the band structure calculations (figure 5), the DOS values at the Fermi level of the end members are in close agreement with 0.90 and 0.84 states per eV and unit cell, respectively, and there is no reason for a significant decrease in DOS at intermediate compositions. Nevertheless, in our very recent experimental study, we found that NaAlGe is not a simple metallic conductor as expected from the band structure calculation but has a unique electronic state with a pseudogap [23]. The Sommerfeld coefficients determined from heat capacity measurements were 2.17 mJ K^{-2} mol^{-1} for NaAlSi and 0.44 mJ K^{-2} mol^{-1} for NaAlGe, indicating that the DOS of NaAlGe is enormously reduced to 20% that of NaAlSi [20, 23].

The composition dependence of T_c, calculated by assuming that DOS linearly decreases with Ge composition x, is plotted in figure 4. In this calculation, following the previously reported evaluation of T_c for NaAlSi [19], the T_c of the solid solution was calculated using the Allen–Dynes modification of the McMillan formula, [24, 25] as given below:

$$T_c = \frac{\omega_{\ln}}{1.2} \exp\left(-\frac{1.04(1+\lambda)}{\lambda-\mu^*(1+0.62\lambda)}\right), \quad (1)$$

The parameters used for NaAlSi were the electron–phonon interaction strength $\lambda = 0.68$, the screened Coulomb pseudo-potential parameter $\mu^* = 0.10$, the logarithmically-averaged phonon frequency $\omega_{\ln} =$ 216 K [19]. The λ of the solid solution is assumed to decrease in proportion to the decrease in the DOS. Here, the change of electron–phonon interaction and the decrease of $\omega_{\ln}$ due to the Ge substitution are ignored, and μ^* is taken to be always 0.10. The calculated x_c, above which T_c becomes less than 1.8 K, is 0.42, which is consistent with the experimental results. However, the initial slope of the change in T_c is much larger than the experimental value. This discrepancy suggests that the change in DOS is not linear with x, but decreases gradually in the region of small x values, and drops sharply in the vicinity of the middle composition. Therefore, the composition dependence of T_c for $NaAlSi_{1-x}Ge_x$ can be explained by taking into account a suitable decrease in DOS.

The incompatibility of the electronic state of NaAlGe with the band structure calculation results remains a mystery. A larger spin–orbit interaction (SOI) than that of NaAlSi may be contributing to the electronic state of NaAlGe with Ge heavier than Si. However, even when SOI is considered, the calculated change in the band structure of NaAlGe is small [17, 18], and the cause of the decrease in DOS and pseudo gap formation is unknown. It is expected that some interactions, which are not taken into account in the band structure calculation, play an important role. The system appears to be a simple quasi-two-dimensional *sp* electron system, but it may hide important physics beyond our current understanding.

4. Conclusions

Single crystals of the $NaAlSi_{1-x}Ge_x$ solid solution were synthesized using Na–Ga flux. The crystal structure parameters of the compounds of each composition were determined using the single crystal X-ray diffraction method. The refined crystal structure parameters varied continuously with x. The superconducting transition temperature determined by magnetization measurements decreases gradually with increasing Ge substitution for Si from 6.9 K for $x = 0$ to 5.6 K for $x = 0.33$, and then decreases rapidly to below 1.8 K for $x = 0.45$. Single crystals of NaAlGe with $x = 1.0$ also did not exhibit superconductivity above 1.8 K. The suppression of superconductivity in the solid solution cannot be explained by the decrease of phonon frequency but may be related to an unusual electronic state on the Ge rich side.

Acknowledgments

The authors would like to thank Yumi Oikawa, Reina Kusaka, and Chikako Nagahama for their help in the sample preparation. This research was financially supported by JSPS KAKENHI Grants (20H05150 and 20H02820).

References

[1] Hasan M Z and Kane C L 2010 *Rev. Mod. Phys.* **82** 3045
[2] Qi X -L and Zhang S -C 2011 *Rev. Mod. Phys.* **83** 1057
[3] Burkov A A 2016 *Nat. Mater.* **15** 1145
[4] Schoop L M, Pielnhofer F and Lotsch B V 2018 *Chem. Mater.* **30** 3155
[5] Nandkishore R 2016 *Phys. Rev.* B **93** 020506
[6] Ali M N, Gibson Q D, Klimczuk T and Cava R J 2014 *Phys. Rev.* B **89** 020505
[7] Bian G, Chang T -R, Sankar R, Xu S -Y, Zheng H, Neupert T, Chiu C -K, Huang S -M, Chang G, Belopolski I, Sanchez D S, Neupane M, Alidoust N, Liu C, Wang B, Lee C -C, Jeng H -T, Zhang C, Yuan Z, Jia S, Bansil A, Chou F, Lin H and Hasan M Z 2016 *Nat. Commun.* **7** 10556
[8] Guan S-Y, Chen P -J, Chu M. W, Sankar R, Chou F, Jeng H -T, Chang C -S and Chuang T -M 2016 *Sci. Adv.* **2** e1600894.
[9] Yamakage A, Yamakawa Y, Tanaka Y and Okamoto Y 2015 *J. Phys. Soc. Jpn.* **85** 013708.
[10] Okamoto Y, Inohara T, Yamakage A, Yamakawa Y and Takenaka K 2016 *J. Phys. Soc. Jpn.* **85** 123701
[11] Okamoto Y, Saigusa K, Wada T, Yamakawa Y, Yamakage A, Sasagawa T, Katayama N, Takatsu H, Kageyama H and Takenaka K 2020 *Phys. Rev.* B **102** 115101
[12] Westerhaus W and Schuster H U 1979 *Z. Naturforsch.* B **34** 352
[13] Kuroiwa S, Kawashima H, Kinoshita H, Okabe H and Akimitsu J 2007 *Physica* C **466** 11
[14] Rhee H B, Banerjee S, Ylvisaker E R and Pickett W E 2010 *Phys. Rev.* B **81** 245114
[15] Muechler L, Guguchia Z, Orain J C, Nuss J, Schoop L M, Thomale R and von Rohr F O 2019 *APL Mater.* **7** 121103
[16] Jin L, Zhang X, He T, Meng W, Dai X and Liu G 2019 *J. Mater. Chem.* C **7** 10694
[17] Yi X, Li W Q, Li Z H, Zhou P, Ma Z S and Sun L Z 2019 *J. Mater. Chem.* C **7** 15375
[18] Wang X, Ding G, Cheng Z, Suruc G, Wang X L and Yang T 2020 *Adv. Res.* **23** 95
[19] Tütüncü H M, Karaca E and Srivastava G P 2016 *Philos. Mag.* **96** 1006.
[20] Yamada T, Hirai D, Yamane H and Hiroi Z 2021 *J. Phys. Soc. Jpn.* **90** 034710
[21] Schoop L, Müchler L, Schmitt J, Ksenofontov V, Medvedev S, Nuss J, Casper F, Jansen M, Cava R J and Felser C 2012 *Phys. Rev.* B **86** 174522
[22] Tinkham M 1996 *Introduction to superconductivity* (New York: Dover Publications, inc)
[23] Yamada T, Hirai D, Yamane H and Hiroi Z (submitted)
[24] McMillan W L (1968) *Phys. Rev.* **167** 331
[25] Allen P B and Dynes R C (1975) *Phys. Rev.* B **12** 905

Electrical Transport and Magnetic Properties of High-Quality $Sr_xBi_2Se_3$ Single Crystals Obtained by Annealing

T Matsumae, S Pyon and T Tamegai

Department of Applied Physics, The University of Tokyo, 7-3-1 Hongo, Bunkyo-ku, Tokyo 113-8656, Japan

E-mail: matsumae-takumi817@g.ecc.u-tokyo.ac.jp

Abstract. Topological insulator Bi_2Se_3 becomes topological superconductors by doping Cu or Sr. These materials attract much interest because of the nematic state showing two-fold symmetry when the magnetic field is applied parallel to *ab*-plane. In this study, single crystals of $Sr_xBi_2Se_3$ ($0.1 \leq x \leq 0.3$) were grown by the self-flux method followed by annealing process, which improved superconducting properties. Using these high-quality samples, in-plane magnetic-field-angle dependence of resistivity was measured. The result showed two-fold symmetry and indicated the existence of nematic domains. In addition, magnetoresistance as a function of in-plane field direction was measured and is discussed with regard to crystal inhomogeneity. These results help us to perform future experiments of direct visualization of nematic domains.

1. Introduction

Topological insulator Bi_2Se_3 becomes topological superconductors by doping Cu or Sr [1, 2]. The crystal structure of topological superconductors $A_xBi_2Se_3$ (A = Cu, Sr) have three-fold rotational symmetry in its *ab*-plane (Fig. 1). However, their physical quantities such as resistivity, upper critical field, and specific heat exhibit two-fold symmetry below T_c when magnetic field is applied parallel to *ab*-plane [3, 4]. This spontaneous symmetry lowering is called nematicity, the mechanism of which have not been elucidated yet. A scenario is suggested that the nematicity is attributed to the possible superconducting gap symmetry Δ_{4x} or Δ_{4y} [5, 6]. Since these superconducting states are related to topological superconductivity due to their odd parity, the nematicity should be a consequence of the topological feature in this scenario. Another scenario is that the symmetry lowering is caused simply by the crystal structure distortion [7]. In fact, crystal axis elongation was confirmed by a high-resolution XRD measurements, although the degree of the distortion was not large enough to explain the anisotropy with $\gamma \sim 4$. However, these measurements were performed at room temperature, and thus the possibility of huge distortion around 2 K is not fully excluded. The third scenario is that Sr ordering causes the nematicity, although it was not seen in STM observations [8]. For experimental explorations of these scenarios, the sample quality is quite important. However, we should point out that the existence of large inhomogeneity of dopants can be an obstacle for future research. For example, the direct visualization of nematic domains which show different nematic orientations is one of the fascinating topics because it may tell us what happens at the interface between the domains. Such a measurement will require high homogeneity of crystals because we have to distinguish two kinds of domains originating from different selection of nematic orientations and from mixture of crystallites with different crystal axis directions. It is the former that is physically interesting, and thus highly homogeneous single crystals should be prepared.

Here, Sr is chosen as a dopant and the following experiments (I) and (II) were performed focusing on Sr inhomogeneity. (I) We investigated crystal growth condition dependence of T_c. (II) We explored the presence of current path bending observed in the in-plane magnetic-field-angle dependence of resistivity, which is relevant to visualization of nematic domains. The existence of nematic domains is expected and has been suggested in some experiments [7, 9]. As mentioned above, direct visualization of nematic domains will be the next goal since their presence is a natural consequence of nematic superconductivity if the observed two-fold symmetry is not an artifact due to extrinsic effects such as distortion of the crystal [7]. One of the feasible ways of the visualization of nematic domains is to use magneto-optical (MO) imaging. In this method, magnetic field distribution can be imaged by utilizing the Faraday effect of an indicator film. A specific magnetic field distribution is expected to be generated by the current distribution following the nematic domain pattern. Alternatively, nematic domain distribution can be imaged by detecting transport current paths. However, current path bending was pointed out to explain anomalous in-plane magnetic-filed-angle dependence of magnetoresistance above T_c [7]. This current bending is one of the obstacles to MO imaging. Here, the origin of this anomalous effect was explored and discussed.

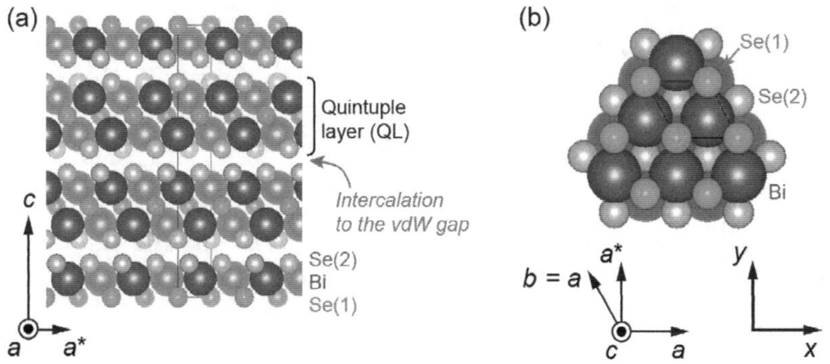

Figure 1. Crystal structure of Bi_2Se_3. (a) View from the a-axis direction. The intercalated metallic ions most likely sit in the van der Waals (vdW) gap between the quintuple layers (QL). (b) View from the c-axis direction [10].

2. Experimental Methods

2.1. Crystal growth

Single crystals $Sr_xBi_2Se_3$ ($x = 0.17$) were obtained by direct reaction of constituent elements. Sr pieces and Bi_2Se_3 powder (prepared in advance) were sealed into a quartz tube and heated in an electrical furnace. After reaching to the maximum temperature 1000°C, it was cooled down to 700°C (sample #1) or 650°C (sample #2 and #3) at a rate of 3°C/h. Samples #1 and #2 were quenched into water and sample #3 was slowly cooled down from 650°C to room temperature after the furnace was shut down (furnace cooling). Typical dimensions of obtained crystals are 1 mm × 1 mm × 0.3 mm. These crystals were additionally annealed in vacuum (< 0.05 Torr) at 560°C for 24 h, 48 h, and 96 h.

2.2. Measurements

Magnetization of each sample was measured by a superconducting quantum interference device (SQUID) magnetometer (MPMS-XL5, Quantum Design). Crystal axes were determined by X-ray diffractometer (Smartlab, RIGAKU). In-plane magnetic-field-angle dependences of sample #3-5 were measured by AC four-probe method using physical property measurement system (PPMS, Quantum Design). In this measurement system, the crystal was set on a rotator which can be rotated around the horizontal axis, while the direction of magnetic field was fixed perpendicular to the rotation axis. The crystal was mounted on the rotator so that its plane is perpendicular to the rotational axis. The angle between a-axis and transport current was 4° (almost parallel) and the current was applied along the longer edge of the sample (2.0 mm × 1.2 mm × 0.28 mm). In this measurement, the applied AC current amplitude was set to 1 mA, and its frequency was 17 Hz.

3. Results and Discussion

3.1. Magnetization

Temperature dependence of magnetization is shown in Fig. 2 (a). The critical temperature (T_c) of samples #1 (T_c = 2.4 K) are lower than that of samples #3 (T_c = 3.0 K). The latter T_c is as high as the highest value reported in Ref. [11] (T_c = 3.1 K), which is against the conventional idea that quenching process is essential for superconductivity as was demonstrated for Cu-doped Bi_2Se_3 [12]. The difference between Cu-doping and Sr-doping might be the consequence of different ionic radii. Ionic radius of Sr (1.18 Å) is larger than that of Cu (0.73 Å). This indicates that Sr ions cannot move easily in a crystal compared with Cu ions. As a result, quenching, which fixes the position of ions in the crystal, is not necessary for the Sr-doping case. In fact, Cu-doped Bi_2Se_3 loses its superconductivity when it is kept in air [1], which means that Cu atoms tend to escape from the sample due to their mobility. In Sr-doped samples, improvement of Sr homogeneity by slow cooling positively affects to increase T_c, while fixing the position of Cu ions by quenching is more important in Cu-doped samples.

Temperature dependence of magnetization after the annealing is shown in Figs. 2 (b), (c), and (d). Annealing for ∼ 96 h effectively improved the superconducting transition width, although T_c did not change so much. This should have been caused by Sr distribution that became more homogeneous after annealing. Superconducting volume fraction of the sample #3-4 was 73%. Here, diamagnetization coefficient of n = 0.70 calculated by the sample geometry was used. Since this superconducting volume fraction is as high as those reported in [2], the quality of samples #3 is good in terms of T_c, superconducting transition width, and volume fraction. From these results, it is concluded that furnace cooling followed by annealing at 500°C - 600°C for ∼ 96 h is effective to obtain high-quality samples of $Sr_xBi_2Se_3$.

3.2. In-plane magnetic-field-angle dependence

The main panel of Fig. 3 (a) shows magnetic-field dependence of resistivity at 2.0 K, and the inset shows temperature dependence of resistivity for sample #3-4 (T_c = 2.6 K) at zero field. Based on the magnetic-filed dependence, upper critical field H_{c2} is evaluated using a criterion $\rho/\rho_n = 0.9$, where ρ_n = 0.45 mΩcm is used (Fig. 3 (b)). It is confirmed that H_{c2} exhibited a clear two-fold symmetry. The principal axis of nematicity, where H_{c2} takes maximum value, is along a-axis ($\varphi = 0°$ and 180°). This direction is identical to those reported in [4,7,13–15], while others reported the direction perpendicular to a-axis [7,14,16]. These two directions correspond to which of Δ_{4x} or Δ_{4y} occurs as a superconducting gap symmetry. This issue is important to topological superconducting features but has not been solved yet. Figure 4 (a) shows polar plot of resistivity at 1 T and at different temperatures. Two-fold symmetry is evident, although small dips indicated by blue arrows are also observed. These dips indicate that the crystal is inhomogeneous, i.e., crystallites with different crystal orientations are likely to be present in

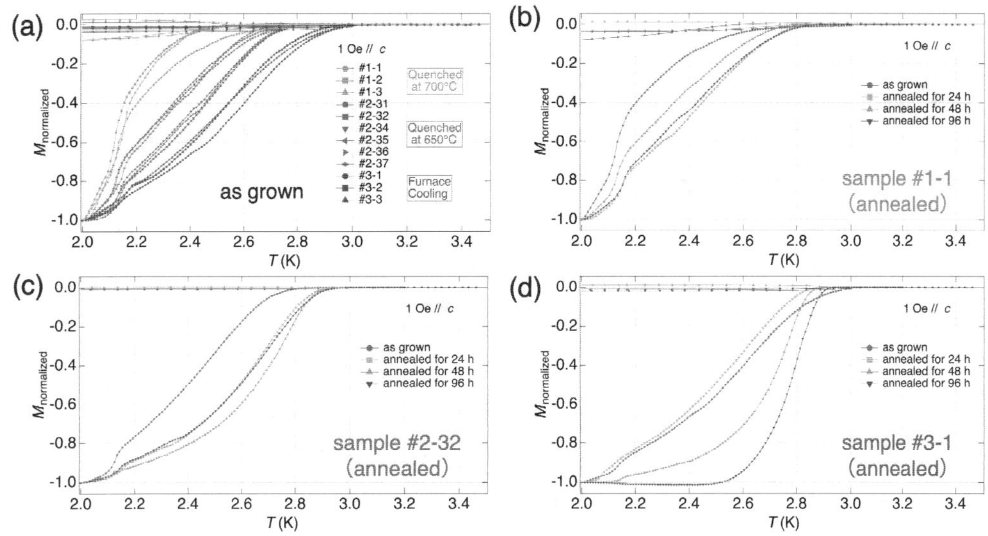

Figure 2. (a) Temperature dependence of magnetization of as-grown samples. Temperature dependence of magnetization of annealed samples (b) #1-1, (c) #2-32, and (d) #3-1. Applied field is 1 Oe and magnetization are normalized by the value at 2.0 K.

the sample. These small dips might have been observed in angular dependence of H_{c2} as well. However, it is difficult to confirm them due to lack of data points. Figure 4 (b) shows an angular dependence of resistivity by subtracting a constant value of $\tilde{\rho} = 0.4$ mΩcm from ρ in order to emphasize the behavior close to the normal state resistivity. The existence of dips at $\varphi \sim 300°$ (shown by purple arrows) are confirmed. This angle is rotated by 60° from $\varphi = 0°$, which may suggest the existence of nematic domains. Thus, this particular sample piece can be a candidate for future measurement of nematic domains.

Angular dependence of magnetoresistance at 2.0 K and 8.5 T is shown by green triangles in Fig. 5 (a). Here, $\Delta\rho$ is $\rho - \rho_0$ and ρ_0 is resistivity at zero field evaluated from fitting (described below). This peculiar behavior is explained by Fourier series expansion; the total resistivity is decomposed into a constant, the first harmonic component, and the second harmonic component as shown by Eq. (1) [7].

$$\rho(B, \varphi) = \rho_0 + \rho_1(B)\cos(\varphi + \alpha_1) + \rho_2(B)\cos(2\varphi + \alpha_2) \quad (1)$$

The second harmonic component represents a usual magnetoresistance behavior. On the other hand, the first harmonic component can be interpreted as a term originating from the Hall effect. Since each wire for I-V measurement was contacted on the surface of the sample, Hall resistivity should not be observed ideally. However, the sample is inhomogeneous to some extent as pointed out above. So, the transport current may flow along c-axis locally due to the presence of defects. In this picture, current, magnetic field, and the direction of voltage are perpendicular to each other, and Hall signal can be generated. In order to examine this scenario, magnetic field dependences of ρ_1 and ρ_2 are plotted in Fig. 5 (b). The fitting lines are $\rho_1 = \beta_1 B^{p_1}$ and $\rho_2 = \beta_2 B^{p_2}$, respectively. The fact that $\rho_1(B)$ is proportional to B ($p_1 = 1.0$) strongly supports the interpretation that ρ_1 is originated from Hall resistivity. On the other hand, $p_2 = 1.6$, which is close to 2, is expected for usual magnetoresistance. The presence of unintentional c-axis current strongly suggests that the sample is still not homogeneous enough. Further improvements of sample quality may be achieved by some methods other than self-flux method. For example, floating zone method was tried for Cu-doped Bi_2Se_3 [12], although the sample quality was worse than the self-flux method. However, it may not be the case with Sr-doped Bi_2Se_3 because of the different sizes of dopant ions.

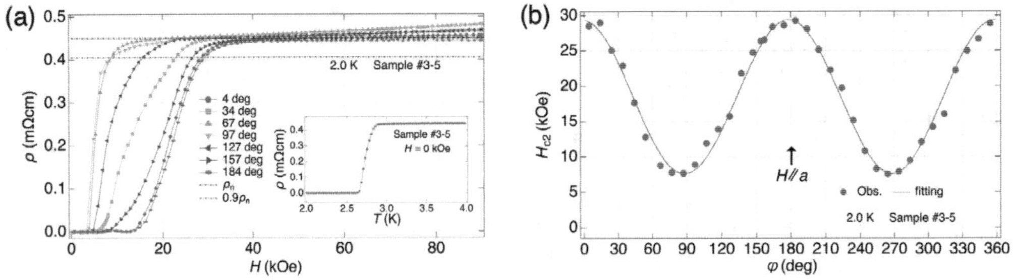

Figure 3. (a) Magnetic field dependence of resistivity at 2.0 K for different directions of in-plane magnetic field. $\varphi = 0°$ and $180°$ corresponds to $H \parallel a$. Inset of (a) shows temperature dependence of resistivity at 0 Oe, showing $T_c^{zero} = 2.6$ K. (b) In-plane magnetic-field-angle dependence of upper critical field H_{c2}. H_{c2} is defined at 90% of normal state resistivity. The fitting curve is $A\cos(2\varphi) + B\sin(2\varphi) + C$.

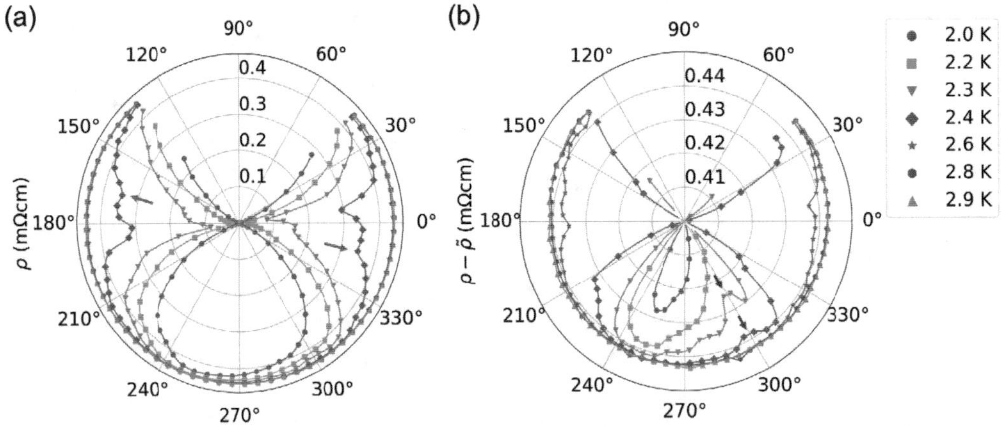

Figure 4. (a) In-plane magnetic-field-angle dependence of resistivity at 1 T. Dips shown by blue arrows are caused by crystallites with different crystal orientation. (b) The same data as (a) plotted after subtracting a constant value of $\tilde{\rho} = 0.4$ mΩcm from ρ to emphasize small dips at $\varphi = 300°$ (shown by purple arrows).

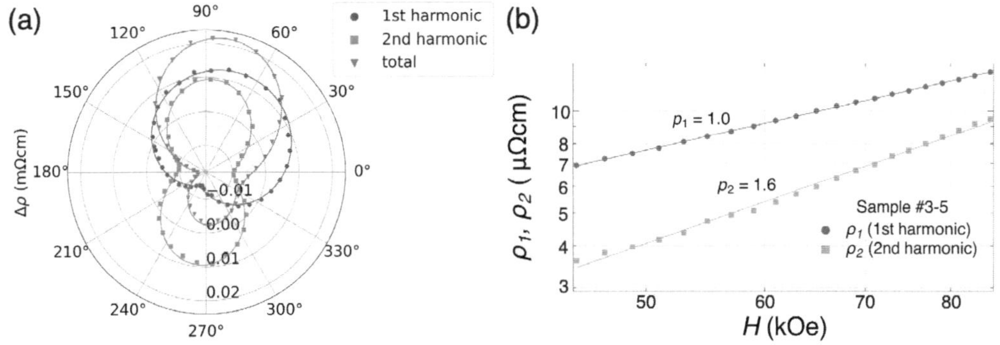

Figure 5. (a) Polar plot of in-plane magnetic-field-angle dependence of magnetoresistance $\Delta\rho = \rho - \rho_0$ at 8.5 T. (b) Magnetic field dependence of ρ_1 and ρ_2, where p_1 and p_2 represent the power exponents of ρ_1 and ρ_2, respectively ($\rho_1 \propto B^{p_1}$ and $\rho_2 \propto B^{p_2}$).

4. Summary

$Sr_xBi_2Se_3$ ($x = 0.17$) single crystals were grown by self-flux method. Quenching from high temperature, which was essential for $Cu_xBi_2Se_3$ was not effective for $Sr_xBi_2Se_3$. We interpret this in terms of different sizes of ions. Additional annealing at 500°C - 600°C for $\sim$ 96 h also effectively made Sr distribution homogeneous and improved the superconducting transition width. Using these high-quality samples, in-plane magnetic-field-angle dependence of resistivity and H_{c2} were measured, and clear nematic superconductivity was confirmed. We also confirmed the presence of minor nematic domains, accompanied by large inhomogeneities. In addition, deviation of angular dependence of magnetoresistance from two-fold symmetry is interpreted by the presence of c-axis current generating Hall voltage. Such inhomogeneous current flow should be removed for future experiments of direct visualization of nematic domains by MO imaging.

References

[1] Hor Y S, Williams A J, Checkelsky J G, Roushan P, Seo J, Xu Q, Zandbergen H W, Yazdani A, Ong N P and Cava R J 2010 *Phys. Rev. Lett.* **104** 57001
[2] Liu Z H, Yao X, Shao J F, Zuo M, Pi L, Tan S, Zhang C J and Zhang Y H 2015 *J. Am. Chem. Soc.* **137** 10512
[3] Matano K, Kriener M, Segawa K, Ando Y and Zheng G Q 2016 *Nat. Phys.* **12** 852
[4] Pan Y, Nikitin A M, Araizi G K, Huang Y K, Matsushita Y, Naka T and de Visser A 2016 *Sci. Rep.* **6** 28632
[5] Fu L and Berg E 2010 *Phys. Rev. Lett.* **105** 97001
[6] Hashimoto T, Yada K, Yamakage A, Sato M and Tanaka Y 2013 *J. Phys. Soc. Jpn.* **82** 44704
[7] Kuntsevich A Y, Bryzgalov M A, Prudkoglyad V A, Martovitskii V P, Selivanov Y G and Chizhevskii E G 2018 *New J. Phys.* **20** 103022
[8] Du G, Shao J F, Yang X, Du Z G, Fang D L, Wang J H, Ran K J, Wen J S, Zhang C J, Yang H, Zhang Y H and Wen H 2017 *Nat. Commun.* **8** 14466
[9] Kostylev I, Yonezawa S, Wang Z W, Ando Y and Maeno Y 2020 *Nat. Commun.* **11** 4152
[10] Yonzezawa S 2019 *Condens. Matter* **4** 2
[11] Sun Y, Kittaka S, Sakakibara T, Machida K, Wang J H, Wen J S, Xing X Z, Shi Z X and Tamegai T 2019 *Phys. Rev. Lett.* **123** 27002
[12] Schneeloch J A, Zhong R D, Xu Z J, Gu G D and Tranquada J M 2015 *Phys. Rev. B.* **91** 144506
[13] Nikitin A M, Pan Y F, Huang Y K, Naka T and de Visser A 2016 *Phys. Rev. B* **94** 144516
[14] Du G, Li Y F, Schneeloch J, Zhong R D, Gu G D, Yang H, Lin H and Wen H H 2017 *Sci. China Phys. Mech. Astron.* **60** 37411
[15] Smylie M P, Willa K, Claus H, Koshelev A E, Song K, Kwok W K, Islam Z, Gu G D, Schneeloch J A, Zhong R D and Welp U 2018 *Sci. Rep.* **8** 7666

[16] Willa K, Willa R, Song K W, Gu G D, Schneeloch J A, Zhong R, Koshelev A E, Kwok W K and Welp U 2018 *Phys. Rev. B* **98** 184509

Surface electronic structure of possible high-T_c surface superconductor WO_3 based on the first-principles calculation

Takuya Sekikawa, Hiroyuki Kawai, Yoshiaki Ōno

Department of Physics, Niigata University, Niigata, 950-2181, Japan

E-mail: sekikawa.t.phys@gmail.com

Abstract. The first-principles calculation is performed to investigate the surface electronic structure of WO_3 in which high-T_c surface superconductivity up to T_c=120 K is observed when alkali metals are alightly doped. To take into account of the surface effect, we employ the tetragonal 14-layers slab model and find that a remarkable in-gap band with a narrow band width of 0.13 eV appears at about 0.6 eV below the conduction bands. Detailed analysis of the wave functions reveal that the in-gap band mainly consists of orbitals in the first and second layers on the surface. Therefore, the in-gap band is considered to play crucial role for the surface superconductivity in the slightly electron-doped WO_3.

1. Introduction

Tungsten oxide WO_3 is a band insulator with a band gap of 3 eV. When electron carriers are introduced in WO_3 by alkali metal doping, A_xWO_3 (A=Na, K, Rb, Cs) the so-called tungstens bronze become metallic and show the superconductivity for $0.1 \lesssim x \lesssim 0.35$, where the superconducting transition temperature T_c is found to increase with decreasing x up to 7 K [1, 2]. At further low doping $x\sim 0.05$, remarkable two dimensional superconductivities on the surface of A_xWO_3 are reported at $T_c = 91$ K for A=Na [3] and $T_c = 120$ K for A=H [4], although the bulk systems are non-metallic due to the effect of the Anderson localization [5]. Theoretically, the recent first-principles calculations have well accounted for the experimental x-dependence of T_c for $x \gtrsim 0.1$ by using the McMillan equation with including the plasmon effect [6, 7]. The surface electronic structure and the possible surface superconductivity, however, were not discussed there. The purpose of this paper is to clarify the surface electronic structure of WO_3 as a starting point to investigate the surface superconductivity.

2. Method

The first-principles DFT calculations are performed by using the OpenMX code [8], and the GGA-PBE [9] is used for the electron exchange and correlation. Core electrons were replaced by norm-conserving pseudopotentials [10] in their fully non-local formulation [11]. A uniform mesh with a planewave cutoff of 200 Ry was used for integrations in real space . The basis sets used are made of numerical atomic orbitals of finite range and are explicitly described below. We also employ the (GGA)+U with $U = 8.25$ eV to reproduce the experimental value of the band gap of 3 eV.

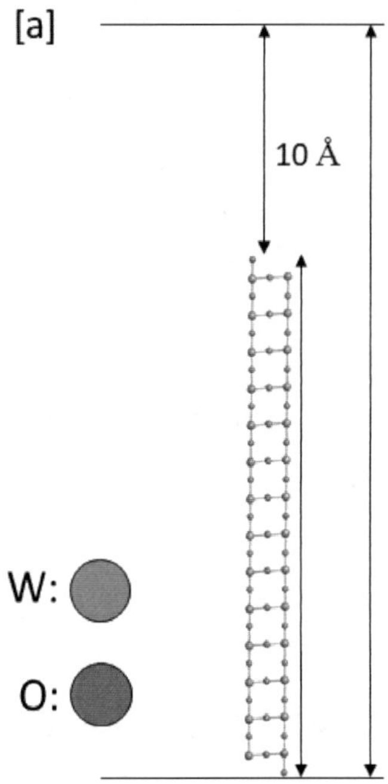

Figure 1. Schematic structures of the 14-layers slab model before optimized (a) and after optimized (b).

The WO_3 surface is modeled by a periodic array of 14-layers slabs separated by a vacuum spacing of 10Å as shown in Fig. 1. The plain-direction lattice constants are taken to be those calculated for tetragonal WO_3(4/nmm). All atomic positions are fully optimized using atomic force calculations. The structures before optimized and after optimized are shown in Figs. 1 (a) and (b), respectively.

We adopt norm-conserving pseudopotentials with an energy cutoff of 200 Ry for charge density including the $5p$, $5d$ and $6s$-states as valence states for W; $1s$ for H; $2s$ and $2p$ for O in both cases of the bulk and the surface systems. We use $7 \times 7 \times 7$ regular k-point mesh for the bulk and $7 \times 7 \times 1$ for the surface. The numerical pseudo atomic orbitals are used as follows: the numbers of the s-, p-, d- and f-characterorbitals for W are 3, 2, 2 and 1, respectively, and those for O are 2, 2, 1 and 0, respectively. The cutoff radii of W and O are 7.0 and 7.0, respectively, in units of Bohr. The cutoff radii for the atomic orbitals of each element were obtained for an energy shift of 50 meV.

3. Result

Figures 2 (a) and (b) show the band dispersions of the tetragonal WO_3 in the cases of the bulk system and the 14-layers slab model, respectively. In the 14-layers slab case, a remarkable in-gap band with a narrow band width of 0.13 eV appears at about 0.6eV below the conduction bands. The detailed analysis of the wave functions revel that the in-gap (LUMO) bands mainly consist of orbitals in the first and second layers on the surfaces as shown in Fig. 3 (a) and the 2nd

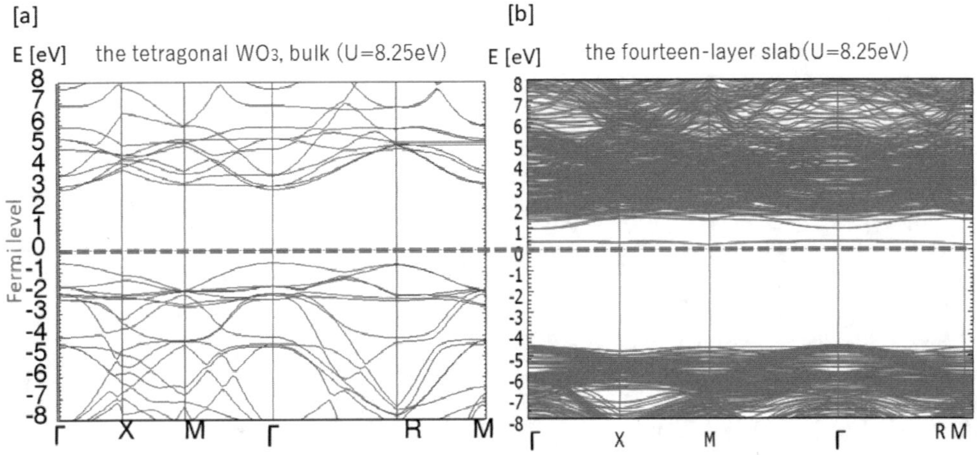

Figure 2. Band dispersions of the tetragonal WO_3 in the cases of the bulk (a) and the 14-layers slab (b), respectively.

LUMO bands mainly consist of orbitals in the first layers on the surfaces. We note that the energy bands in the slab model are 2-fold degenerate due to the inversion symmetry along the *c*-axis as shown in Fig. 3.

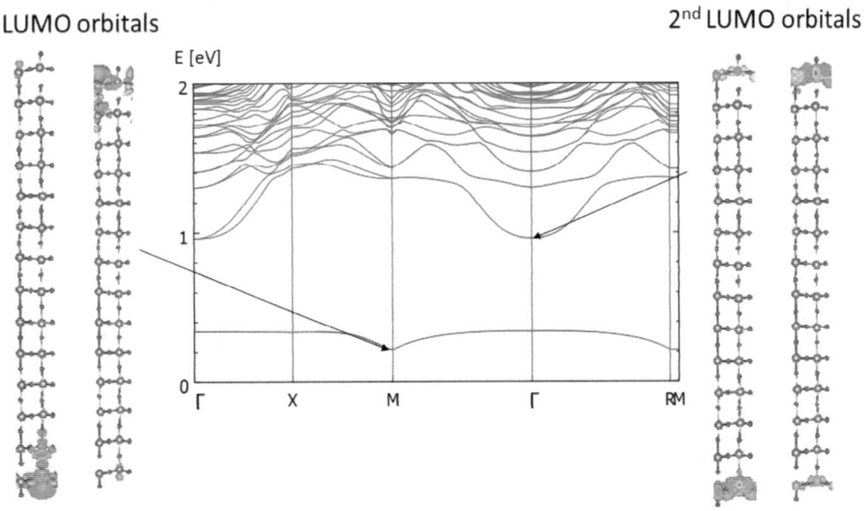

Figure 3. Schematic figures of the wave functions for the in-gap (LUMO) bands and the 2nd LUMO bands, respectively.

4. Conclusion

We have investigated the surface electronic structure of WO_3 on the basis of the first-principles calculation with the use of the 14-layers slab model and have found to exsit a remarkable narrow in-gap bands mainly consists of orbitals in the first and second layers on the surface. When the electron carriers are slightly doped, the in-gap bands are partially occupied and then is considered to play crucial role for the surface superconductivity which is observed in the slightly Alkali-doped WO_3 with a high-T_c. To clarify the mechanism of the High-T_c surface superconductivity, we need further calculations including the surface phonon states and the electron correlation effects in the narrow in-gap bands. Such calculations are now under way.

Acknowledgments

This work was partially supported by JSPS KAKENHI Grant Numbers JP 21K03399 and 21J15788.TS was supported by a JSPS Research Fellowship for Young Scientists. Numerical calculations were performed partially using OFP at the CCS, University of Tsukuba and the MASAMUNE-IMR, Tohoku University.

References

[1] H. R. Shans, *Solid State Commun.* **15**, 753 (1974).
[2] K. Stanley, R. Morris and W. Moulton, *Phys. Rev. B* **20**, 1903 (1979)
[3] S. Reich and Y. Tsabba, *Eur. Phys. J. B* **9**, 1 (1999).
[4] S. Reich, G. Leitus, R. Popovitz -Biro, A. Goldbourt and S. Vega, *J Supercond. Nov. Magn.* **22**, 343 (2009).
[5] S. Raj *et al.*, *Phys. Rev. B* **75**, 155116 (2007).
[6] T. Sekikawa, R. Watabe, J. Ishizuka, Y. Nitta, K. Sano, Y. Ōno, *JPS Conf. Proc.* **30**, 011043 (2020).
[7] K. Sano, Y.Nitta and Y. Ōno, *J. Phys. Soc. Jpn.* **89**, 113704 (2020).
[8] OpenMX. Open source package for Material eXplorer. http://www.openmx-square.org .
[9] J. P. Perdew and K. Burke and M. Ernzerhof, *Phys. Rev. Lett.* **77**. 3685 (1996) .
[10] N.Troullier and J.L. Martins, *Phys. Rev. B* **43**.1993 (1991) .
[11] L.Kleinman and D. M. Bylander, *Phys. Rev. Lett.* **48**,1425 (1982) .

Journal of Physics: Conference Series **2323** (2022) 012007 doi:10.1088/1742-6596/2323/1/012007

Magnetic properties of Hf(P, Se)$_2$ single crystal by high-temperature solution method using CdSe flux

H Kitô[1], K Kawashima[1,2], S Ishida[1], H Fujihisa[1], Y Gotoh[1], A Iyo[1], H Ogino[1], H Eisaki[1] and Y Yoshida[1]

[1]National Institute of Advanced Industrial Science and Technology (AIST), 1-1-1 Umezono, Tsukuba, Ibaraki 305-8568, Japan

[2] IMRA JAPAN Co., Ltd, 2-36 Hachiken-cho, Kariya-shi, Aichi 448-8650, Japan

E-mail: h.kito@aist.go.jp

Abstract. By the high pressure and high temperature solution method mixing CdSe fluxes, layered PbFCl-type $HfP_{2-x}Se_x$ (x = 0.45) and HfSiSe single crystals were successfully grown and obtained under a high-pressure for the first time. It was clarified that the melting temperature of the solid solution was lowered by adding a small amount of CdSe. Typical sizes of ~ 180 × ~ 180 × ~ 20 μm^3 for $HfP_{2-x}Se_x$ (x = 0.45) and ~ 120 × ~ 100 × ~ 20 μm^3 for HfSiSe were obtained and characterized by scanning electron microscopy (SEM) and energy-dispersive electron spectroscopy. X-ray diffraction analysis showed that $HfP_{2-x}Se_x$ (x = 0.45) and HfSiSe were grown oriented c - axis. Magnetic susceptibility versus temperature measurement were performed for $HfP_{2-x}Se_x$ (x = 0.45). Onset of the T_c was showed at around ~ 5.36 K for $HfP_{2-x}Se_x$ (x = 0.45) single crystal. Here, the crystal growth and magnetic properties for $HfP_{2-x}Se_x$ (x = 0.45) single crystal were presented.

1. Introduction

The layered PbFCl-type $AP_{2-x}X_x$ (A=Zr, Hf, X=S and Se) intermetallic superconducting materials [1] are composed A-(P, X) blocks and a two-dimensional square P planar nets. This crystal structure is shown in Figure 1 and is an isostructural to topological Dirac ZrSiS semiconductor [2] in a non-symmorphic space group ($P4/nmm$).

Using polycrystalline sample study, superconducting transition temperature (T_c) for $ZrP_{2-x}Se_x$ has been observed to exhibit below T_c = 6.31 K at x = 0.75 [1, 3]. From this result, T_c was tuned by (P, X) site for $AP_{2-x}X_x$. The T_c versus S (or Se) substitution for X site in $AP_{2-x}X_x$ are dome-shaped. Replaced by rare earth element in A site of $AP_{2-x}X_x$, the T_c versus substitution y of $AP_{2-x}X_x$ was also dome-shaped. From this result, in $AP_{2-x}X_x$, it was clarified that T_c was also tuned by the rare earth substitution of A site in $AP_{2-x}X_x$ [4, 5], similar as the doping concentration of high T_c superconducting cuprates [6] and iron-based compounds [7].

We had reported growth the partial substitution by the rare earth in Zr site in ZrPSe [8] and T_c was reached at 8.36 K for (Zr, Sc)PSe single crystal [9].

Understanding the mechanism of superconductivity in this $AP_{2-x}X$ system leads to the development

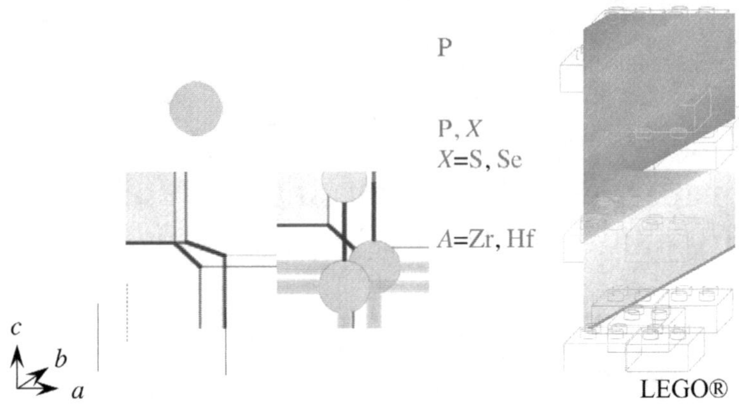

Figure 1. The PbFCl-type structure of the $AP_{2-x}X_x$ (A = Zr, Hf; X = S, Se). This structure is a tetragonal layered structure composed A-(P, X) blocks and a two-dimensional square P planar nets, like as a stacking of LEGO® blocks.

of new superconductivity and is extremely important. The high pressure and high temperature synthesis in earth science research is very popular tools to synthesize new material and grown single crystals. Using this technique, the crystal growth of MgB_2 superconductor [10], Yb_2Cu_2In [11] and CoP_3 Skutterudite [12] had already reported.

Since the melting point of CdSe is ~ 1,270 °C [13], it can be expected to act as a flux material during the synthesis of refractory compounds like as $HfP_{2-x}Se_x$. Growth single crystals for $HfP_{2-x}Se_x$ and HfSiSe using CdSe flux at the condition of the high temperature and high pressure were reported.

2. Experiment and results

2.1. Experiment

Single crystals of $HfP_{2-x}Se_x$ (x = 0.45) and HfSiSe were grown using a flux method at the condition of high pressure by a cubic anvil press. As a first step, the powder of Hf (powder, 99.9 %, including Zr < 1 %), P (grain, 99.9999%) (or Si (powder, 99.999 %)) and Se (grain, 99.999%) were weighted in the nominal composition ($HfP_{2-x}Se_x$ (x = 0.45) and HfSiSe) well mixed in a nitrogen filled glove box. As a second step, 0.25 mol of CdSe was added to these nominal compositions and mixed. Total mass ~ 0.55 g of the pellet was sealed in BN crucibles, placed in a high-pressure cell and heated at ~ 1,680 °C and the applied pressure at ~ 2.0 GPa, keeping for 2.0 h and rapidly quenched to room temperature.

The microstructure of these samples was obserevd by means of scanning electron microscopy (SEM Hitachi High-Technologies TM3000) and chemical composition of obtained $HfP_{2-x}Se_x$ (x = 0.45) and HfSiSe was analyzed by an energy dispersive X-ray spectrometry (EDX Oxford Instruments SwiftED 3000).

The X-ray diffraction data of the samples were collected using Cu Kα radiation under 40 kV and 40 mA on Rigaku Ultima-IV diffractometer. Collecting conditions were 2θ range from 5° to 65° of 0.020° step width and a counting rate of 8.0° / minutes.

The magnetic susceptibility χ (T) of $HfP_{2-x}Se_x$ (x = 0.45) (3.16 mg) sample was investigated from 2 K to 10 K with the applied magnetic field of 20 Oe using a superconducting quantum interference device magnetometer (Quantum Design MPMS). The magnetization hysteresis (M - H) major loops in fields from -7 to 7 T were also measured at 2K and 4K.

2.2. Results

Using SEM observation, the microstructure of obtained samples was shown in Figure 2. The size and morphology of these samples have been analyzed by SEM. The typical size of single crystals was ~ 180 × ~ 180 × ~ 20 μm^3 for $HfP_{2-x}Se_x$ (x = 0.45) and was ~ 120 × ~ 100 × ~ 20 μm^3 for HfSiSe, respectively. The chemical compositions for $HfP_{2-x}Se_x$ (x = 0.45) and HfSiSe single crystals were investigated elemental analysis by EDX. EDX analysis clearly indicated that the chemical composition of $HfP_{2-x}Se_x$ (x = 0.45) (or HfSiSe) were the different from the bulk composition. The metal ratios of Hf : Zr : P : Se were examined to be 0.9900(12) : 0.0100(15) : 1.612(14) : 0.347(7) for any 4 points of $HfP_{2-x}Se_x$ (x = 0.45) and the metal ratios of Hf : Zr : Si : Se were examined to be 0.9293(13) : 0.0000(0) : 1.214(2) : 0.785(2) for any 14 points of HfSiSe, respectively. Impurity HfP and $HfSe_2$ phases were observed in $HfP_{2-x}Se_x$ (x = 0.45) and $HfSe_2$ were also observed in HfSiSe, respectively.

X-ray diffraction for obtained $HfP_{2-x}Se_x$ (x = 0.45) and HfSiSe single crystals were shown in Figure 3. There exist only sharp 0 0 l peaks attributed to $HfP_{2-x}Se_x$ (x = 0.45) and HfSiSe. The peaks on pattern were indexed as the PbFCl phase. This indicates three dimensional regular and perfect atomic arrangements in the crystal.

Magnetic susceptibility versus temperature for $HfP_{2-x}Se_x$ (x = 0.45) with the applied magnetic field

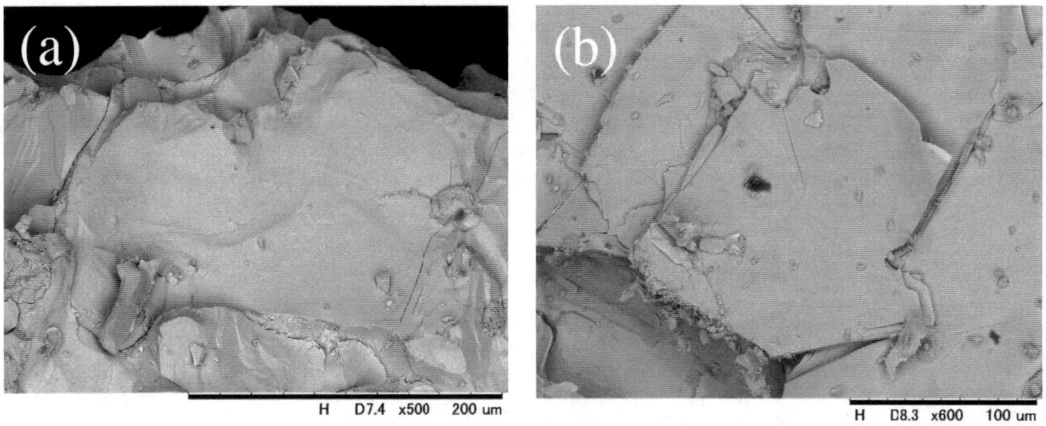

Figure 2. SEM image of (a) $HfP_{2-x}Se_x$ crystal and (b) HfSiSe crystal.

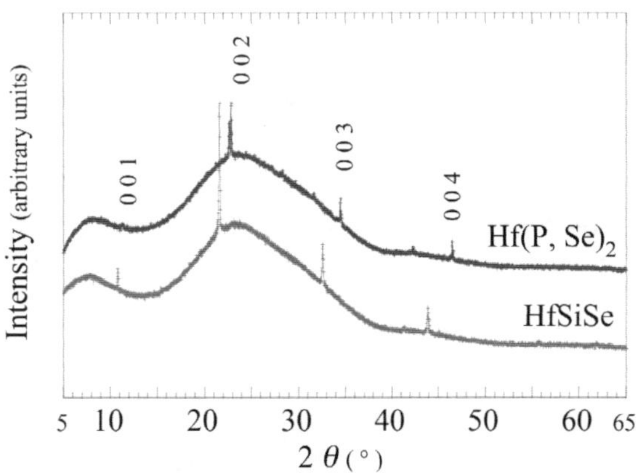

Figure 3. XRD patterns at room temperature for typical $HfP_{2-x}Se_x$ (x = 0.45) and HfSiSe single crystals.

of 20 Oe were shown in Figure 4. With decreasing temperature, $T_{c\ onset}$ was exhibited around 5.36 K at the applied field *H* parallel to *c*-axis and perpendicular to *c*-axis for $HfP_{2-x}Se_x$ ($x = 0.45$). T_c(onset) are ~ 6.31 K for $ZrP_{1.25}Se_{0.75}$ and ~ 3.5 K for $ZrP_{1.54}S_{0.46}$ (A = Zr, X = S, x = 0.46) had already reported [14, 15]. Without shape correction, the Meisner volume fraction were roughly estimated 65.69 % (*H* parallel to *c*-axis) and 48.96 % (*H* perpendicular to *c*-axis) under the Zero-field-cooled (Z.F.C) process. The diamagnetic shielding volume fraction were also estimated to be 36.43 % (*H* parallel to *c*-axis) and 18.62 % (*H* perpendicular to *c*-axis) at T = 2.0 K under field-cooled (F.C) process, respectively.

Figure 5(a) shows $M - H$ curves at 2K for $HfP_{2-x}Se_x$ ($x = 0.45$). From the results of this magnetic hysteresis loops, the lower critical field, $Hc_1(0)$ was determined in the same manner as in reference [16]. The value of $Hc_{1\ H\ \mathrm{parallel\ to}\ c\text{-axis}}$ (0) was estimated to be 0.0014 T and the value of $Hc_{1\ H\ \mathrm{perpendicular\ to}\ c\text{-axis}}$ (0) was also estimated to be 0.0011 T, respectively.

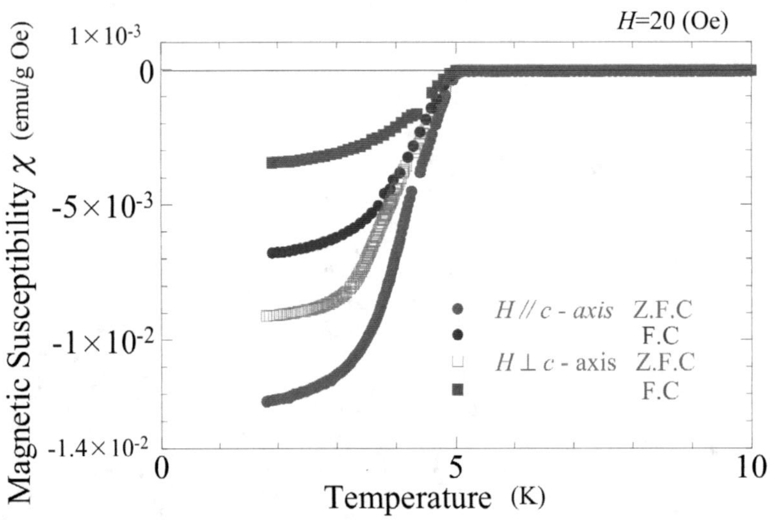

Figure 4. Magnetic susceptibility χ versus temperature at *H* parallel and *H* perpendicular to *c*-axis with a applied magnetic field H = 20 Oe for $HfP_{2-x}Se_x$ ($x = 0.45$) single crystal.

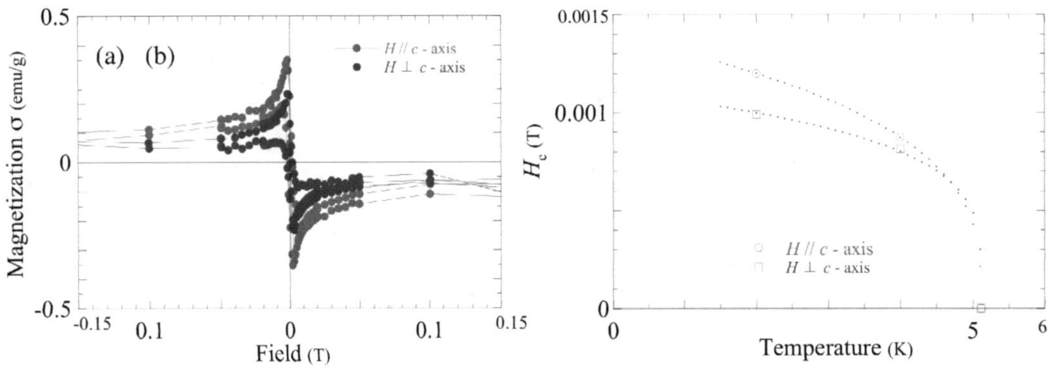

Figure 5 (a) $M - H$ curves at 2K for $HfP_{2-x}Se_x$ ($x = 0.45$) single crystal. **Figure 5 (b)** Lower critical field $H_{c1}(T)$ versus temperature for $HfP_{2-x}Se_x$ ($x = 0.45$) single crystal.

3. Summary

By CdSe flux, by the high-pressure apparatus, HfSiSe single crystal were obtained and in particularly, $HfP_{2-x}Se_x$ single crystal was successfully grown for the first time. From the EDX analysis, element concentration can be slightly different from the nominal concentrations of $HfP_{2-x}Se_x$ (x = 0.45) and HfSiSe. A sharp transition T_c(onset) ~ 5.36 K for $HfP_{2-x}Se_x$ (x = 0.45) was observed from the magnetic susceptibility measurements. The value of $Hc_{1\ H \text{ parallel to } c\text{-axis}}$ (0) and $Hc_{1\ H \text{ perpendicular to } c\text{-axis}}$ (0) for $HfP_{2-x}Se_x$ (x = 0.45) were also estimated from the results of $M - H$ curves each temperature.

References

[1] Kitô H, Yanagi Y, Ishida S, Oka K, Gotoh Y, Fujihisa H, Yoshida Y, Iyo A, and Eisaki H, 2014 *J. Phys. Soc. Jpn.* **83** 074713
[2] L. M. Schoop, M. N. Ali, C. Strasser, A. Topp, A. Varykhalov, D. Marchenko, V. Duppel, S. S. P. Parkin, B. V. Lotsch1 and C. R. Ast, 2016 *Nature Commun.* **7** 11696
[3] Ishida S, Fujihisa H, Hase I, Yanagi Y, Kawashima K, Oka K, Gotoh Y, Yoshida Y, Iyo A, Eisaki H, and Kito H, 2016 *Supercond. Sci. Technol.* **29** 055004
[4] Iwakiri K, 2016 Graduation Thesis, Tokyo Univ. of Science
[5] Iwakiri K, Nishio T, Kawashima K, Ishida S, Oka K, Fujihisa H, Gotoh Y, Iyo A, Ogino H, Eisaki H, Yoshida Y, and H Kitô, 2018 *J. Phys. Conf. Series* **1054** 012002
[6] Squire R H, March N H 2010 *Inter. J. Quantum Chem.* **110** 2808
[7] Sun J P, Matsuura K, Ye G Z, Mizukami Y, Shimozawa M, Matsubayashi K, Yamashita M, Watashige T, Kasahara S, Matsuda Y, Yan J -Q, Sales B C, Uwatoko Y, Cheng J -G, and Shibauchi T, 2016 *Nature Communications* **7** 12146
[8] Kitô H, Iwakiri K, Nishio T, Kawashima K, Ishida S, Oka K, Fujihisa H, Gotoh Y, Iyo A, Ogino H, Eisaki H and Yoshida Y, 2018 *J. Phys. Conf. Series* **1054** 012003.
[9] Kitô H, Kawashima K, Iwakiri K, Nishio T, Ishida K, Fujihisa H, Gotoh Y, Iyo A, Ogino H, Eisaki H and Yoshida Y, 2021 *Materials Science Forum* (Trans Tech Publications Ltd. ISBN(softcover): 978-3-0357-1630-6, ISBN(eBook): 978-3-0357-3630-4) **1016** 708
[10] Tsuda S, Yokoya T, Takano Y, Kito H, Matsushita A, Yin F, Itoh J, Harima H, and Shin S, 2003 *Phys. Rev. Lett.* **91** 127001
[11] Tsujii N, Kitô H, Kitazawa H, and Kido G, 2001 *J. Alloys and Compunds* 322 74
[12] Lee C.H, Kito H, Ihara H, Akita K, Yanase N, Sekine C, and Shirotani I, 2004 *J. Cryst. Growth* **263** 358
[13] Romesh C. Sharma, and Y. Austin Chang, 1989 *J. Electrochern.* Soc. **136** 1536
[14] Cichorek T, Bochenek L, Czulucki A, Schmidt M, Niewa R, Auffermann G, Prots Y, Wawryk R, Baenitz M, Steglich F, and Kniep R, 2011 *Scientific Report MPG/CPfS* 103 (www.cpfs.mpg.de/2081022/B_2009-2010.pdf)
[15] Baenitza M, Lüders K, Kniep R, Steglich F, and Schmidt M, 2016 *Physics Procedia* **81** 65
[16] Ishikawa Y, Mori K, Kobayashi K, and Sato K, 1988 *J. J. Appl. Phys. Lett.* **27** L403

Characterizations of Anisotropic Superconductivity in $(BiSe)_{1+\delta}NbSe_2$

S Matsuzawa, S Pyon and T Tamegai

Department of Applied Physics, The University of Tokyo, 7-3-1 Hongo, Bunkyo-ku, Tokyo 113-8656, Japan

Email: creator-m318@g.ecc.u-tokyo.ac.jp

Abstract. Anisotropy of superconductivity in misfit layered compounds $(BiSe)_{1+\delta}NbSe_2$ was studied. We synthesized single crystals of $(BiSe)_{1+\delta}NbSe_2$ using CsCl/KCl flux method. There existed two phases with different T_c depending on the starting composition δ, although their structures were almost unaffected by changing δ. Anisotropy of the upper critical field was determined by magneto-resistance measurements with various out-of-plane and in-plane field angles. The out-of-plane anisotropy of H_{c2} showed a cusp-like behavior for fields close to *ab*-plane similar to that predicted by 2D Tinkham model. Furthermore, a clear two-fold symmetry was observed in the in-plane angular dependence of H_{c2}.

1. Introduction

Misfit layered compounds (MLCs) are layered chalcogenides with incommensurate stacking of two sublattices with different symmetries and have been studied for decades due to very unique structures and various interesting physical properties [1]. The chemical formula of MLCs is generally represented as $(MX)_{1+x}(TX_2)_n$, where M is Sn, Sb, Pb, Bi, or rare earth metals, T is transition metals, and X is S or Se. The structure of MLCs is composed of van-der-Waals coupling of two sublattice layers, *i.e.*, rocksalt-type monochalcogenide MX and trigonal-prisms dichalcogenide TX_2. The naming “misfit” comes from a special mismatch of lattice constants of their two subsystems, which match perfectly in the *b*-axis direction but a mismatch occurs in the *a*-axis direction. In this way, the structure of MLCs is so complicated, and is usually approximated by an averaged orthorhombic system. Assuming this average structure, MLCs have two-fold symmetry in *ab*-plane as a whole.

Among various MLCs family, $(BiSe)_{1.10}NbSe_2$, sometimes referred to as $BiNbSe_3$, is known to show superconductivity with the critical temperature T_c of 2.3 ~ 3.3 K [2,3]. Recently, there are some reports claiming that the superconductivity of $Nb_xBi_2Se_3$, a material that Nb atoms are doped into a topological insulator Bi_2Se_3, is identical to the misfit phase $(BiSe)_{1.10}NbSe_2$ [4,5]. In addition to $Nb_xBi_2Se_3$, Cu and Sr doped ones also have been actively investigated, since they are expected to be candidates for topological superconductors, and they show rotational-symmetry breaking superconductivity, so called nematic superconductivity [6-8]. However, the origin of superconductivity in $Nb_xBi_2Se_3$ remains still controversial.

In this study, we report the growth of single crystals of $(BiSe)_{1+\delta}NbSe_2$ and its anisotropy in both out-of-plane and in-plane directions by the upper critical field (H_{c2}) measurements.

2. Experimental Methods

Five kinds of $(BiSe)_{1+\delta}NbSe_2$ single crystals were prepared as shown in table 1. Each sample can be labeled by the nominal ratio parameter δ of BiSe to $NbSe_2$. (For example, if the ratio of BiSe:$NbSe_2$ = 1.10:1.00, δ becomes 0.10.) Single crystals of $(BiSe)_{1+\delta}NbSe_2$ were synthesized by CsCl/KCl flux method as described in ref. [3]. For the raw materials, high purity fine powders of Bi (99.999%), Se (99.999%), and Nb (99.5%) were used. The molar ratio of the flux was CsCl:KCl = 5:3. The mixture of the raw materials (0.8 g) and the flux (5.0 g) were sealed in an evacuated quartz tube (10 mm o.d., 200 mm length). The quartz tube was heated at 700°C for 10 h and cooled slowly down to 600°C at a rate of 0.5°C/h followed by furnace cooling to room temperature. After that, the product was washed with distilled water to remove the flux. Single crystals with typical dimensions of ~1×1×0.02 mm^3 were obtained.

The structure and compositional ratio of the crystal were evaluated by X-ray diffraction (XRD) with Cu Kα radiation and energy dispersive X-ray spectroscopy (EDX), respectively.

The temperature dependence of magnetization (*M-T*) was measured using a superconducting quantum interference device (SQUID) magnetometer (MPMS-XL5, Quantum Design). The critical temperature (T_c) was determined as a merging temperature of zero field-cooled (ZFC) and field-cooled (FC) *M-T* curve.

The transport measurement was performed by a four-probe method using a physical property measurement system (PPMS, Quantum Design). The ac electrical resistivity along *ab*-plane was measured. The onset of the superconducting transition temperature (T_c^{onset}) was defined as the inflection point in the temperature dependence of resistivity (ρ–T), and the zero resistivity (T_c^{zero}) was determined considering the criterion of a resistivity of 0.1 μΩcm in the ρ–T curve. The magnetic field dependence of resistivity (ρ-H) was also measured at various out-of-plane and in-plane field angles.

3. Results and Discussion

Table 1 shows the nominal and analyzed compositions of prepared $(BiSe)_{1+\delta}NbSe_2$ crystals grown from each nominal composition normalized by Nb = 1.00. Analyzed compositions are close to the ratio of Bi:Nb:Se ~ 1:1:3 for all crystals. Figure 1 shows XRD patterns of $(BiSe)_{1+\delta}NbSe_2$ with different δ. Inspite of different nominal ratios δ, all (00*l*) peaks are observed at the same angles for all samples. Detailed inspections of diffraction data show the presence of minor impurity phase of Bi_2Se_3. The *c*-axis lattice constants are calculated to be ~24.18 Å for all crystals. These facts suggest that the starting composition for synthesis is not important for the stacking structure of $(BiSe)_{1+\delta}NbSe_2$ even with δ as large as 1.0.

Figure 2(a) shows the temperature dependence of normalized magnetization of $(BiSe)_{1+\delta}NbSe_2$ with different δ. The applied field was 5 Oe parallel to *c*-axis. The evaluated T_c of sample A, B, and C is about 2.6 K, while that of D and E is about 3.0 K. There is no sample showing a transition with T_c

Table 1. Nominal and analyzed chemical compositions of prepared $(BiSe)_{1+\delta}NbSe_2$ crystals grown from each nominal composition normalized by Nb = 1.00.

Sample	δ	Nominal		EDX	
		Bi	Se	Bi	Se
Sample A	0.10	1.10	3.10	1.05(2)	3.0(2)
Sample B	0.00	1.00	3.00	1.02	2.91
Sample C	0.20	1.20	3.20	1.06	2.95
Sample D	0.50	1.50	3.50	1.08	2.90
Sample E	1.00	2.00	4.00	1.16	3.03

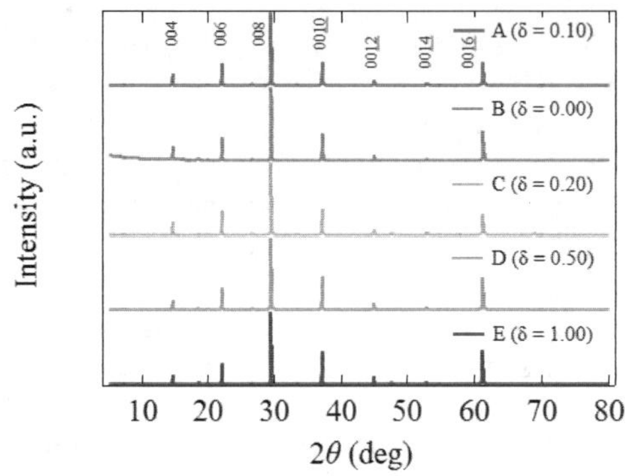

Figure 1. X-ray diffraction patterns of $(BiSe)_{1+\delta}NbSe_2$.

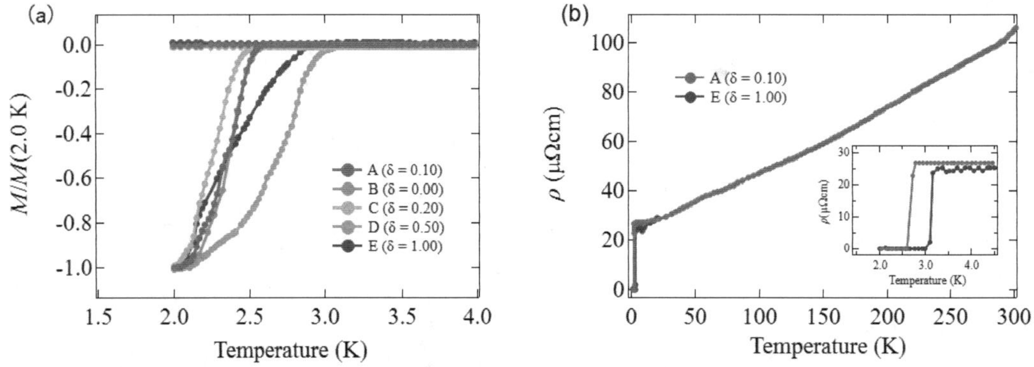

Figure 2. (a) Temperature dependence of normalized magnetization of $(BiSe)_{1+\delta}NbSe_2$ with different δ. (b) Temperature dependence of in-plane resistivity of $(BiSe)_{1+\delta}NbSe_2$ with δ = 0.10 and 1.00. (The sample with δ = 1.00 was measured only below 25 K.) The inset shows ρ-T curve near the transition temperature.

between 2.6 K and 3.0 K, which implies that $(BiSe)_{1+\delta}NbSe_2$ may have two different phases. In the low-T_c phase, starting ratios of BiSe to $NbSe_2$ are close to 1.10, while in the high-T_c phase they are over 1.50. Figure 2(b) shows the temperature dependence of resistivity of $(BiSe)_{1+\delta}NbSe_2$ with δ = 0.10 and 1.00. The superconducting transition starts at 2.8 K and 3.2 K (corresponding to T_c^{onset}), and zero resistivity is observed at 2.6 K and 3.0 K (T_c^{zero}), respectively. T_c values examined from our magnetization and resistivity measurements almost agree with those in the previous reports [2,3]

Figure 3(a) shows the field dependence of resistivity of $(BiSe)_{1+\delta}NbSe_2$ (δ = 0.10) with different out-of-plane field angles at T = 2.0 K. Note that the angle θ is defined as an angle between the field direction and ab-plane of the crystal. The magnetic field was applied along a-axis. ρ-H curves show broad superconducting transition at angles close to 0°, indicating that the superconducting region still remain in the sample under higher fields than the irreversibility field close to $H \parallel ab$. On the other hand, for $\theta \geq$ 60° the transition remains sharp, while ρ-H curves shift toward higher fields as the angle approaches to 90°. We determined the upper critical field, H_{c2}, by the criterion $\rho = 0.9\rho_n$ (ρ_n: normal state resistivity), since the characteristic behavior seen at the end of the superconducting transition is considered to be intrinsic. The anisotropy parameter in the out-of-plane direction defined by the ratio of H_{c2} for two principal directions is estimated to be $\gamma = 1.13$, indicating that the anisotropy of $(BiSe)_{1+\delta}NbSe_2$ is small. The out-of-plane angular dependence of H_{c2} is shown in Fig. 3(b). $H_{c2}(\theta)$ shows a cusp-like behavior close to 0°.

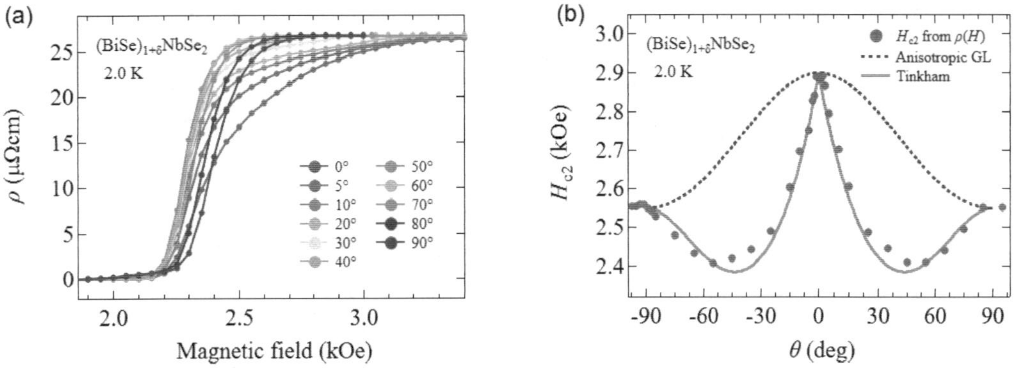

Figure 3. (a) Magnetic field dependence of resistivity of $(BiSe)_{1+\delta}NbSe_2$ (δ = 0.10) with different out-of-plane angles at 2.0 K. (b) Out-of-plane anisotropy of the upper critical field H_{c2} of $(BiSe)_{1+\delta}NbSe_2$ at 2.0 K. $\theta = 0°$ and 90° correspond to *ab*-plane and *c*-axis, respectively. The solid and dotted lines are Tinkham model and the anisotropic GL model, respectively.

The angular dependence of H_{c2} in anisotropic superconductors is usually explained by two theoretical models. For 3D anisotropic superconductors, $H_{c2}(\theta)$ is well-fitted by the anisotropic Ginzburg-Landau (GL) model represented by the following equation;

$$\left(\frac{H_{c2}(\theta)\sin\theta}{H_{c2}^{\perp}}\right)^2 + \left(\frac{H_{c2}(\theta)\cos\theta}{H_{c2}^{\parallel}}\right)^2 = 1, \quad (1)$$

where $H_{c2}^{\parallel}$ and $H_{c2}^{\perp}$ are upper critical fields corresponding to $\theta = 0°$ and 90°, respectively. On the other hand, for 2D film superconductors $H_{c2}(\theta)$ obeys the following equation;

$$\left|\frac{H_{c2}(\theta)\sin\theta}{H_{c2}^{\perp}}\right| + \left(\frac{H_{c2}(\theta)\cos\theta}{H_{c2}^{\parallel}}\right)^2 = 1, \quad (2)$$

which is generally referred to as Tinkham model. Generally speaking, Tinkham model can be only applied when the thickness of the film is thinner than the coherence length perpendicular to the film plane. As shown in Fig. 3(b), the cusp-like behavior of $H_{c2}(\theta)$ of $(BiSe)_{1+\delta}NbSe_2$ is reproduced by Tinkham model in spite of its small anisotropy. This is considered to be related to the existence of stacking faults in the crystal. In some layered superconductors, it has been known that $H_{c2}(\theta)$ can be fitted well by the Tinkham model despite bulk nature of crystals, such as CaAlSi [9] and YbC_6 [10].

We also measured the field dependence of resistivity for $(BiSe)_{1+\delta}NbSe_2$ (δ = 0.10) with different in-plane field angles as show in Fig. 4(a). The angle φ is defined as an angle between the field direction and *a*-axis direction of the BiSe sublattice. In other words, 0° and 90° correspond to the *a*-axis and *b*-axis, respectively. The variation of zero resistivity field is more pronounced than that of out-of-plane, indicating that the in-plane anisotropy is larger than that for the out-of-plane direction. This is consistent with the intrinsic misfit structure of MLCs. In the range between $\varphi = 0°$ and 40°, the curves show plateau-like structures close to saturation. For clarifying the origin of this behavior, further research will be needed. The upper critical field is determined by the criterion $\rho = 0.5\rho_n$ to avoid anomalies related to the presence of the plateau. The anisotropy parameter in the in-plane direction is estimated as $\gamma_{ab} = H_{c2}^{\parallel b}/H_{c2}^{\parallel a} = 1.92$. The in-plane anisotropy of $H_{c2}(\varphi)$ is shown in Fig. 4(b). There are two peaks separated by ~180° reflecting two-fold symmetry of the whole crystal system. $H_{c2}(\varphi)$ can be also fitted by the anisotropic GL model for in-plane direction represented as

$$H_{c2}(\varphi) = \frac{H_{c2}(0°)}{\sqrt{\cos^2\varphi + \gamma_{ab}^2\sin^2\varphi}}. \quad (3)$$

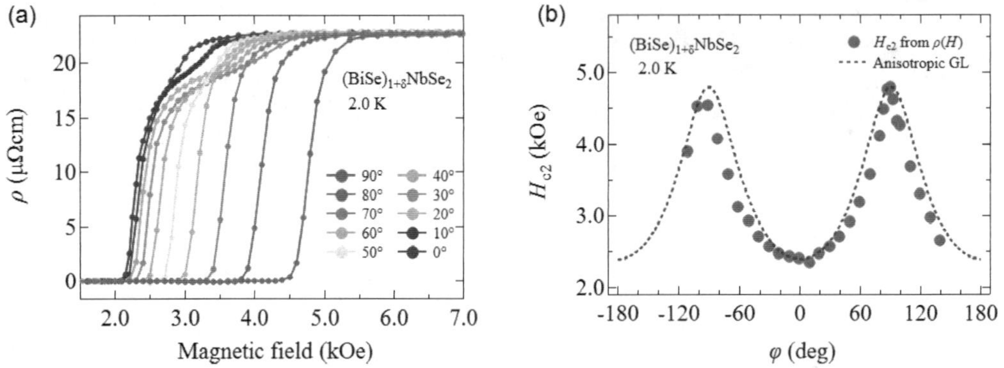

Figure 4. (a) Magnetic field dependence of resistivity for $(BiSe)_{1+\delta}NbSe_2$ ($\delta = 0.10$) with different in-plane field angles at 2.0 K. (b) In-plane anisotropy of H_{c2} of $(BiSe)_{1+\delta}NbSe_2$ at 2.0 K. The broken line is the anisotropic GL model for in-plane anisotropy. $\varphi = 0°$ and 90° correspond to *a*-axis and *b*-axis, respectively.

Experimentally measured $H_{c2}(\varphi)$ is very similar to that reported for $Nb_xBi_2Se_3$ [11]. This result strongly supports the claim that the origin of superconductivity in $Nb_xBi_2Se_3$ is actually misfit layered compound $(BiSe)_{1.10}NbSe_2$.

4. Summary

We have grown single crystals of $(BiSe)_{1+\delta}NbSe_2$ starting with different δ. We find that there are two phases with distinct T_c values depending on δ values; $T_c = 2.6 \sim 2.8$ K for $\delta \sim 0.1$, and $T_c = 3.0 \sim 3.2$ K for $\delta > 0.5$. However, XRD measurements show that crystal structures for the two phases are almost identical. Resistivity measurements at different field angles show quite different behaviors between out-of-plane and in-plane directions. The out-of-plane field angle dependence of H_{c2} is found to be well described by Tinkham model, despite the fact that its anisotropy parameter is very small. In the case of in-plane direction, the angular dependence of H_{c2} shows two-fold symmetry, which resembles to that observed in $Nb_xBi_2Se_3$, supporting the claim that superconductivity in $Nb_xBi_2Se_3$ originates from $(BiSe)_{1+\delta}NbSe_2$.

References

[1] Wiegers G A 1996 *Prog. Solid State Chem.* **24** 1
[2] Nader A, Briggs A and Gotoh Y 1997 *Solid State Commun.* **101** 149
[3] Nagao M, Miura A, Horibe Y, Maruyama Y, Watauchi S, Takano Y and Tanaka I 2020 *Solid State Commun.* **321** 114051
[4] Kamminga M E, Batuk M, Hadermann J and Clarke S J, 2020 *Commun. Mater.* **1** 82
[5] Dalgaard K L, Kevy S M, Wollesen L, Ma Q, Wiedmann S, Clauss K G V S and Bremholm M 2021 *Phys. Rev.* B **103** 184103
[6] Asaba T, Lawson B J, Tinsman C, Chen L, Corbae P, Li G, Qiu Y, Hor Y S, Fu L and Li L 2017 *Phys. Rev.* X **7** 011009
[7] Matano K, Kriener M, Segawa K, Ando Y and Zheng G 2016 *Nat. Phys.* **12** 852
[8] Pan Y, Nikitin A M, Araizi G K, Huang Y K, Matsushita Y, Naka T and deVisser A 2016 *Sci. Rep.* **6** 28632
[9] Ghosh A K, Tokunaga M and Tamegai T 2003 *Phys. Rev.* B **68** 054507
[10] Kawai N F and Fukuyama H 2008 *Physica* C **468** 2403
[11] Shen J, He W, Yuan N F, Huang Z, Cho C, Lee S H, Hor Y S, Law K T and Lortz R 2017 *npj Quant. Mater.* **2** 59

Journal of Physics: Conference Series **2323** (2022) 012009 doi:10.1088/1742-6596/2323/1/012009

Bulk-Combinatorial Method for Exploring New Superconductors and its Mathematical Aspect

I Hase[1,*], A Iyo[1], Y Higashi[1], T Yanagisawa[1] and K Kawashima[2]

[1] National Institute of Advanced Industrial Science and Technology (AIST), Tsukuba, Ibaraki 305-8568, Japan
[2] IMRA JAPAN Co. Ltd., Kariya, Aichi 448-8650, Japan

*E-mail: i.hase@aist.go.jp

Abstract. In this paper we introduce our method to explore new superconductors: bulk-combinatorial method. Using this method, we have succeeded in finding many new superconductors so far. We also show that the problems that arise when identifying new superconducting phases in this way are related to the famous mathematical puzzle, the so-called "1000 poisoned wine bottles", under some assumptions.

1. Introduction

In recent years, the development of new superconductors has become more and more important from both academic and applied viewpoints. However, despite the effort of many researchers, the general guiding principle for exploring new superconductors has not yet been established. Especially in the field where human resources are limited, a method for efficiently discovering superconductors with high throughput is desired.

In this paper we introduce our method to explore new superconductors efficiently: bulk-combinatorial (BC) method. We have succeeded in finding many new superconductors so far [1-6]. This name comes from the combinatorial chemistry method in the field of chemistry [7]. Most of the combinatorial chemistry methods are used for thin film samples. On the other hand, our method is used for ordinary *bulk* samples and does not require expensive equipment. The name BC method was named for this feature.

Usually, crystal synthesis in the field of solid-state physics is performed separately for each compound. In the BC method shown below, various compounds can be synthesized at one time, so that the throughput is very high. In addition, since many elements are mixed in this method, it may be compatible with the synthesis of high entropy alloys [8]. However, it takes time and effort to identify the superconducting phase from the synthesized composite. In this paper, we also present an efficient and systematic algorithm for making this identification.

The BC method can be regarded as a kind of group test. Group testing itself is a very profound mathematical problem [9], but it can be considerably simplified in this case. In this paper, when the mixture contains one new superconductor or even multiple superconductors with different T_c, a method for identifying the new superconductor at the lowest cost is concretely provided. In addition, the cost for the case of including two new superconductors having the same T_c is estimated based on combinatorics theory. These problems are related to a famous mathematical puzzle so called "1000 poisoned wine bottles".

This paper is organized as follows: The experimental details of the BC method are described in Section 2. An algorithm for identifying new superconductor and the discussions is shown in Section 3. Finally, we give some conclusions in Section 4.

2. Bulk combinatorial method

Bulk combinatorial method (BCM) is quite simple.

1. Choose several (typically five or six) kinds of elements, typically at random.
2. Mix the raw materials roughly for example on a piece of aluminum foil in a glove box filled with inert gas. Do not overmix to avoid a homogeneous mixture.
3. Make a pellet by pressing the mixture.
4. Heat the pellet in a short time (typically few minutes) at 800-1000°C in a sealed quartz tube /metal tube/a high-pressure cell.
5. Measure temperature dependent magnetic susceptibility of the pellet.

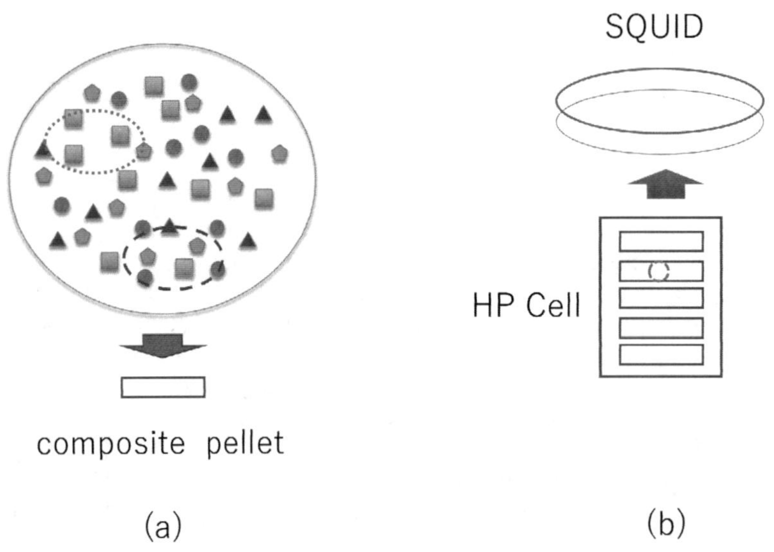

Figure 1. (a) Schematic picture of a roughly mixed composite. Note that this composite is inhomogeneous. For example, the dotted area includes almost one element while the dashed area includes four elements. (b) This composite (a) is pressed into a pellet, and the pellet is heated in a short time. Several pellets can be heated simultaneously in the case of high-pressure (HP) cell. Magnetic susceptibility is measured using SQUID magnetometer. If even a small part of the pellet is superconducting (shown in dashed circle), it can be detected by SQUID.

Steps 2 and 3 are the most important part of this method. If the raw materials are mixed too much using a mortar and pestle, it will result in a homogeneous mixture and less number of different compounds formed in the pellet. By roughly mixing the raw materials and heating it for a short time, the pellet that are expected to contain a large number of different compounds can be obtained as shown in Figure 1(a).

If a superconductor is produced in the pellet, even if only by a small amount (0.01%), it can be detected by a SQUID magnetometer and its T_c can be determined. We can know the known superconductors that can be produced by the combinations of used elements and their T_c using the database [https://mits.nims.go.jp/en/]. If its T_c does not coincide with any known superconductor, then a new superconductor may have formed in the pellet. See Figure 1(b). More details of BCM have been explained in ref. 10.

This method is very efficient in that it can produce a large number of different compounds in a single synthesis. For example, if we mix six elements, see {A,B,C,D,E,F}, then there are maximally 2^6=64 possibilities for the new superconductors. Namely, six simple substances A, …, F, fifteen binary alloys AB, AC, …, EF, twenty ternary alloys ABC, ABD, …, DEF, fifteen quaternary alloys ABCD, ABCE, …, CDEF, six quintuple alloys ABCDE, …, BCDEF, one sextuple alloys ABCDEF, and we can consider empty set {}, which may mean that there is a mistake in measurement.

One might think that it is not easy to identify the compound that exhibit superconductivity in the pellet. We will show how to identify the superconducting phase in the next section.

3. Mathematical Aspect of the Bulk-Combinatorial Method

This section discusses the mathematical aspects of BCM. To simplify the problem, we first make some assumptions:

(a) Only one compound can be made by reacting specific elements.
(b) BCM-synthesized composites include all possible compounds.
(c) The sensitivity of the measurement is very good, and it can be detected if even one superconductor is contained.
(d) There is only one new superconductor contained in the composite.

Assumption (a) is for simplicity, and you don't really have to make this assumption. For example, if two compounds AB and AB_2 are generated from two elements {A,B} and only AB exhibits superconductivity, then the composite consisting of the elements {A,B,…,F} also shows superconductivity because it includes AB. Therefore, it can be treated in the same way as if only AB is generated from the two elements {A,B}.

Assumption (b) is not always guaranteed in practice. Sometimes there is no compound that can be made from a particular combination of elements. However, this combination of elements does not cause superconductivity, so there is no problem anyway. Note that if the number of elements N becomes too large, the number of possible combinations (=2^N) explodes, and the possibility arises that some compounds will not be produced. This is also related to assumption (c).

Regarding assumption (c), high-sensitivity measurement can be performed by using SQUID to measure the magnetic susceptibility. We have empirically found that adding one more element to a composite reduces the volume percentage by about an order of magnitude. Considering the sensitivity of SQUID, it is possible to measure up to 6 element composites without any problem.

Assumption (d) is rather controversial. If the N is increased, the number of compounds contained in the composite will be 2^N. Therefore, it is highly possible that multiple unknown superconductors will be found for a very large N. However, since the T_c of superconductors usually differs depending on the compound, it is possible to identify the two superconductors individually by measuring the magnetic susceptibility. In this case, the superconductor can be identified by N measurements for the composite containing the compound of 2^N, as in the case where the assumption (d) holds. This is shown in subsection 3.1.

If these T_c are the same or very close, it will be more difficult to identify the superconducting phase, as shown in subsection 3.2.

3.1. Order-N algorithm for determining superconducting phase

In this subsection, we show how to identify a superconductor in N measurements for the case where the above assumption (d) holds, i.e., there is only one superconductor in a composite containing 2^N compounds with N elements.

To make the problem more concrete, we will consider the case of N=6 as before. The composite {A,B,C,D,E,F} contains 2^6=64 compounds. If, for example, only the compound ACEF is a superconductor, how many measurements would be needed to identify this compound?

The answer is 6 times. Make six composites of five elements as shown below, and measure the magnetic susceptibility of each of them.

{B,C,D,E,F} (without A)
{A,C,D,E,F} (without B)
{A,B,D,E,F} (without C)
{A,B,C,E,F} (without D)
{A,B,C,D,F} (without E)
{A,B,C,D,E} (without F)

For example, if the superconductor is the compounds ACEF, the result would be:

{B,C,D,E,F} (without A) X
{A,C,D,E,F} (without B) ✔
{A,B,D,E,F} (without C) X
{A,B,C,E,F} (without D) ✔
{A,B,C,D,F} (without E) X
{A,B,C,D,E} (without F) X

Where X means no superconducting signal and ✔ means superconducting signal in the magnetic susceptibility measurement. It is because the composite {B,C,D,E,F} does not include the element A, so the compound ACEF is also not included, and so on.

In this case, read the part marked with a X from the top, ACEF is the superconductor.

Let us clarify this method. Assign a number to each of the 64 = 2^6 compounds in the composite {A,B,C,D,E,F} and display them in binary.

i.e. ACEF = $[101011]_2$.

In this way, all the compounds can be displayed in 6 digits. If you decide whether each "digit" as 0 or 1, you can see which of the 64 compounds it is.

It is straightforward to generalize this result to the N-elements case. We can identify the superconductor by measuring N times using composites of (N-1) elements.

In fact, this is a surprising result. The number of elements in existence is about 100, so if the T_c's were different from each other, we could identify all superconductors with as few as $N \sim 100$ measurements! However, life is not always what one likes. Recall assumptions (b) and (c) above: if N is too large, these assumptions do not hold. Realistically, N should be no more than 10 at most.

So far, we have dealt with the case where hypothesis (d) holds, that is, the case where only one superconductor is contained in the composite. Even if two or more superconductors are included, there is no problem if their T_c are different from each other. However, the problem becomes very difficult if these T_c are the same or very similar. This problem belongs to the combination theory of mathematics and is essentially the same problem as the famous mathematical puzzle called "1000 poisoned wine bottle puzzles".

3.2. *"1000 poisoned wine bottle problem"*

There is a famous mathematical puzzle as follows: [11]

"*You have 1000 bottles of wine, one of which is poisoned. You also have a supply of rats (say). You want to determine which bottle is the poisoned one by feeding the wines to the rats. The poisoned wine takes exactly one hour to work and is undetectable before then. How many rats are necessary to find the poisoned bottle in one hour?*"

In our problem to identify a superconductor within a N-element composite, "wine bottles" correspond to all possible (2^N) compounds, "poisoned wine" to superconductors, and "rats" to magnetic susceptibility measurements. Now $1000 < 1024 = 2^{10}$, so the answer to this puzzle is 10 rats, as shown in the previous subsection.

The above puzzle can be generalized. That is, "one of which (wine bottle) is poisoned" is generalized as " k of which (wine bottles) are poisoned". As far as we know, the general solution to this problem is not yet known. However, in the case of $k = 2$, the following argument holds: [11]

"*Each bottle of wine corresponds to the set of rats who tasted it. Let $\mathscr{F}$ be the family of the resulting sets. If bottles corresponding to sets A and B are poisoned then $A \cup B$ is the set of dead rats. Therefore we can identify the poisoned bottles as long as for all $A,B,C,D \in \mathscr{F}$ such that if $A \cup B = C \cup D$ then we have $\{A,B\}=\{C,D\}$* [12]. *Families with this property are called* (strongly) union-free *and the maximum possible size f(n) of a union free family $\mathscr{F} \subset 2^{[n]}$ has been studied in extremal combinatorics. In the question context, f(n) is the maximum number of bottles of wine which can be tested by n rats.*"

The lower bound of the maximum possible size *f(n)* of a union free family $\mathscr{F}$ is estimated as [13]

$$2^{(n-3)/4} \leq f(n)$$

In the 1000 wine bottles problem, we can take lower bound of *f(n)* as $1000 < 2^{10} = 2^{(n-3)/4}$, which reads $n = 43$ if we use Eqn.(1) in ref.13.. Similarly, in our problem to identify two superconductors within a N-element composite, $n \leq 4N + 3$. In case of $N = 6$ we need 27 times measurements, which is much larger than the one-superconductor case.

4. Conclusion

We explained the bulk combinatorial method that we developed. This method can synthesize many compounds at the same time, so is very efficient and high throughput for exploring new superconductors. The problem of identifying superconducting phases in synthesized composites results in an interesting problem of combinatorics in mathematics, though in most realistic cases there is a simple solution. This method may be used not only for the search for superconductors, but also for the search for other functional materials.

Acknowledgments

This work was partially supported by KAKENHI (Grant No. JP19K03731) from Japan Society for the Promotion of Science (JSPS).

References

[1] Iyo A *et al.* 2014 *Supercond. Sci. Technol.* **27** 025005; *ibid.* 072001
[2] Kinjo T *et al.* 2016 *Supercond. Sci. Technol.* **29** 03LT02
[3] Iyo A *et al.* 2015 *Scientific Reports* **5** 10089
[4] Iyo A, Hase I, Fujihisa H, Gotoh Y, Takeshita N, Ishida S, Ninomiya H, Yoshida Y, Eisaki H and Kawashima K 2019 *Phys. Rev. Mater.* **3** 124802
[5] Iyo A *et al.* 2017 *Inorg. Chem.* **56** 8590
[6] Iyo A *et al.* 2021 *J. Alloys and Compounds* **882** 160733
[7] Nikolaou K C, Hanko R and Hartwig W 2005 *Handbook of Combinatorial Chemistry: Drugs, Catalysts and Materials* (Wiley Online Library)
[8] Yeh J W, Chen S K, Lin S J, Gan J Y, Chin T S, Shun T T, Tsau C H and Chang S Y 2004 *Adv. Eng. Mater.* **6** 299
[9] Du D and Hwang F 2000 *Combinatorial Group Testing and Its Applications* (Singapole: World Scientific)
[10] Iyo A and Kawashima K 2020 *Materials Informatics: Kaihatsu Jirei Saizensen* (in Japanese) ed. Ito S (Tokyo NTS) ISBN: 978-4-86043-708-4 C3058 chapter 7.2 pp.255-266
[11] URL: https://mathoverflow.net/questions/59939/identifying-poisoned-wines
[12] We changed part of the expression in ref.11 for clarity. The original expression is " … such that $A \cup B = C \cup D$ we have {A,B} = {C,D}".
[13] Frankl P and Füredi Z 1984 *Europ. J. Combinatrics* **5** 127

STM/STS study of the local density of states in Cu doped Fe(Se,Te) superconductors

Akira Sugimoto[1], Satoru Ishimitsu[1], Hironori Ohtsubo[1], Kaito Matsumoto[1], Takeshi Saito[1], Toshikazu Ekino[1], Alexander M. Gabovich[2]

[1] Graduate School of Advanced Science and Engineering, Hiroshima University, Higashi-Hiroshima, 739-8521, Japan
[2] Institute of Physics, National Academy of Sciences of Ukraine, 46 Nauka Ave., Kyiv 03028, Ukraine

e-mail: asugimoto@hiroshima-u.ac.jp

Abstract. The scanning tunnel microscopy/scanning tunnel spectroscopy (STM/STS) measurements of the local quasiparticle density of states (LDOS) in Cu_x-$Fe_{1.01-x}Se_{0.4}Te_{0.6}$ (x = 0.01-0.03) superconductors were carried out. Critical temperatures T_c decrease with increasing Cu-impurity doping level x. STM images show clear regular square arrangements of spots with the lattice spacing ~ 0.38 nm as well as the randomly distributed large spots, which are most possibly due to excess Fe atoms. While the STM image resembles that of non-doped Fe(Se,Te), the LDOS map shows various extra spot structures with bright (with higher magnitudes of dI/dV) and dark (with lower magnitudes of dI/dV) regions. With increasing x, the number of such spots grows proportionally, indicating that these spots most probably correspond to the dopant Cu atoms.

1. Introduction

Recent development of the research on iron chalcogenide (11-type) superconductors led to the deeper understanding of their properties, in particular, the new aspects of the nematic order and its influence on superconductivity [1-4]. At the same time, the investigation of the critical temperature (T_c) dependence for those compounds on the artificial impurity atom doping level also remains the useful method to probe the superconducting order parameter [5-7]. Previous studies of impurity effect on the 11-type superconductor Fe(Se,Te) revealed the order parameter of $\pm$s-wave symmetry for up to 5 % of Mn, Co, Ni, Cu and Zn impurities in polycrystalline specimens [6], and the various contents (< 5 %) of Zn, Co, Ni in Fe(Se,Te) single crystals [7].

Scanning tunnel microscopy/scanning tunnel spectroscopy (STM/STS) is a powerful tool to study such impurity effect, because it can measure the local density of state (LDOS) directly in the vicinity of the impurity atoms at the atomic scale. Actually, the STS measurements of Ni-doped cuprate Bi-based superconductors clarified that they possess d-wave order parameters [8]. In addition, we have previously carried out the STS measurements of the local quasiparticle density of states (LDOS) distributions in MgB_2 and found the spot structures of the zero-bias peak being due to certain impurities [9] and the LDOS spots of multi-layered apical F cuprate superconductors due to the irregularities of apical O/F ratios [10-13]. On the other hand, in the case of iron superconductors, the impurity signal at the atomic scale was widely observed as the spot structures created by the excess Fe in pristine FeTe [14], various shapes of spot structures with multiple peaks in iron 111-type LiFeAs [15], and unidirectional LDOS structures in 122 type $CaFe_2As_2$ samples demonstrating the nematic order [16]. The observed peaks were considered to be generated at random by impurity or excess atoms. It shows the usefulness of the LDOS studies in the vicinity of the impurities. Among various metals, copper atoms can be considered as effective dopants bearing in mind the existence of chalcopyrite $FeCuS_2$, despite valence and magnetic differences between Fe and Cu atoms. Hence, we suggested that LDOS measurements in the vicinity of Cu impurity may be fruitful in order to better understand the nature of superconductivity in the materials concerned. In this paper, we present the results of the STM/STS studies of the LDOS distributions in Cu_x-Fe(Se,Te) (x = 0.01 - 0.03) superconductors.

2. Experimental

Cu-doped iron chalcogenide single crystals were synthesized by a self-flux method using the double-sealed quartz tubes. The nominal compositions were Cu_x-$Fe_{1.01-x}$ $Se_{0.4}Te_{0.6}$ (x = 0.01, 0.02, 0.025, 0.03). The tube was heated at 1000 °C and cooled to 400 °C at the rate of – 3 °C/h, then kept 100 hours at 400 °C [17,18], followed by furnace cooling to room temperature. Details of the synthesis procedures were described elsewhere [19,20]. The resistivity measurements were done by a standard dc four probe method. The samples were cooled by liquid helium. STM apparatus used in the experiment was upgraded Omicron LT-UHV-STM system [21-23]. STM/STS measurements were carried out at T = 4.9 K under the condition of 10^{-8} Pa, the chamber being evacuated by ion pump. A constant current mode was applied to observe STM images. The dI/dV curves were obtained by numerical differentiation of the measured I-V characteristics with the spatial interval of ~ 0.20 nm and ~ 0.05 nm. The voltage V refers to the sample bias.

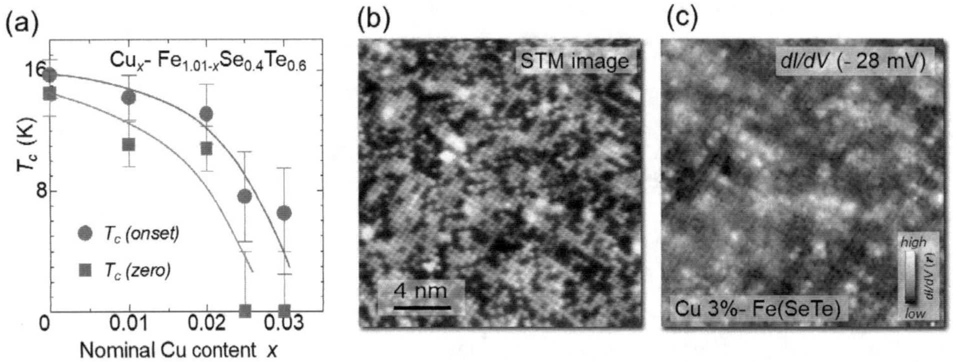

Figure 1. (a) The dependence of the critical temperature T_c on the nominal Cu content x; (b) the STM image (topography) of $Cu_{0.03}$-$Fe_{0.98}Se_{0.4}Te_{0.6}$ single crystal at T = 4.9 K; (c) the spatial distribution of LDOS (dI/dV map) at V = – 28 mV on the area of Fig.1 (b).

3. Results and discussion

Figure 1 (a) shows the x (~nominal Cu doping level) dependence of the critical temperature T_c. Both the onset critical temperature T_c (onset) and zero-resistivity temperature T_c (zero) are presented. Both critical quantities decrease with x. At x = 0.025 and 0.03, the zero resistivity was not observed because of the T_c (zero) fell below 4.2 K. Hence, we denoted it as T_c (zero) = 0. According to the previous data [7], superconductivity was destroyed by 2 - 3 % of Ni doping of the $FeSe_{0.3}Te_{0.7}$ single crystal, while it still survived 3% doping by Co and Zn. Our results demonstrate that the suppression of superconductivity by Cu is somewhat between that for Ni and Co. From the previous data for polycrystalline samples [6], it comes about that the superconductivity disappears for 5 % of Cu doping of $FeSe_{0.5}Te_{0.5}$ sample, which agrees with our results. Assuming that the doped Cu atom works as the non-magnetic impurity, the T_c reduction by the small concentration of Cu might lead to exclude a conventional s-wave Cooper pairing due to the Anderson theorem. On the other hand, the experiments of the neutron scattering of Cu-doped Fe(Se,Te) showed the Cu atoms work as the weak magnetic impurity at T = 100 K [24,25]. The possibility of the magnetism of Cu atom in this compound should not be excluded at present. However, taking into account the fact that the T_c reduction by the "magnetic" impurity of Ni in Fe(Se,Te) was stronger than that by the Cu and the "non-magnetic" Zn impurities[6,7], the mechanism of the T_c reduction is considered to be different from that of the cuprate superconductor, which showed smaller reduction of T_c by the magnetic impurity of Ni than by that of Zn [26]. Hence, other kinds of pairings proposed for iron based superconductors seem preferable [27,28].

Figure 1 (b) shows the typical cleaved surface of the STM image (topography) of the $Cu_{0.03}$-$Fe_{0.98}Se_{0.4}Te_{0.6}$ single crystal. The STM image reveals clear regular square arrangements with the lattice spacing ~ 0.37 nm. The brighter lattice spots correspond to Te atoms and darker ones to Se atoms [19].The fraction of brighter spots is about 60% that agrees with the nominal Te concentration. Moreover, the randomly distributed large spots were also seen, which are most possibly due to the excess Fe atoms. The number of those spots is very small as compared to the Cu dopant level x ~ 0.03. Figure 1 (c) demonstrates the spatial distribution of LDOS, (*dI/dV* map), describing the same area as the STM image Fig. 1 (b). While the STM image resembles those for non-impurity doped Fe(Se,Te) samples with a few excess Fe spots [19], the *dI/dV* map shows extra amount of various spot structures with bright (having higher magnitudes of *dI/dV*) regions. Those spot structures show various kinds of shapes and bias voltage dependences.

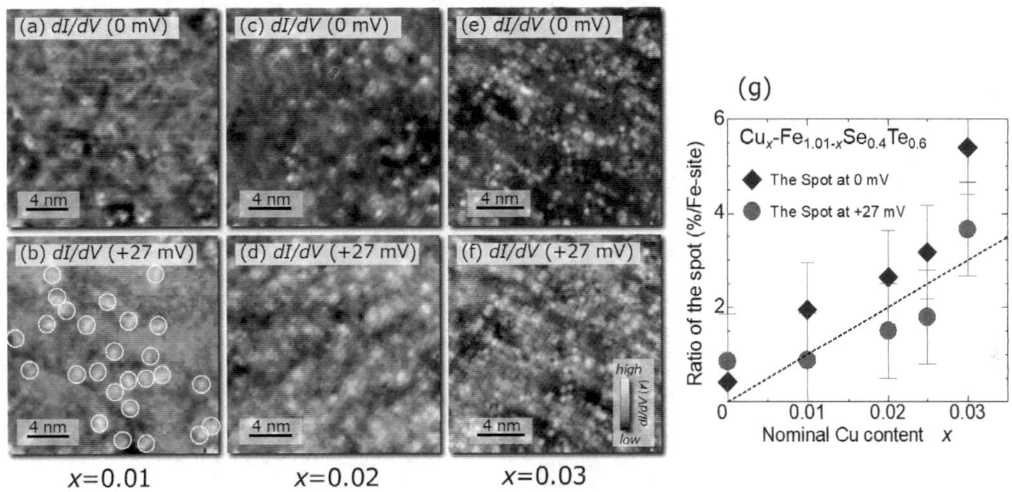

Figure 2. (a) –(f) *dI/dV* maps for various Cu content x (x = 0.01, 0.02, 0.03), at the sample bias voltages V = 0 mV and + 27 mV; (g) the x-dependence of the ratio of the spot number per regular Fe site.

In Fig. 2 (a) - (f), *dI/dV* maps for various Cu content x (x = 0.01, 0.02, 0.03) are depicted at the sample bias voltages of V = 0 mV and + 27 mV. With increasing x, the number of the brighter spots gradually grows. Figure 2 (g) shows x dependence of the ratio of the spot number per Fe site obtained by counting the number of the brighter spots in *dI/dV* maps for V = 0 mV and + 27 mV, including the data for the non-Cu doped sample (x = 0) [19]. Such spots are defined as the brighter region compared with the background and are indicated by white circles in Fig. 2 (b). The number of the Fe sites was estimated as follows. At first, we calculated the number of the basic lattice cells in the observed field of the *dI/dV* map found based on the observed basic constant ~ 0.37 nm of the chalcogenide lattice. Because there are two Fe sites in one basic lattice cell of the chalcogenide layer, the number of the Fe sites is estimated as twice the number of the basic lattice cells in the observed field. The dashed line in the Fig. 2 (g) is the guideline describing the studied ratio. With increasing the Cu doping level x, the number of such spots grows proportionally. Those spot structures are considered to originate from the doped Cu and excess Fe atoms. The ratio of the spot of the non-Cu doped case (x = 0) is roughly ~ 0.5 %/regular Fe-site. Therefore, the contribution from the excess Fe was not the dominant origin of the spot structure, but the latter was mainly due to the doping Cu atoms. The multiple peaks of the spot structures were also observed, especially at $V = 0$ mV (shown as small red dots in Fig. 2 (c)). There were more such spots than it had to be according to the nominal Cu composition. At the same time, in the case of $V = +$ 27 mV, the x dependence of the spot ratio nearly corresponds to the nominal Fe/Cu ratio line. In addition, almost all the spots in the *dI/dV* map at $V = +$ 27 mV are smeared. Taking into account that the spot number correlates the nominal Fe/Cu ratio line, those large smeared spots seems to originate from one Cu dopant each.

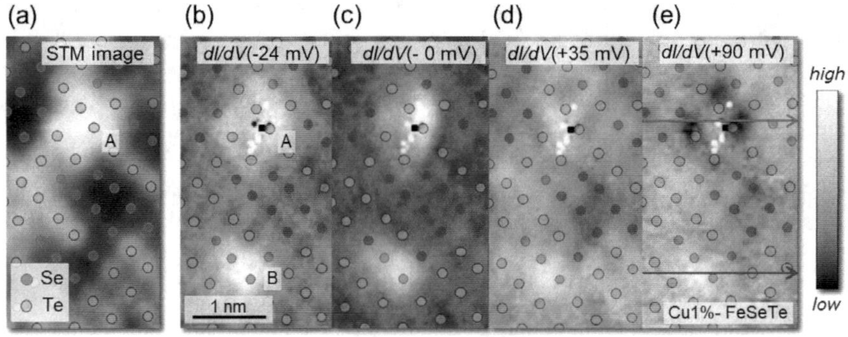

Figure 3. The precise STS *dI/dV* images with the intervals of ~0.05 nm on the $Cu_{0.01}$-$Fe_{1.00}Se_{0.4}Te_{0.6}$ single crystal ($T = 4.9$ K). (a) The simultaneously obtained STM topography ($I = 0.4$ nA, $V = -200$ mV); (b) *dI/dV*(– 24 mV) map; (c) *dI/dV*(0 mV) map; (d) *dI/dV*(+ 35 mV) map; (e) *dI/dV*(+90 mV) map.

The details of such multiple peaks and their location in the crystal are very interesting because the impurity spots in *dI/dV* maps may serve as a guide to elucidate some details of the superconducting state including the symmetry of the order parameter. To clarify them, we also performed the precise measurements of STS with the intervals of ~0.05 nm.

Figure 3 demonstrates the precise *dI/dV* maps measured by STS simultaneously with the STM topography in the vicinity of some spot structures on the surface of the $Cu_{0.01}$-$Fe_{1.00}Se_{0.4}Te_{0.6}$ sample. In the topography shown in Fig. 3 (a), the regular atomic arrangement of the chalcogenides (bright Te and dark Se) can be observed. The Se (darker) and Te (brighter) atoms were indicated by red and light blue dots, respectively. Also, one noticeable bright spot structure was observed (denoted as "A"). In Fig. 3 (b) to (e), the *dI/dV* maps are shown for the same area as in Fig. 3 (a), with red and light blue marks obtained from topography of Fig. 3 (a). In Fig 3 (b) obtained for $V = -24$ mV, two representative kinds of spot structures (noted by the label A and B) are seen. Spot A (which is at the same position as spot A in the topography) has the square shape at $V= -24$ mV. Then, at 0 mV and +35 mV (Fig. 3 (c), (d)), the brighter spot A area becomes narrower. At higher voltage of $V= +$ 90mV, the darker (lower *dI/dV* magnitudes) spots were formed with the four fold symmetry. On the other hand, in contrast to the type A spot, the B one remained bright (higher *dI/dV* magnitude compared with the background) for all observed bias voltages. We note that almost all type-A spots were located near Te atoms of the chalcogenide layer, while B spots were located near Se atoms.

To demonstrate the continuous energy dependence of the characteristic spot structures, we displayed in Fig. 4 (a) and 4 (b) the normalized *dI/dV* spectra obtained along the red and blue lines in Fig. 3 (e), respectively. The dependences are represented by the red-colored contrast. The magnitude of the normalized *dI/dV* was found by dividing over the averaged *dI/dV* curve obtained from whole of the STS area, $dI/dV\,(y,V)/dI/dV_{allave}\,(V)$. In Fig. 4 (a) and (b), the labels A and B indicate the peak of "spot A" and "spot B" on *dI/dV* maps from Fig. 3. Two dashed ovals in Fig. 4 (a) show the suppression of the density of states at the positive (unoccupied) bias side that corresponds to the inversed contrast spot (darker spot) structure shown in Fig. 3 (e). Furthermore, we note that the *dI/dV* magnitude on the Te site is suppressed and 4-fold symmetry structures are formed in the vicinity of spot A of Fig. 3 (e), while that of the other ordinary Te site is enhanced at $V = +90$ mV. On the other hand, the *dI/dV* curves of spot B in Fig. 4 (b) show continuously higher *dI/dV* magnitude as compared to other areas.

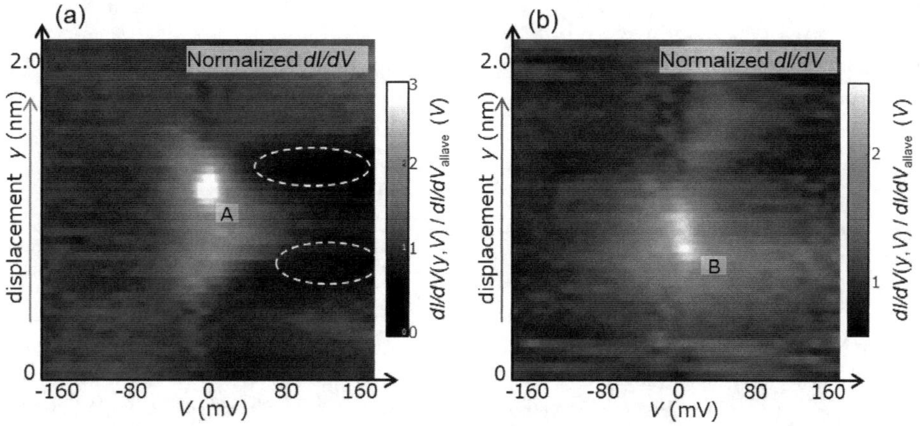

Figure 4. The normalized *dI/dV* spectra represented by the red-colored contrast, (a) along the red and (b) blue line of Fig. 3 (e). The magnitude of the normalized *dI/dV* was obtained by dividing over the averaged *dI/dV* curve from whole of the STS area: $dI/dV(y,V)/dI/dV_{allave}(V)$.

As has been mentioned above, the spot structures are considered to originate from Cu dopants and excess Fe atoms as a minor element. However, there are many kinds of spots with different bias dependences, spatial shapes, contrast inversion etc., indicating that Cu impurity (and/or Fe one) may generate various kinds of spot structures. Those features of the spot structures are not directly correlated to the dopant atom positions. The observations suggest that those observed characteristic contrast changes and/or multiple peak and anti-peak structures on *dI/dV* maps are mainly due to the relationship between impurity position and topmost Se/Te (chalcogenide) arrangement above the Fe layers. Therefore, to elucidate the symmetry of the order parameter or find the nematicity effects by observing the impurity induced LDOS modulation seems to be rather difficult, because of the diversity of the impurity spot structures. One should elaborate a technique, which allows to carry out the direct replacement of Fe atoms at their sites by impurity atoms, to reduce the number of possible additional structures. Then, definite conclusions could be obtained.

4. Summary

STM/STS measurements of the LDOS in doped Cu_x-$Fe_{1.01-x}Se_{0.4}Te_{0.6}$ ($x = 0.01 - 0.03$) superconductors were reported. The critical temperatures T_c linearly decrease with increasing doping level x. While the STM topography reveals clear regular square arrangements, which are similar to those of non-doped Fe(Se,Te), the LDOS map shows extra energy-dependent spot structures with bright and dark regions. With increasing x, the number of such spots grows proportionally, indicating that those spots correspond to copper dopant atoms. From precise measurements of STS, various kinds of the spot structures were observed including the multiple peaked spots and/or inversed contrast ones, which are induced by impurities.

Acknowledgement
We would thank the Natural Science Center for Basic Research and Development, Hiroshima University for supplying liquid helium and the element analysis. This research was supported by Grant-in-Aid for Scientific Research (No. 19540370) and (No. 245403770) from JSPS, Japan, the Project grant N23 of the 2022-2024 Scientific Cooperation Agreement between Poland and Ukraine, as well as Grants 1.4.B-197 of the NASU.

References

[1] Böhmer A E and Kreisel A 2018 *J. Phys.: Condens. Matt.* **30** 023001

[2] Zhang P, Yaji K, Hashimoto T, Ota Y, Kondo T, Okazaki K, Wang Z, Wen J, Gu G D, Ding H, Shin S 2018 *Science* **360** 182

[3] Machida T, Sun Y, Pyon S, Takeda S, Kohsaka Y, Hanaguri T, Sasagawa T and Tamegai T 2019 *Nature Materials* **18** 811

[4] Wang Z, Rodriguez J O, Jiao L, Howard S, Graham M, G. D. Gu G D, Hughes T L, Morr D K and Madhavan V 2020 *Science* **367** 104

[5] Abdel-Hafiez M, Pu Y J, Brisbois J, Peng R, Feng D L, Chareev D A, Silhanek A V, Krellner C, Vasiliev A N and Chen X J 2016 *Phys. Rev. B* **93** 224508

[6] Zhang A M, Xia T L, Kong L R, Xiao J H and Zhang Q M 2010 *J. Phys.: Condens. Matt.* **22** 245701

[7] Inabe T, Kawamata T, Noji T, Adachi T and Koike Y 2013 *J. Phys. Soc. Jpn.* **82** 044712

[8] Hudson E W, Lang K M, Madhavan V, Pan S H, Eisaki H, Uchida S, Davis J C 2001 *Nature* **411** 920

[9] Sugimoto A, Yanase Y, Ekino, Muranaka T and Gabovich A M 2019 *Low Temp. Physics* **45** 1209

[10] Sugimoto A, Shohara K, Ekino T, Watanabe Y, Harada Y, Mikusu S, Tokiwa K and Watanabe T 2009 *Physica C* **469** 1023

[11] Sugimoto A, Shohara K, Ekino T, Watanabe Y, Harada Y, Mikusu S, Tokiwa K and Watanabe T 2010 *Physica C* **470** S160

[12] Sugimoto A, Ekino T, Gabovich A M, Sekine R, Tanabe K and Tokiwa K, 2017 *Physical Review B* **95** 1745081

[13] Ekino T, Sugimoto A, Sekine R, Tanabe K and Tokiwa K 2018 *J. Energ. Chall. and Mech.* **5** 1

[14] Machida T, Kogure K, Kato T, Nakamura H, Takeya H, Mochiku T, Ooi S, Mizuguchi Y, Takano Y, Hirata K and Sakata H 2013 *Phys. Rev. B* **87** 214508

[15] Hanaguri T, Kitagawa K, Matsubayashi K, Mazaki Y, Uwatoko Y and Takagi H 2012 *Phys. Rev.B* **85** 214505

[16] Chuang T-M, Allan M P, Lee J, Xie Y, Ni N, Bud'ko S L, Boebinger G S, Canfield P C, Davis J C 2010 Science **327** 181

[17] Sun Y, TsuchiyavY, Taen T, Yamada T, Pyon S, Sugimoto A, Ekino T, Shi Z, Tamegai T 2014 *Sci. Rep.* **4** 4585

[18] Sun Y, Pan Y, Zhou N, Xing X, Shi Z, Wang J, Zhu Z, Sugimoto A, Ekino T, Tamegai T and Kitano H 2021 *Phys. Rev. B* **103** 224506

[19] Ekino T, Sugimoto A and Gabovich A M 2013 *Low Temp. Phys.* **39** 265

[20] Ekino T, Sugimoto A and Gabovich A M 2018 *IOP Conf. Ser.: Mat. Sci. Eng.* **369** (1) 012024

[21] Sugimoto A, Iwano M, Ishimitsu S, Ekino T and Gabovich A M 2020 *Supercond. Sci. Tech.* **33** 095011

[22] Sugimoto A, Ohtsubo H, Matsumoto K, Ishimitsu S, Iwano M, Ekino T and Gabovich A M 2021 *J. Phys.:Conf. Ser.***1975** 012015

[23] Ekino T, Iwano M, Sugimoto A, Jun Akimitsu J and Gabovich A M 2021 *Phys. Rev. B* **104** 054514

[24] Wen J, Li S, Xu Z, Zhang C, Matsuda M, Sobolev O, Park J T, Christianson A D, Bourret-Courchesne E, Li Q, Gu G, Lee D H, Tranquada J M, Xu G, Birgeneau R J 2013 *Phys. Rev. B* **88** 144509

[25] Tranquada J M, Xu G, Zaliznyak I A 2020 *J. Phys.: Condens. Matter* **32** 374003

[26] Sidis Y, Bourges P, Fong H F, Keimer B, Regnault L P, Bossy J, Ivanov A, Hennion B, Gautier-Picard P, Collin G, Millius D L, Aksay I A 2000 *Phys. Rev. Lett.* **84** 5900

[27] Li J, Guo Y F, Zhang S B, Yuan J, Tsujimoto Y, Wang X, Sathish C I, Sun Y, Yu S, Yi W, Yamaura K, Takayama-Muromachiu E, Shirako Y, Akaogi M and Kontani H 2012 *Phys. Rev. B* **85** 214509

[28] Chen X, Dai P, Feng D, Xiang T and Zhang F-C 2014 *Nat. Sci. Rev.* **1** 371

^{75}As NMR studies on iron-based La1111 superconductors in lightly and heavily electron doped states

T Kouchi[1], K Yoshinaga[1], S Nishioka[1], T Shiota[1], M Yashima[1], H Mukuda[1], K Kuroki[2], S Miyasaka[2], and S Tajima[2]

[1]Graduate School of Engineering Science, Osaka University, Osaka 560-8531, Japan
[2]Graduate School of Science, Osaka University, Osaka 560-0043, Japan

E-mail: kouchi@nmr.mp.es.osaka-u.ac.jp

Abstract. We report ^{75}As NMR experiments in heavily electron-doped LaFePnO$_{0.75}$H$_{0.25}$ (Pn=As$_{1-x}$Sb$_x$ and As$_{1-x'}$P$_{x'}$) compounds with the maximum $T_c^{\max}$ ($\sim$33.1 K), and compare with the previous results in lightly electron-doped LaFePn(O,F) compounds. The T_c of this series can be sensitively controlled by the pnictogen height (h_{Pn}) through the substitution at Pn site, and the electron doping level through the substitution at the O site with H or F. In heavily electron-doped LaFePnO(La1111) compounds, we found that spin fluctuations at low-energies were moderately suppressed upon cooling in the Sb-substituted high T_c compounds with high h_{Pn} ($x \geq 0$), although they are completely suppressed in P-substituted non-superconducting compounds ($x' \geq 0.2$) with lower h_{Pn}. This feature is largely different from that in the lightly doped La1111 compounds with the well-nested Fermi surfaces, where the spin fluctuations are critically enhanced upon cooling. Here, we present the characteristics of spin fluctuations over wide doping region of La1111-based compounds, and discuss the relationship with the superconductivity.

1. Introduction

High-T_c superconductivity (SC) has been observed in heavily electron-doped (HED) states of FeSe-based compounds ($T_c \sim 60$ K)[1, 2] and LaFeAsO$_{1-x}$H$_x$($T_c \sim 36$ K)[3]. These electronic states are dominated by relatively large electron Fermi surfaces (FSs) and a tiny or no hole FS, which differs significantly from the typical Fe-based SCs with well nested hole and electron FSs in the lightly electron doped (LED) regimes. The intrinsic factor for the high T_c SC transition in HED regimes is still not identified, because the universality of the magnetic and orbital order phases near the SC phases is unclear so far. To clarify the universality or diversity of the SC mechanism in the Fe-pnictides, it is necessary to elucidate the evolution of the electronic states over wide doping regions in a same family. In order to discuss the possible origin of re-enhanced high-T_c state in the HED Fe-pnictides, we have performed ^{75}As NMR measurements on the HED La1111-compounds, LaFePnO$_{1-y}$H$_y$ (Pn = As$_{1-x}$Sb$_x$ and As$_{1-x'}$P$_{x'}$)[3, 4, 5, 6]. We note that the electron doping level of this system is widely tuned by F or H substitution (y) from the LED SC phase to the HED SC phase. Moreover, the T_c values in the HED SC state markedly rise up through the substitution of the Pn site with Sb, when $x > 0$, namely, the pnictogen height from the Fe plane (h_{Pn}) increases[6]. On the other hand, the substitution at the Pn site with

P ($x' > 0$) totally suppresses the SC in association with the decreases of h_{Pn} (see Fig. 2(a)). Theoretically, it is expected that the energy level of the tiny hole Fermi pocket dominated by Fe-d_{xy} orbital is quite sensitive to the variation in h_{Pn}, which increases (decreases) when the h_{Pn} is relatively high (low)[7], together with the non-negligible effect of the electron correlation that should be strong (weak)[8, 9].

In previous NMR studies on the HED states, the low energy-spin fluctuations have not been observed, which is different from many cases of the LED SC states[10, 11, 12], where the low energy-spin fluctuations are critically enhanced upon cooling and correlated with the appearance of the SC state. In this report, we show the recent ^{75}As NMR studies on the HED LaFePn($O_{0.75}H_{0.25}$) where the h_{Pn} is controlled by the substitution at Pn site, and compare with the previous studies on the LED states of La1111-based series[11].

2. Experimental

^{75}As NMR experiments were carried out on the polycrystalline powder samples of LaFe$Pn$$O_{0.75}H_{0.25}$ at x = 0, 0.2, 0.3, and 0.4 (Pn=$As_{1-x}Sb_x$), and x' = 0.2, and 0.4 (Pn=$As_{1-x'}P_{x'}$), which were synthesized by solid state reaction method[6]. The T_c values were evaluated by an onset of diamagnetic response in susceptibility and zero-resistivity, which are summarized in Fig. 2(a)[6]. The Knight shift (K) was determined at center peak of ^{75}As NMR spectra indicated in Fig. 1 (a). The nuclear spin-lattice relaxation rate ($1/T_1$) measured at ^{75}As site was determined by fitting a recovery curve for nuclear magnetization to a multiple exponential function $m(t) = 0.1\exp(-t/T_1) + 0.9\exp(-6t/T_1)$.

3. Results and Discussion

Figure 1 (a) shows field-swept ^{75}As NMR spectra at a fixed frequency f = 58.634 MHz. The nuclear quadrupole frequency ($^{75}\nu_Q$) at ^{75}As site for each sample was evaluated from the spectra by measuring the resonance field (H_{res}) as a function of several fixed frequencies (f)[13, 14]. As shown in Fig. 1(b), the linear relation between $^{75}\nu_Q$ and the content $x(x')$ was roughly observed, which is due to the monotonic change in h_{Pn} caused by Sb or P-substitution, because the $^{75}\nu_Q$ is in proportion to an electric field gradient (EFG) at the ^{75}As site[15]. This gives a microscopic evidence that h_{Pn} can be tuned continuously by contents of Sb/P-substitution. Here, the subtle variation in $^{75}\nu_Q$ for $x \geq 0.3$ is owing to that in h_{Pn}[6]. Figure 1 (c) shows the temperature(T) dependences of K. The monotonic decrease of K upon cooling is observed in these compounds, which is universally seen in many electron-doped Fe-based systems, not only in La1111 compounds but also in FeSe-based compounds[16]. Figure 1(d) shows the T dependence of $(1/T_1T)$ in these samples, which also exhibits the decrease from high temperatures to low temperatures. Here, we point out that the absolute values of $(1/T_1T)$ are especially enlarged by the increase of Sb contents, i.e., $x > 0$. According to the previous studies[11, 17, 18], we assume that the $(1/T_1T)$ is given by the sum of $(1/T_1T)_{AFM}$ and $(1/T_1T)_0$. Here, the first term $(1/T_1T)_{AFM}$ represents the component of antiferromagnetic spin fluctuations (AFMSFs) at low energies ($\omega \to 0$) and finite wave vector ($\vec{q}$), as given by,

$$\left(\frac{1}{T_1T}\right)_{AFM} \propto \lim_{\omega\to 0}\sum_{\vec{q}} A_{hf}(\vec{q})^2 \frac{\chi"(\vec{q},\omega)}{\omega}, \quad (1)$$

where $A_{hf}(\vec{q})$ is the hyperfine coupling constant at $\vec{q}$, and $\chi"(\vec{q},\omega)$ is dynamical spin susceptibility at energy ω and at $\vec{q}$. The second term, $(1/T_1T)_0$, is related to the square of $N(E_F)$ that is proportional to the spin-part of K ($\equiv K_s$). Here K_s is given by subtracting K_{chem} ~0.11 % from K[14]. To discuss the characteristics of $(1/T_1T)_{AFM}$, the component of AFMSFs in $(1/T_1T)$, i.e., $(1/T_1T)_{AFM}$, is extracted by evaluating the T dependence of $(1/T_1T)_0$ ($\propto K_s^2$).

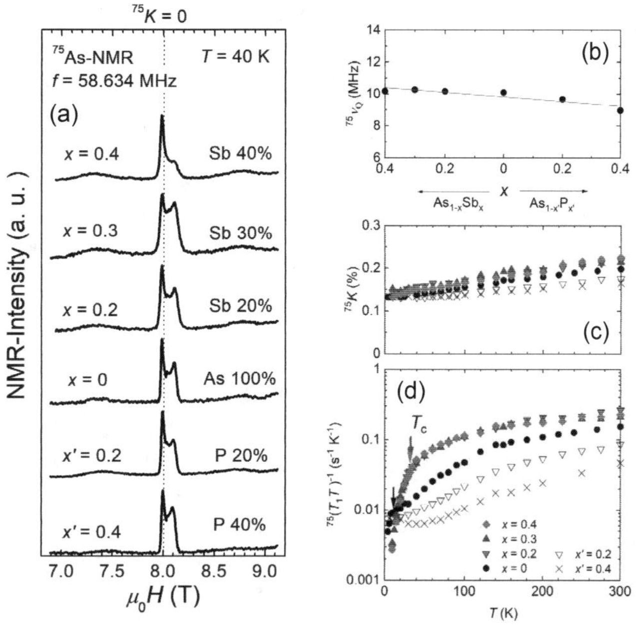

Figure 1. (Color online) (a) ^{75}As NMR spectra (T = 40 K) in LaFePnO$_{0.75}$H$_{0.25}$ for $0 \leq x \leq 0.4$ (Pn=As$_{1-x}$Sb$_x$), and $0 \leq x' \leq 0.4$ (Pn=As$_{1-x'}$P$_{x'}$). (b) The $x(x')$ dependence of $^{75}\nu_Q$ at ^{75}As site, which exhibits a linear dependence for Sb (P) substitutions at As site. The $^{75}\nu_Q$ tends to increase as h_{Pn} becomes higher in many iron-pnictides[10, 15, 26, 27, 28, 29, 30]. (c) T dependences of ^{75}K and (d) $(1/T_1T)$ for all the compounds. It should be noted that, although the T-dependence of ^{75}K is similar, the $(1/T_1T)$ is more enlarged by the Sb substitution ($x > 0$),

Figure 2(b) shows the contour plot of $(1/T_1T)_{\rm AFM}$ as functions of content $x(x')$ and T, which is useful to see the relationship between the h_{Pn}, the AFMSFs, and T_c, This figure clearly shows that the enhanced spin fluctuations at $T > 200$ K are more remarkable in Sb-substituted compounds with higher h_{Pn} and higher T_c. On the other hand, the characteristic spin fluctuations totally disappear in the non-SC P-substituted samples with lower h_{Pn}. These results suggest that such spin fluctuations characterized by a gap-like behavior at low-energies may have some relations with the emergence of high-T_c state in the HED SC3 phase. In this context, these results imply that at least the spin fluctuations may play an indispensable role for re-enhancement of T_c even in the HED regimes of La1111-based compounds. However, we point out that this feature differs from the results on the LED states of SC(=SC1/SC2) phases[11]. The low energy-spin fluctuations significantly develop upon cooling, which have a correlation with the emergence of high T_c state, as displayed in Fig. 2(c). Here the corresponding phase diagram of the SC phases and antiferromagnetic (AFM) phases for LaFePnO$_{1-y}$H$_y$ is also presented by the contour plot in Fig. 2(a)[6]. The presented NMR results in the heavily and lightly electron doped regimes correspond to the compositions in the line B and line C of Fig. 2(a), respectively. The significant difference should be due to the topology of FSs: The FS topology at SC3 phase is characterized by the large electron FS and tiny hole Fermi pocket, which differs from the well-nested FSs expected in the LED SC phases. In this FS topology without the nested FSs,

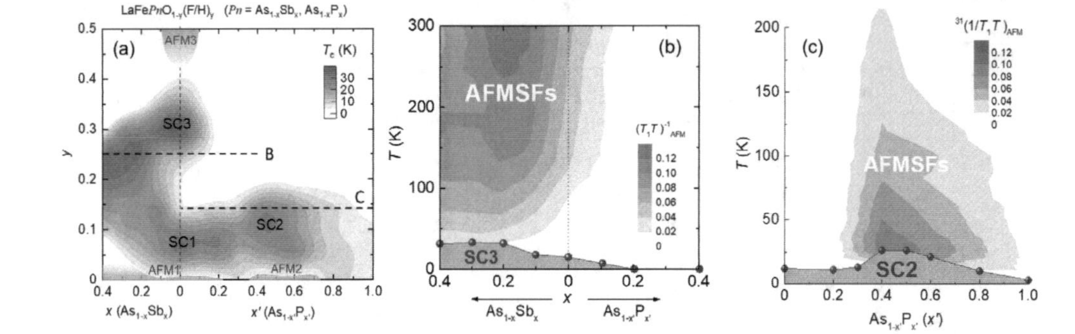

Figure 2. (Color online) (a) Unique phase diagram of SC and AFM states of LaFePnO$_{1-y}$(F/H)$_y$ cited from Kawashima et al[6]. The line B is the composition ($y = 0.25$) of this work in the heavily doped region. The line C is the composition ($y = 0.14$) of our previous NMR studies in the lightly doped region[11]. (b) Contour plot of $(1/T_1T)_{\rm AFM}$ as functions of $x(x')$ and T at the line B. The AFMSFs exhibit a broad peak at T >200 K and are suppressed upon cooling. This feature is remarkable in the Sb-substituted compounds ($x \geq 0$) with larger x or higher h_{Pn}, while it disappears when $x' \geq 0.2$ with lower h_{Pn}. (c) Contour plot in the LED SC1/SC2 phases at line C[11]. The AFMSFs increases toward T_c, which is more significant at higher T_c region[11].

the theoretical studies suggest that the spin fluctuations at finite energies derived mainly from d_{xy} orbital play a rather positive role in the HED SC states[19, 20], even when the low energy spin fluctuations are suppressed. In addition, it has been suggested theoretically that, in the case of unconventional SCs caused by the effects of strong electron correlations, the low energy spin fluctuations will act negatively for SC[19, 20, 21, 22, 23, 24]. Thus, it is suggested that one possible factor for the enhancing T_c in the HED SC3 phase can be attributed to the presence of the finite energy spin fluctuations with gap-like feature at at low-energies. On the other hand, the electronic states of the LED SC states are characterized by the well-nested FSs, which are dominated by d_{xz}/d_{yz} orbitals. The contribution of the d_{xy} orbital is less important in the case of the LED regimes, since the $h_{\rm Pn}$ is low[7], and hence the well-nested FSs derived mainly from d_{xz}/d_{yz} orbitals bring about the low energy spin fluctuations developed critically upon cooling[10, 11, 18, 25]. The remaining problem is to elucidate the possible contribution of nematic fluctuations derived from the orbital degrees of freedom, because the structural transition with the other AFM order phase is present in the vicinity of the SC3 phase, as seen in Fig. 2(a). However, there is no information of some other low energy local fluctuations except for AFMSFs in La1111-based families, since the large single crystals is not available. The further experiments that probes the possible orbital fluctuations are desired to discuss the SC mechanism of Fe-based superconductors.

4. Summary

Systematic ^{75}As NMR studies in HED LaFePnO$_{0.75}$H$_{0.25}$ revealed that the finite energy spin fluctuations with a gap at low-energies are enhanced in the Sb-substituted compounds with high T_c and high h_{Pn}. On the other hand, such characteristics of the spin fluctuations disappears in non-SC P-substituted compounds. Our findings suggest that this type of AFMSFs is one of the indispensable factors for re-emergence of hight T_c state in HED regime. This feature is different from that in the LED states of La1111, where the low energy- spin fluctuations developed critically upon cooling. Although the types of AFMSFs is different in doping levels, it is suggested that the spin fluctuations may play an indispensable role on the La1111-based compounds in wide electron-doping regimes. Recently, the similar spin fluctuations suppressed moderately at low-energies was reported in the HED state of the intercalated FeSe-based compounds[16]. Toward understanding of the SC mechanism over various Fe-based families, further spectroscopic experiments across the Fe-based compound families should be performed to clarify possible roles of nematic fluctuations due to the orbital degrees in the future.

Acknowledgements

This work was partially supported by KAKENHI (18K18734), the Mitsubishi Foundation, the Tanigawa Fund, and the Murata Science Foundation. One of the author (T.K.) is supported by a JSPS Fellowship (21J14053), and the Kato Foundation (KS-3227).

References

[1] Wang Q -Y, Li Z, Zhang W -H, Zhang Z -C, Zhang J -S, Li W, Ding H, Ou Y -B, Deng P, Chang K, Wen J, Song C -L, He K, Jia J -F, Ji S -H, Wang Y -Y, Wang L -L, Chen X, Ma X -C and Xue Q -K 2012 *Chin. Phys. Lett.* **29** 037402

[2] He S, He J, Zhang W, Zhao L, Liu D, Liu X, Mou D, Ou Y, Wang Q, Li Z, Wang L, Peng Y, Liu Y, Chen C, Yu L, Liu G, Dong X, Zhang J, Chen C, Xu Z, Chen X, Ma X, Xue Q -K and Zhou X -J 2013 *Nat. Mater.* **12** 605

[3] Iimura S, Matsuishi S, Sato H, Hanna T, Muraba Y, Kim S W, Kim J E, Takata M and Hosono H 2012 *Nat. Commun.* **3** 943

[4] Hiraishi M, Iimura S, Kojima K M, Yamamura J, Hiraka H, Ikeda K, Miao P, Ishikawa Y, Torii S, Miyazaki M, Yamaguchi I, Koda A, Ishii K, Yoshida M, Mizuki J, Kadono R, Kumai R, Kamiyama T, Otomo T, Murakami Y, Matsuishi S and Hosono H 2014 *Nat. Phys.* **10** 300

[5] Miyasaka S, Uekubo M, Tsuji H, Nakajima M, Tajima S, Shiota T, Mukuda H, Sagayama H, Nakao H, Kumai R and Murakami Y 2017 *Phys. Rev.* B **95** 214515

[6] Kawashima T, Miyasaka S, Tsuji H, Yamamoto T, Uekubo M, Takemori A, Lai K -T and Tajima S 2021 *Sci. Rep.* **11** 10006

[7] Kuroki K, Usui H, Onari S, Arita R and Aoki H 2009 *Phys. Rev.* B **79** 224511

[8] Miyake T, Nakamura K, Arita R and Imada M 2010 *J. Phys. Soc. Jpn.* **79** 044705

[9] Hirayama M, Misawa T, Miyake T and Imada M 2015 *J. Phys. Soc. Jpn.* **84** 093703

[10] Oka T, Li Z, Kawasaki S, Chen G -F, Wang N -L and Zheng G -Q 2012 *Phys. Rev. Lett.* **108** 047001

[11] Shiota T, Mukuda H, Uekubo M, Engetsu F, Yashima M, Kitaoka Y, Lai K -T, Usui H, Kuroki K, Miyasaka S and Tajima S 2016 *J. Phys. Soc. Jpn.* **85** 053706

[12] Sakano F, Nakamura K, Kouchi T, Shiota T, Engetsu F, Suzuki K, Horikawa R, Yashima M, Miyasaka S, Tajima S, Iyo A, Guo Y -F, Yamaura K, Takayama-Muromachi E, Yogi M and Mukuda H 2019 *Phys. Rev.* B **100** 094509

[13] Shimizu S, Tabata S, Iwai S, Mukuda H, Kitaoka Y, Shirage P M, Kito H and Iyo A 2012 *Phys. Rev.* B **85** 024528

[14] Kouchi T, Nishioka S, Suzuki K, Yashima M, Mukuda H, Kawashima T, Tsuji H, Kuroki K, Miyasaka S and Tajima S submitted.

[15] Mukuda H, Terasaki N, Kinouchi H, Yashima M, Kitaoka Y, Suzuki S, Miyasaka S, Tajima S, Miyazawa K, Shirage P -M, Kito H, Eisaki H and Iyo A 2008 *J. Phys. Soc. Jpn.* **77** 093704

[16] Nishioka S, Kouchi T, Suzuki K, Yashima M, Mukuda H, Kodani M, Mita K, Kakuto T, Lee J -H, Fujii T and Kambe T 2021 *J. Phys. Soc. Jpn.* **90** 124709

[17] Ning F -L, Ahilan K, Imai T, Sefat A S, McGuire M A, Sales B C, Mandrus D, Cheng P, Shen B and Wen H -H 2010 *Phys. Rev. Lett.* **104** 037001

[18] Mukuda H, Engetsu F, Yamamoto K, Lai K -T, Yashima M, Kitaoka Y, Takemori A, Miyasaka S and Tajima S 2014 *Phys. Rev.* B **89** 064511

[19] Nakata M, Ogura D, Usui H and Kuroki K 2017 *Phys. Rev.* B **95** 214509

[20] Matsumoto K, Ogura D and Kuroki K 2020 *J. Phys. Soc. Jpn.* **89** 044709

[21] Millis A J, Sachdev S and Varma C M 1988 *Phys. Rev.* B **37** 4975

[22] Kuroki K, Higashida T and Arita R 2005 *Phys. Rev.* B **72** 212509

[23] Matsumoto K, Ogura D and Kuroki K 2018 *Phys. Rev.* B **97** 014516

[24] Rodriguez J P 2021 *Phys. Rev.* B **103** 184513

[25] Kitagawa S, Iye T, Nakai Y, Ishida K, Wang C, Cao G -H and Xu Z -A 2014 *J. Phys. Soc. Jpn.* **83** 023707
[26] Grafe H J, Lang G, Hammerath F, Paar D, Manthey K, Koch K, Rosner H, Curro N J, Behr G, Werner J, Leps N, Klingeler R, Klauss H H, Litterst F J and Büchner B 2009 *New J. Phys.* **11** 035002
[27] Mukuda H, Furukawa S, Kinouchi H, Yashima M, Kitaoka Y, Shirage P M, Eisaki H and Iyo A 2012 *Phys. Rev. Lett.* **109** 157001
[28] Kinouchi H, Mukuda H, Yashima M, Kitaoka Y, Shirage P M, Eisaki H and Iyo A 2011 *Phys. Rev. Lett.* **107** 047002
[29] Li Z, Ooe Y, Wang X -C, Liu Q -Q, Jin C -Q, Ichioka M and Zheng G -Q 2010 *J. Phys. Soc. Jpn.* **79** 083702
[30] Kitagawa K, Mezaki Y, Matsubayashi K, Uwatoko Y and Takigawa M 2011 *J. Phys. Soc. Jpn.* **80** 033705

The interplay between the incommensurate spin correlations and superconductivity in Mn substituted $La_{1.90}Sr_{0.10}CuO_4$

Masaki Fujita, Masanori Enoki[1]

Institute for Materials Research, Tohoku University, Sendai, Miyagi 980-8577, Japan

E-mail: fujita@imr.tohoku.ac.jp

Abstract. The Mn substitution effect on superconductivity and spin correlations in $La_{1.90}Sr_{0.10}Cu_{1-y}Mn_yO_4$ was investigated by means of magnetic susceptibility and neutron scattering measurements. The superconducting (SC) transition temperature (T_c) and the SC volume fraction are reduced by Mn substitution with $y = 0.01$. It was clarified that the peak-splitting of the incommensurate (IC) low-energy spin excitations narrows by 30% and the magnetic ordering temperature is significantly enhanced in the $y = 0.01$ sample. Thus, the SC state and the magnetic ordering state coexist at low temperature. Furthermore, incommensurability (δ), corresponding to half the distance between the IC peaks, and T_c ($\sim$ 18 K) of $y = 0.01$, both follow the linear T_c–δ relation established for the pristine $La_{2-x}Sr_xCuO_4$ (LSCO). Therefore, although T_c varies as a result of Mn substitution, spin correlations are closely coupled to superconductivity as is the case for LSCO.

1. Introduction

The relationship between spin correlations and superconductivity in cuprate oxide has been studied for a long time. It is known that underdoped (UD) $La_{2-x}Sr_xCuO_4$ (LSCO) exhibits low-energy magnetic fluctuations centered at four incommensurate (IC) positions ((0.5$\pm\delta$, 0.5, 0) and (0.5, 0.5$\pm\delta$, 0)) in the reciprocal lattice space [1, 2]. The incommensurability (δ) linearly increases upon hole-doping, and the superconducting (SC) transition temperature (T_c) in the UD region is proportional to the value of δ. Furthermore, the low-energy spin excitation in the optimally-doped LSCO opens the energy gap below T_c, and IC spin fluctuations concomitantly diminish with the decrease of T_c in the overdoped region [3]. Thus, according to the hole concentration (p), spin fluctuations are connected with superconductivity in the entire SC phase.

To better understand the characteristics of spin correlations, the impurity substitution effects have been intensively studied. In contrast to the rapid suppression of both T_c and the SC volume fraction (V_c) by Ni and Fe substitution, T_c in LSCO with fixed p is weakly suppressed by Mn substitution, while V_c decreases as in Fe substitution [4, 5]. Huang and co-workers claimed that the suppression of superconductivity as a result of Mn substitution is caused by carrier localization rather than the magnetic pair breaking effect [6]. Therefore, the magnetic

[1] Present address: Institute of Multidisciplinary Research for Advanced Materials, Tohoku University, Sendai, Miyagi 980-8577, Japan

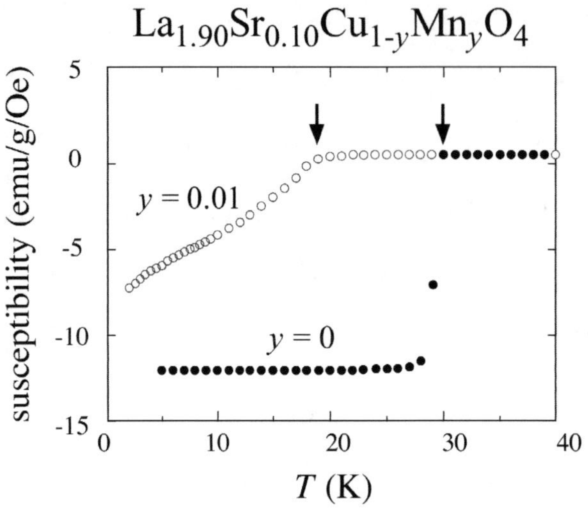

Figure 1. Temperature dependence of magnetic susceptibility in $La_{1.90}Sr_{0.10}Cu_{1-y}Mn_yO_4$ with $y = 0$ and 0.01. The arrows indicate the superconducting transition temperature.

impurity effect on spin correlations differs from other 3*d* transition metals and the relevance to superconductivity could be investigated in Mn-substituted LSCO.

In this study, low-energy spin fluctuations at 1–2 meV and static spin correlations were examined for a single crystal of UD $La_{0.19}Sr_{0.10}Cu_{1-y}Mn_yO_4$ (LSCMO) with $y = 0$ and 0.01 by neutron scattering measurements.

2. Experimental Details

2.1. Sample preparation

For the crystal growth, we first prepared the powder sample of LSCMO with $y = 0$ and 0.01 by a standard solid-state reaction. The dried powder of La_2O_3, $SrCO_3$, CuO and MnO_2 were mixed with an appropriate molar ratio and the mixed powder was sintered at 960 °C for 24 hours with intermediate re-grindings. The sintered sample was pressed in the shape of a rod and fired at 1200°C for 12 hours. The crystal was grown using a traveling-solvent floating-zone method [2, 7]. A 100 mm-long crystalline rod with a diameter of ~8 mm was successfully obtained. We cut the rod into 35 mm-long pieces and annealed them at 900°C for 72 hours under oxygen gas flow.

2.2. Neutron scattering measurement

For the neutron scattering measurements, we co-aligned two crystalline pieces of LSCMO with $y = 0$ and 0.01 individually, using neutron beams on the triple-axis spectrometer AKANE installed at the JRR-3 research reactor operated by the Japan Atomic Energy Agency. Inelastic neutron scattering (INS) and elastic neutron scattering (ENS) measurements were performed on the triple-axis spectrometers TOPAN and HER at JRR-3, respectively. We employed an E_f-fix mode with a neutron energy of 13.5 meV and the collimator sequences (CS) of 50′-30′-30′-150′ at TOPAN. (E_f is the energy of scattering neutrons.) Sapphire and pyrolytic graphite filters were placed in the beam to achieve a better signal to noise ratio by reducing higher-order neutrons. For the ENS measurement at HER, incident neutron energy of 3.5 meV and CS of 32′-40′-80′-80′ were selected. The sample was mounted to a closed cycle refrigerator so that the *c*-axis was perpendicular to the scattering plane.

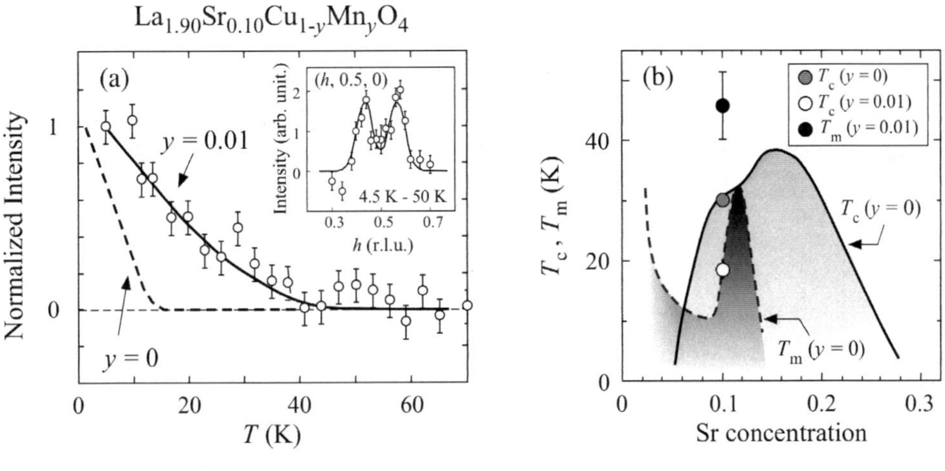

Figure 2. (a) Temperature dependence of intensity of elastic incommensurate magnetic peak in $La_{1.90}Sr_{0.10}Cu_{1-y}Mn_yO_4$ with $y = 0.01$. The dashed line is the result for $y = 0$ taken from Ref. 9. The inset shows the deferent spectra obtained by subtracting data at 50 K from that at 4.5 K. (b) Magnetic and superconducting phase diagram for $La_{1.90}Sr_{0.10}Cu_{1-y}Mn_yO_4$. The present results are plotted with circles.

3. Results and discussion

3.1. *Superconducting property*

To investigate the Mn substitution effect on superconductivity, we measured the magnetic susceptibility after the zero-field-cooling process (χ) by using a superconducting quantum interference device. The sample used for the susceptibility measurement was taken from the same crystalline rod prepared for the INS measurements. As shown in Fig. 1, we observed the diamagnetic signal corresponding to the appearance of the SC state for both the LSCMO with $y = 0$ and 0.01 samples at low temperatures. T_c for $y = 0$ (30 K) is consistent with the previously reported value [8] and is reduced by 1% Mn substitution. χ in the $y = 0$ sample drops rapidly upon cooling, indicating the sharp SC transition. Conversely, χ decreases almost linearly below ~18 K in the $y = 0.01$ sample. Thus, Mn substitution induces an SC state which develops gradually with decreasing temperature. The sizable $|\chi|$ value at the lowest temperature measured for $y = 0.01$ suggests that the bulk SC state is realized in the Mn substituted LSCO.

3.2. *Static magnetism*

The appearance of a magnetic ordered state with the IC correlation was confirmed, as shown by the subtracted data in Fig. 2(b). The temperature dependence of the peak intensity indicates a magnetic ordering temperature (T_m) of ~45 K in the 1% Mn substituted sample. For reference, we present a plot of magnetic intensity as a function of temperature for pristine LSCO with $x = 0.10$ [9]. The intensity was normalized by the value at the lowest temperature measured. Since T_m in the pristine sample is ~15 K, T_m is markedly enhanced by a tiny amount of Mn substitution. Furthermore, the magnetic order develops rather gradually in $y = 0.01$, suggesting that the magnetism in the $y = 0.01$ sample is inhomogeneous. The evaluated T_c and T_m for the present samples are plotted in the magnetic and SC phase diagram. (See Fig. 2(b).) The present results suggest that the bulk SC state in the 1%-substituted sample coexists with the static IC magnetic state, which is strengthened by Mn substitution.

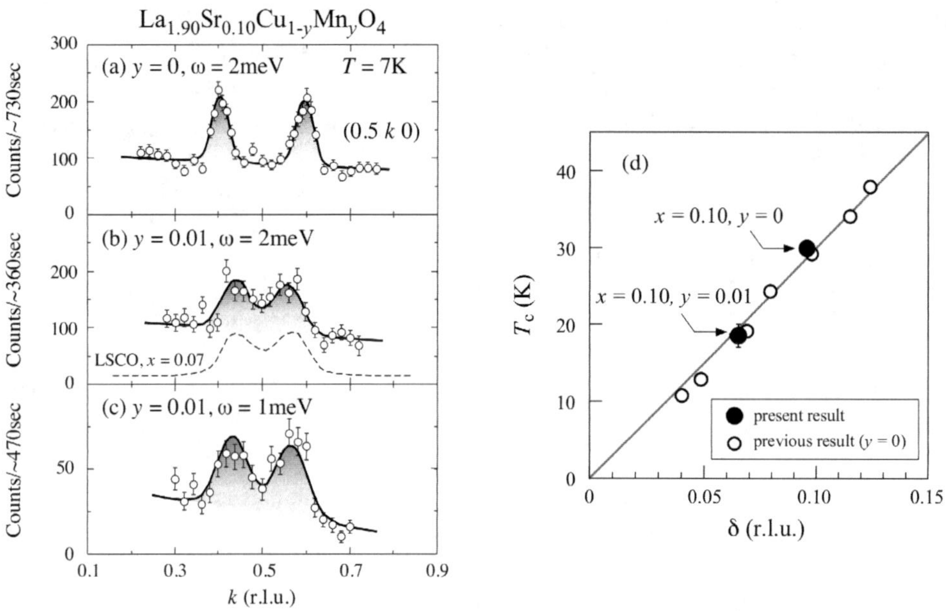

Figure 3. Inelastic neutron scattering spectra in $La_{1.90}Sr_{0.10}Cu_{1-y}Mn_yO_4$ (a) with $y = 0$ at $\omega = 2$ meV, (b) $y = 0.01$ at $\omega = 2$ meV , and (c) $y = 0.01$ at $\omega = 1$ meV. The spectra were measured at 7 K along the k-direction through the (0.5, 0.5. 0) reciprocal position. The result for $La_{1.93}Sr_{0.07}CuO_4$ were taken from Ref. 10. (d) Superconducting transition temperature (T_c) as a function of incommensurability (δ). The closed and open symbols indicate the present and previous results, respectively. The previous results for $La_{2-x}Sr_xCuO_4$ are from Kimura *et al.* [9]

3.3. Spin fluctuations

We next present the results of our INS measurements. The low-energy INS spectra measured at constant energies of 1–2 meV at 7 K are shown in Fig. 3. Compared to the spectrum for the pristine sample, those for the 1% Mn-substituted sample exhibit broader IC peaks with narrower peak splitting. The incommensurability (δ) is defined as half of the distance between two IC peaks. At 2 meV, δ is evaluated as 0.065 ± 0.003 reciprocal lattice units (r.l.u.) for $y = 0.01$ and 0.096 ± 0.001 r.l.u. for $y = 0$ using two-gaussian-peak fitting.

The value of δ for $y = 0.01$ is close to the result for pristine LSCO with $x = 0.07$. We note that the peak profile for LSCMO with $y = 0.01$ is quite similar to that for LSCO with $x = 0.07$, shown by a dashed line in Fig. 3(b) [10]. This similarity recalls the effect of carrier capture associated with the difference in the valences between Cu^{2+} and doped $3d$ transition metals [11]. Because the valence of Mn ions is higher than that of Cu^{2+}, the substitution of Mn induces the reduction of p. The coexistence of trivalent and tetravalent states for doped Mn ions was reported as a result of the Mn substitution effect on T_c in double-doping systems $La_{1.85-y}Sr_{0.15+y}Cu_{1-y}Mn_yO_4$ and $La_{1.85-2y}Sr_{0.15+2y}Cu_{1-y}Mn_yO_4$ [4, 5]. In these system, T_c is almost constant with respect to y, meaning that the reduction of p attributed to Mn substitution can be compensated by additional Sr doping.

If all substituted Mn were to form a trivalent state (tetravalent state), the captured hole by Mn ions associated with the difference in the valences between Mn^{3+}(Mn^{4+}) and Cu^{2+} is 0.01(0.02) holes per Cu ion. Therefore, the value of δ, which is identical to p [1], would be expected to decrease by 0.01 (0.02) r.l.u. However, the experimentally determined reduction in δ due to 1% Mn substitution was 0.031 r.l.u. We note that T_c of $\sim$18 K in $y = 0.01$ is close to that for LSCO with $x = 0.07$ ($T_c \sim 17$ K) rather than $x = 0.08$ ($T_c \sim \sim 21$ K) [8, 10]. These results suggest that the Mn substitution effect has other factors that reduce both δ and T_c in addition to the reduction of p. More importantly, the values of T_c and δ in $y = 0.01$ follow the linear T_c–δ relation for pristine LSCO as seen in Fig. 3(d). Therefore, although Mn substitution causes T_c to vary, spin correlations are closely coupled to superconductivity as is the case for LSCO. That is, although the hole concentration is one of the parameters that control both T_c and δ, our present results suggest more direct relation between spin correlations and superconductivity.

4. Summary

The effect of Mn substitution for Cu on spin correlations was studied for $La_{1.90}Sr_{0.10}Cu_{1-y}Mn_yO_4$ via neutron scattering measurements. The magnetic ordering temperature is enhanced by Mn substitution, and the magnetically ordered state coexists with the bulk superconducting state at low temperature. Mn substitution decreases both incommensurability (δ) for the low-energy spin excitations and the superconducting transition temperature (T_c), whereas those values follow the linear T_c–δ relation for pristine $La_{2-x}Sr_xCuO_4$. Therefore, even T_c varies as a result of Mn substitution, spin correlations are closely connected with superconductivity in $La_{1.90}Sr_{0.10}Cu_{0.99}Mn_{0.01}O_4$.

Acknowledgement

We thank K. Iwasa for sharing machine-time for the neutron scattering measurement at TOPAN, and Adachi, H. Kimura, M. Matsuda, and H. Yamase for stimulating discussions. We also thank M. Sakurai for supporting the crystal growth. This work was carried out by the JRR-3 general user program managed by the Institute for Solid State Physics of the University of Tokyo. This work was financially supported by JSPS/MEXT Grants-in-Aids for scientific Research (KAKENHI) Grant Nos. 21H04448 and 21H04987.

References

[1] Yamada K, Lee C H, Kurahashi K, Wada J, Wakimoto S, Ueki S, Kimura H, Endoh Y, Hosoya S, Shirane G, Birgeneau R J, Greven M, Kastner M A, Kim Y J, 1998 *Phys. Rev.* B **57** 6165
[2] Fujita M, Yamada K, Hiraka H, Gehring P M, Lee S H, Wakimoto S, Shirane G, 2002 *Phys. Rev.* B **65** 064505
[3] Wakimoto S, Zhang H, Yamada K, Swainson I, Kim H, Birgeneau R J, 2004 *Phys. Rev. Lett.* **92** 217004
[4] Zhang C, Zhang Y 2003 *Phys. Rev.* B **68** 054512
[5] Wang C, Sun Y, Zhang Y, 2006 *Supercond. Sci. Technol.* **19** 122
[6] Huang H, Tan M, Zhang C, Zhang Y, 2005 *Physica C* **421** 56
[7] Fujita M, Goka H, Yamada K, Tranquada J M, Regnault L P, 2004 *Phys. Rev.* B **70** 104517
[8] Takagi H, Ido T, Ishibashi S, Uota M, Uchida S, Tokura Y, 1989 *Phys. Rev.* B 40, 2254
[9] Kimura H, Hirota K, Matsushita H, Yamada K, Endoh Y, Lee S H, Majkrzak C F, Erwin R, Shirane G, Greven M, Lee Y S, Kastner M A, Birgeneau R J 1999 *Phys. Rev.* B **59**, 6517
[10] Hiraka H, Endoh Y, Fujita M, Lee Y S, Kuluda J, Ivanov A, Birgeneau R J, 2001 *J. Phys. Soc. Jpn.* **70** 853
[11] Hiraka H, Ohta S, Wakimoto S, Matsuda M, Yamada K 2007 *J. Phys. Soc. Jpn.* **76** 074703
[12] Fujita M 2012 *Physica C* **481** 23, and references therein

Dc vortex-flow resistivity in an epitaxially grown MgB_2 film probed by pulsed current measurements

R Ueno[1], K Ienaga[1], S Maegochi[1], H Shishido[2] and S Okuma[1]

[1]Department of Physics, Tokyo Institute of Technology, 2-12-1, Ohokayama, Meguro-ku, Tokyo 152-8551, Japan
[2]Department of Physics and Electronics, Osaka Prefecture University, 1-1 Gakuen-cho, Naka-ku, Sakai, Osaka 599-8531、Japan

E-mail: sokuma@o.cc.titech.ac.jp

Abstract. Using pulsed currents, we measure the dc vortex-flow resistivity ρ_f in a fast-flow regime for an epitaxially grown MgB_2 film. In the vortex-solid phase the current-voltage ($I-V$) characteristics in different magnetic fields B exhibit a crossover from nonlinear to linear behavior with an increase in I, indicating a change from pinning-dominated to viscous-force dominated flow. $\rho_f(B)$ extracted from the slope of the linear part of the $I-V$ data shows an unusual field dependence with strong enhancement at low B similar to that reported from microwave surface impedance. The results suggest that two-band properties can be also detected by dc measurements with pulsed currents in MgB_2 films.

1. Introduction

MgB_2 [1] is a multiband superconductor, in which two distinct bands contribute to superconductivity with different gap sizes [2, 3, 4, 5, 6, 7, 8, 9, 10, 11, 12]. The two-gap nature is reflected in a flux-flow resistivity ρ_f [13, 14, 15]. According to the Bardeen-Stephen theory [16], ρ_f in conventional s-wave superconductors increases proportionally to the magnetic field B, reaching a normal-state resistivity ρ_n at an upper critical field B_{c2}. In MgB_2, however, it has been revealed by microwave surface impedance measurements [13] that $\rho_f(B)$ rises at low fields more steeply than expected by the Bardeen-Stephen theory and approaches slowly ρ_n at B_{c2}. This unusual field dependence of ρ_f has been interpreted in terms of the two-band property that the superconductivity with the smaller π-band gap is suppressed at a field well below B_{c2}.

In conducting the microwave impedance measurements [13, 14, 17, 18, 19], very specialized techniques and analyses are needed. Thus, one may ask whether the unusual behavior of ρ_f reported by microwave impedance can be also detected by conventional dc resistivity in the fast-flow regime where pinning is ineffective. To our knowledge, however, this question has not been answered experimentally. This is probably because the large dc current is needed to realize the ideal flux flow with no pinning effects, in particular, for bulk crystals, which in turn generates large heat. Therefore, it is interesting to clarify to what extent the dc measurements can probe the multiband properties revealed by microwave measurements. It is also of interest to investigate whether the two-band nature of MgB_2 can emerge not only in high-quality single crystals with a transition temperature of $T_c \simeq$39 K but also in epitaxially grown films with reduced T_c.

In this work, we measure the field dependence of the dc vortex-flow resistivity $\rho_f(B)$ in the fast-flow regime for an epitaxially grown MgB_2 film [20, 21] with a thickness of about 100 nm by employing pulsed-current measurements. The current-voltage ($I-V$) characteristics taken in the vortex-solid phase at equilibrium show a crossover from nonlinear to linear behavior with an increase in I, indicating a change from pinning-dominated to viscous-force (driving-force) dominated flow. The values of $\rho_f(B)$ extracted from the slope of the linear part of the $I-V$ characteristics exhibit an unusual field dependence similar to that observed by microwave measurements [13, 14]. The results suggest that the two-band properties of MgB_2 can be also detected by dc measurements with pulsed currents in epitaxially grown films with $T_c \simeq$20 K lower than 39 K.

2. Experimental

The MgB_2 film [20, 21] with a thickness of about 100 nm was grown on a sapphire substrate by molecular beam epitaxy. The dc resistivity was measured using a standard four-probe method with voltage probes separated by 4.2 mm. The zero-field ($B=0$) linear resistivity ρ at 25 K in the normal state is $\rho_n \sim 1\times10^{-8}$ Ωm. The superconducting transition temperature in $B=0$ where ρ falls below the experimental sensitivity is T_c =20.3 K. The magnetic field B up to 5 T was directed along the c-axis, i.e., perpendicular to the plane of the sample. By applying a current in the mixed state, the vortices are driven along the sample width of 1.1 mm. The $I-V$ characteristics were measured in the continuous-current and the pulsed-current mode. In the latter mode, the voltage induced by vortex motion was measured using an oscilloscope (Rohde and Schwarz RTO2024) with resolution of 10 MHz [22, 23]. The pulsed current technique was used successfully for the mode-locking resonance at 0.05 K to detect the moving vortex lattice [24, 25].

3. Results and discussion

First, we measured the temperature dependence of ρ in different B and estimated the critical temperature $T_c(B)$ separating the vortex solid from the vortex liquid, and the crossover temperature $T_{c2}(B)$ between the vortex liquid and normal state. Here, $T_c(B)$ is defined as a temperature at which ρ falls below the experimental resolution and $T_{c2}(B)$ is determined from the intersection of the linear part of the transition curve $\rho(T)$ with a horizontal line corresponding to the normal-state resistivity $\rho_n(B)$ at 15-25 K. Thus, we construct the vortex phase diagram in the $B-T$ plane, as illustrated in the left inset of figure 1(a), where $T_c(B)$ and $T_{c2}(B)$ (i.e., $B_{c2}(T)$) are plotted with solid and open circles, respectively.

In the main panel of figure 1(a), we typically show with black circles the $I-V$ characteristics measured at 11 K in 3 T in the continuous-current mode. The low-current and low-voltage region surrounded with a dashed rectangle is enlarged and shown in the right inset. To test the influence of heating generated by the vortex motion, we measured comparatively the $I-V$ characteristics in the pulsed-current mode. In the pulsed mode, the square current pulse with duration of 50 μs was applied and its repetition time was set to 0.1 s, which made the duty ratio as small as 0.05 %. The voltage signal was averaged over 5 s and recorded. Figures 1(b) and (c), respectively, display the set of waveforms for the voltage pulse $V(t)$ and the current pulse $I(t)$ with different amplitudes in the range $|I|$ =22-162 mA measured at 11 K in 3.0 T. Here, $I(t)$ of a given amplitude and corresponding $V(t)$ are indicated with the same color. Except for the high-current region above about $|I|$ =100 mA, the waveform of $V(t)$ as well as that of $I(t)$ exhibits a plateau-like behavior and hence, the $I-V$ characteristics are readily obtained from the plateau value. For $|I|$ >100 mA, in contrast, a slight increase in the plateau value $|V(t)|$ with respect to time is visible, which is most likely due to the heating effect. In more detail, a small spike or transient signal of $V(t)$ was in some cases observed in the initial part ($0 \leq t <2$ μs) of each pulse. This is considered to be an artifact originating from the electrical circuit. In

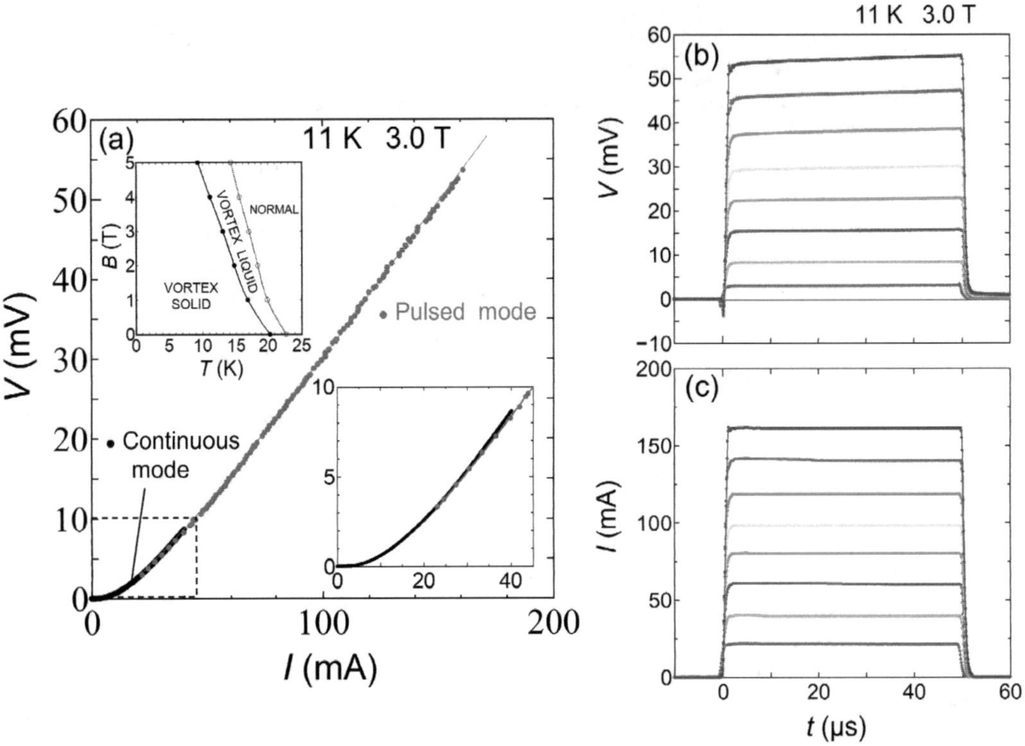

Figure 1. (a) $I - V$ characteristics measured at 11 K in 3.0 T in the continuous mode (black circles) and pulsed mode (red circles). A red straight line is a fit of the data (red circles) taken in the pulsed mode. Left inset: Vortex phase diagram in the $B - T$ plane, where the solid and open circles represent $T_c(B)$ and $T_{c2}(B)$ (or $B_{c2}(T)$), respectively. The lines are guide to the eye. Right inset: The low current region in the main panel is enlarged and shown. (b) The set of selected waveforms for the voltage pulse $V(t)$ generated by vortex motion at 11 K in 3.0 T and (c) the current pulse $I(t)$ with different amplitudes in the range $|I|$ =22-162 mA. Here, $I(t)$ of a given amplitude and corresponding $V(t)$ are indicated with the same color.

the following discussion of $I - V$ characteristics, therefore, we use the plateau value $|V(t)|$ at t =2-4 μs.

In the main panel and the right inset of figure 1(a), the plateau values of $|V(t \simeq 0)|$ thus obtained are plotted against $|I|(\equiv I)$ with red circles. In the broad current range above about 45 mA the data points follow the linear relation, indicating the occurrence of the vortex flow under the very weak pinning. In the low current region, I <30 mA, a good agreement with the data for the continuous mode (black circles) is observed, while with an increase in I, the data points for the continuous mode show a slight upward deviation from the linear relation for the pulsed mode (red circles), as seen in the right inset. This is attributed to the heating effect in the continuous $I - V$ measurements. To avoid a possible damage to the sample from heating, we did not conduct the measurements above 40 mA in the continuous mode. We consider from these results that the pulsed-current method successfully detects the vortex-flow state close to the ideal flux flow, which cannot be attained in the continuous dc $I - V$ measurements.

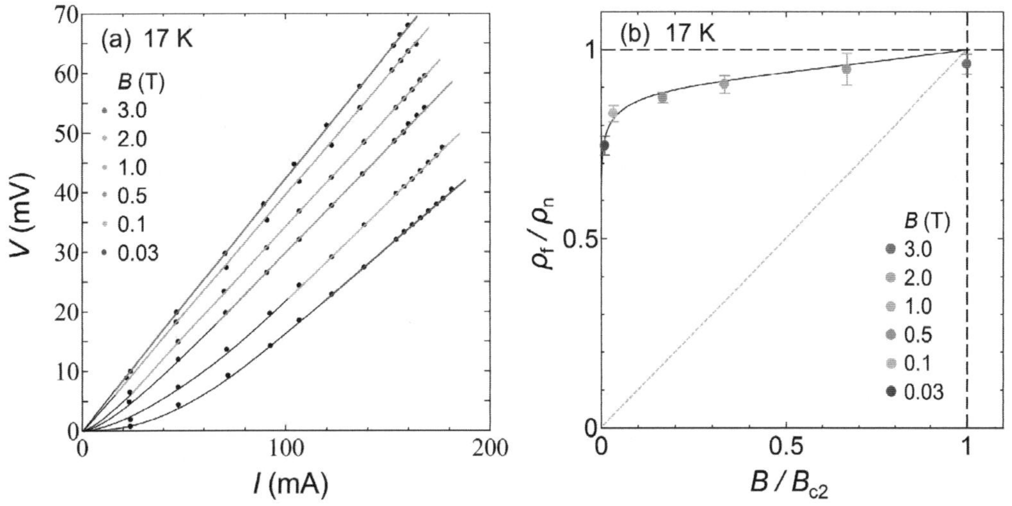

Figure 2. $I-V$ characteristics at 17 K for different B measured in the pulsed mode: $B = 0.03$, 0.1, 0.5, 1.0, 2.0, and 3.0 T from bottom to top. ρ_f is extracted from the slope of the linear part of the $I-V$ data, as indicated with colored lines. The black lines are guides to the eye. (b) ρ_f/ρ_n versus B/B_{c2} at 17 K, where a full line is a guide to the eye. The data points of ρ_f/ρ_n and the corresponding straight lines in (a) are indicated with the same color. A dotted straight line represents ρ_f/ρ_n against B/B_{c2} predicted by the Bardeen-Stephen theory. The horizontal and vertical dashed lines mark the location of ρ_f/ρ_n =1 and B/B_{c2} =1, respectively.

The $I-V$ characteristics thus acquired in the pulsed mode at 17 K in different B are plotted in figure 2(a). Here, we present the data at 17 K instead of 11 K, because the value of B_{c2}, which is necessary for the plot in figure 2(b), is directly obtained from ρ to be 3.0 T at 17 K, whereas at 11 K, B_{c2} exceeds the maximum field available in the present experiment. In high fields B =1-3 T, corresponding to the vortex-liquid phase, V is almost proportional to I in the entire current region. By contrast, in lower fields B <1 T, corresponding to the vortex-solid phase, the $I-V$ data show a crossover from a nonlinear behavior with positive curvature to a linear behavior with an increase in I. This indicates a change from pinning-dominated to viscous-force dominated flow. For all the data presented here, we extract the vortex-flow resistivity ρ_f from the slope of the linear part of the $I-V$ characteristics, as shown with colored lines.

In figure 2(b), we plot ρ_f normalized by ρ_n, ρ_f/ρ_n, as a function of B normalized by B_{c2}, B/B_{c2}, where $B_{c2} = 3.0$ T. According to the Bardeen-Stephen theory, the flux-flow resistivity ρ_f in a conventional s-wave superconductor increases proportionally to B and reaches a normal-state resistivity ρ_n at $B = B_{c2}$, as indicated with a dotted straight line in figure 2(b). The behavior of $\rho_f(B)$ observed in the present MgB_2 film is in marked contrast to that expected for the Bardeen-Stephen theory. At low B in the range B/B_{c2} =0-0.05, ρ_f/ρ_n rises much more steeply than expected by the Bardeen-Stephen theory and then increases very slowly in the range $B/B_{c2} \gtrsim 0.05 (\equiv B^*/B_{c2})$, approaching the normal-state value of 1 at $B = B_{c2}$.

This behavior is similar to that reported from the microwave impedance measurements for MgB_2 single crystals with $T_c \simeq 39$ K [13]. The unusual field dependence of ρ_f/ρ_n with strong enhancement at low fields found in Ref. 13 was interpreted in terms of the two-band superconducting property that the superconductivity with the smaller gap coming from π band is suppressed at a field B_π much lower than B_{c2}. We consider that the unusual dependence of ρ_f/ρ_n on B/B_{c2} presented in figure 2(b) is essentially the same as that observed by the microwave measurements [13]. Thus, our results suggest that the two-band nature of MgB_2 can be observed by dc $I-V$ measurements as long as the vortices are driven fast by large currents and the heat dissipated in the sample is sufficiently reduced. The present results also suggest that the two-band properties can emerge not only in the high-quality single crystals [13] and films [14] with $T_c \sim$39 K but also in the epitaxially grown films with lower $T_c \simeq$20 K.

We expect that the dc measurements of ρ_f based on the pulsed current technique developed here will be used further, e.g., to study the possibility or absence of the two-band nature of amorphous MgB_2 films with much reduced $T_c \simeq$6 K [26].

4. Conclusions

Using pulsed currents, we measure the field dependence of the dc vortex-flow resistivity ρ_f in the fast-flow regime for an epitaxially grown MgB_2 film with a thickness of about 100 nm. The $I-V$ characteristics obtained in the vortex-solid phase exhibit a crossover from nonlinear to linear behavior with an increase in I, indicating a change from pinning-dominated to viscous-force dominated flow. $\rho_f(B)$ extracted from the slope of the linear part of the $I-V$ relation exhibits an unusual B dependence with strong enhancement at low B, which is similar to that reported from microwave impedance for MgB_2 crystals with $T_c \simeq$39 K. The present results suggest that the two-band nature of MgB_2 can be also detected by dc measurements with pulsed currents in epitaxially grown films with reduced $T_c(\simeq$20 K).

While in this work we only show the data of $\rho_f(B)$ at 17 K, it is of great interest to explore how the characteristic field B^* at which a steep increase in ρ_f/ρ_n changes to a slow increase depends on the temperature and to compare $B^*(T)$ with $B_\pi(T)$ reported in the literature [14]. Further measurements of $\rho_f(B)$ at lower temperatures and their analyses are in progress, which will be published elsewhere.

Acknowledgments

We thank S. Kaneko for fruitful discussions. This work was supported by a Grant-in-Aid for Scientific Research on Innovative Areas (KAKENHI Grant No. 20H05266), Challenging Research (KAKENHI Grant No. 21K18598), and Young Scientists (KAKENHI Grants No. 20K14413).

References

[1] Nagamatsu J, Nakagawa N, Muranaka T, Zenitani Y and Akimitsu J 2001 *Nature* **410** 63
[2] Bouquet F, Fisher R A, Phillips N E, Hinks D G and Jorgensen J D 2001 *Phys. Rev. Lett.* **87** 047001
[3] Szabó P, Samuely P, Kačmarčík J, Klein T, Marcus J, Fruchart D, Miraglia S, Marcenat C and Jansen A G M 2001 *Phys. Rev. Lett.* **87** 137005
[4] Iavarone M, Karapetrov G, Koshelev A E, Kwok W K, Crabtree G W, Hinks D G, Kang W N, Choi E M, Kim H J, Kim H J and Lee S I 2002 *Phys. Rev. Lett.* **89** 187002
[5] Gonnelli R S, Daghero D, Ummarino G A, Stepanov V A, Jun J, Kazakov S M and Karpinski J 2002 *Phys. Rev. Lett.* **89** 247004
[6] Eskildsen M R, Kugler M, Tanaka S, Jun J, Kazakov S M, Karpinski J and Fischer Ø 2002 *Phys. Rev. Lett.* **89** 187003
[7] Nakai N, Ichioka M and Machida K 2002 *J. Phys. Soc. Jpn.* **71** 23
[8] Souma S, Machida Y, Sato T, Takahashi T, Matsui H, Wang S C, Ding H, Kaminski A, Campuzano J C, Sasaki S and Kadowaki K 2003 *Nature* **423** 65
[9] Choi H J, Roundy D, Sun H, Cohen M L and Louie S G 2002 *Nature* **418** 758
[10] Kim M S, Skinta J A, Lemberger T R, Kang W N, Kim H J, Choi E M and Lee S K 2002 *Phys. Rev. B* **66** 064511
[11] Koshelev A E and Golubov A A 2003 *Phys. Rev. Lett.* **90** 177002
[12] Blumberg G, Mialitsin A, Dennis B S, Klein M V, Zhigadlo N D and Karpinski J 2007 *Phys. Rev. Lett.* **99** 227002
[13] Shibata A, Matsumoto M, Izawa K, Matsuda Y, Lee S and Tajima S 2003 *Phys. Rev. B* **68** 060501(R)
[14] Sarti S, Amabile C, Silva E, Giura M, Fastampa R, Ferdeghini C, Ferrando V and Tarantini C 2005 *Phys. Rev. B* **72** 024542
[15] Goryo J and Matsukawa H 2005 *J. Phys. Soc. Jpn.* **74** 1394

[16] Bardeen J and Stephen M J 1965 *Phys. Rev.* **140** A1197
[17] Jin B B, Klein N, Kang W N, Kim H J, Choi E M, Lee S I, Dahm T and Maki K 2002 *Phys. Rev. B.* **66** 104521
[18] Okada T, Takahashi H, Imai Y, Kitagawa K, Matsubayashi K, Uwatoko Y and Maeda A 2012 *Phys. Rev. B* **86** 064516
[19] Akutagawa S, Ohashi T, Kitano H, Maeda A, Goryo J, Matsukawa H and Akimitsu J 2008 *J. Phys. Soc. Jpn.* **77** 064701
[20] Shishido H, Yoshida T and Ishida T 2015 *Appl. Phys. Exp.* **8** 113101
[21] Shishido H, Nakagami T, Yoshida T and Ishida T 2017 *J. Phys.: Conf. Ser.* **871** 012036
[22] Maegochi S, Ienaga K, Kaneko S and Okuma S 2019 *Sci. Rep.* **9** 16447
[23] Maegochi S, Ienaga K and Okuma S 2021 *Sci. Rep.* **11** 19280
[24] Ochi A, Sohara N, Kaneko S, Kokubo N and Okuma S 2016 *J. Phys. Soc. Jpn.* **85** 044701
[25] Sohara N, Ochi A, Murakami E, Ienaga K, Kaneko S, Kokubo N and Okuma S 2017 *J. Phys.: Conf. Ser.* **871** 012022
[26] Okuma S, Togo S and Amemori K 2003 *Phys. Rev. B* **67** 172508

Three-Dimensional Molecular Dynamics Method for Vortex pinning by Splayed Columnar Defects

Masaru Kato

Department of Electronics and Physics, Osaka Metropolitan University, 1-1, Gakuencho, Nakaku, Sakai, Osaka 599-8531, Japan

E-mail: mkato76@omu.ac.jp

Abstract. We have developed three-dimensional molecular dynamics method, in which anisotropy of a superconductor and vortex core energy are included. We have applied it to a superconductor with splayed columnar defects. We have found that under a driving force due to an external current, vortices show wandering structures, which comes from the pinning force from splayed columnar defects. Especially, for weak core energy, vortices become soft or flexible. In an isotropic superconductor, these wandering structures disappear.

1. Introduction

Dynamics of vortex, especially pinning of vortex by defects or impurities, is important for applications of superconductivity. In order to investigate such vortex dynamics, we should use some theories. There are at least three levels of theory. Microscopic theory, such as the Bogoliubov-de Gennes equations, the Gor'kov equation, or the Eilenberger equations, can treat several vortices. Phenomenological theory, such as the Ginzburg-Landau theory, can treat hundreds of vortices. The molecular dynamics method (MD) can treat thousands of vortices [1,2]. If we investigate collective motion of vortices, the molecular dynamics method is still useful, we think. So, using the MD, we investigated melting transition of a vortex solid in a slightly dirty nano-sized superconducting plate [3] and in a superconducting plate with a nano-rod lattice [4].

About the pinning of vortices, it was found that splayed columnar defects enhanced the critical current density, recently [5]. Especially, the critical current that is parallel to the splay plane is higher than that for perpendicular current. These splayed columnar defects cannot be treated within usual two-dimensional (2D) MD, because in two-dimension, the columnar defect is just a circular pinning center. Especially, in 2D-MD, directions of the columnar defect cannot be considered. Therefore, in order to apply the MD to the vortex dynamics with splayed columnar defects, we must use the three-dimensional (3D) MD [6,7]. In this study, we include an anisotropy of the superconductor and also the vortex core energy into 3D-MD. Using this 3D-MD, we investigate effects splayed columnar defects on vortex dynamics and also effects of the anisotropy and vortex core energy.

In section 2, we show details of our numerical method. In section 3, we show several simulation results. Section 4 is devoted summary of this study.

2. Numerical Method

In the molecular dynamics method, vortices are considered as particles in two-dimensional space [1,2]. In order to applied the two-dimensional molecular dynamics method to a three-dimensional superconductor, the superconductor should be divided into layers (figure 1) [6,7]. In each layer, we

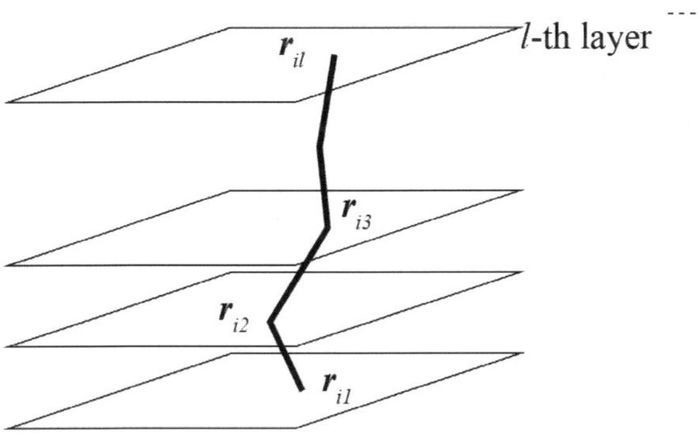

Figure 1. i-th vortex in a layer model for three-dimensional molecular dynamics method.

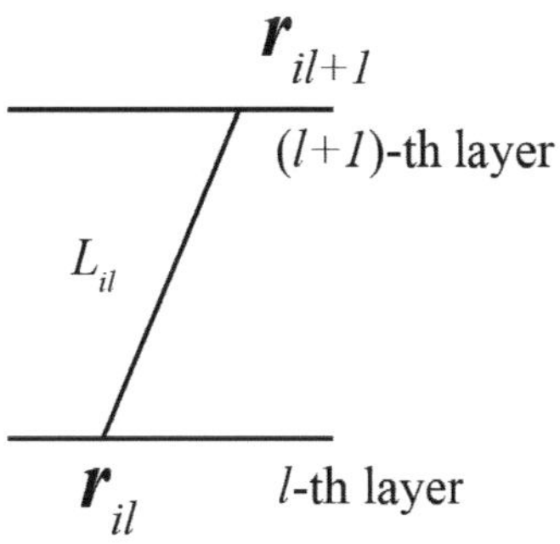

Figure 2. Slanted i-th vortex between l-th and $(l+1)$-th layers.

apply the two-dimensional molecular dynamics method. The equation of motion of i-th vortex in l-th layer is given as,

$$\eta \frac{d\boldsymbol{r}_{il}}{dt} = \boldsymbol{f}_{dil} + \boldsymbol{f}_{pil} + \boldsymbol{f}_{vil} + \boldsymbol{f}_{fil} + \boldsymbol{f}_{bil} + \boldsymbol{f}_{mil} \tag{1}$$

In this equation, we neglect an inertia term, which is relatively small. $\boldsymbol{r}_{il}$ is a position of i-th vortex in l-th layer and η is a viscosity coefficient. Forces on the left-hand side are defined below. First term $\boldsymbol{f}_{dil}$ is a driving force, which comes from an external current. Second term $\boldsymbol{f}_{pil}$ is a pinning force from columnar defects, which is given as,

$$\boldsymbol{f}_{pil} = \sum_{j}^{\text{columnar defects}} \boldsymbol{f}_{ijl}^{p} \tag{2}$$

$$\boldsymbol{f}_{ijl}^{p} = \frac{f_{pj}}{r_{pj}} \left| \boldsymbol{r}_{il} - \boldsymbol{r}_{jl}^{p} \right| \Theta\left(\frac{r_{pi} - \left| \boldsymbol{r}_{il} - \boldsymbol{r}_{jl}^{p} \right|}{\lambda} \right) \hat{\boldsymbol{r}}_{ijl} \tag{3}$$

Here, $\boldsymbol{r}_{jl}^{p}$ is a position of j-th columnar defect in l-th layer, and f_{pj} and r_{pj} are strength and radius of j-th columnar defect, respectively. $\hat{\boldsymbol{r}}_{ijl} = \left(\boldsymbol{r}_{jl}^{p} - \boldsymbol{r}_{il}\right) / \left|\boldsymbol{r}_{jl}^{p} - \boldsymbol{r}_{il}\right|$ is a unit vector along the relative position vector $\boldsymbol{r}_{jl}^{p} - \boldsymbol{r}_{il}$, and $\Theta(x)$ is the step-function. $\lambda = \lambda_0 \left(1 - \frac{T}{T_c}\right)^{-1/2}$ is the penetration depth, and T and T_c are temperature and the critical temperature, respectively. Strictly speaking, shapes of splayed columnar defects in each layer are elliptic. However, in this study we consider only circular shapes. We will consider this point later. Third term $\boldsymbol{f}_{vil}$ is a repulsive interaction from other vortices, which is given as,

$$\boldsymbol{f}_{vil} = f_0 \left(\frac{\lambda_0}{\lambda} \right)^2 \sum_{j \neq i} K_1\left(\frac{r_{ijl}}{\lambda} \right) \hat{\boldsymbol{r}}_{ijl} \tag{4}$$

Here, $K_1(x)$ is the first order modified Bessel function, $f_0 = \Phi_0 / 8\pi^2 \lambda_0^2$ is the strength of vortex-vortex repulsive interaction, and $\hat{\boldsymbol{r}}_{ijl} = \left(\boldsymbol{r}_{il} - \boldsymbol{r}_{jl}\right) / \left|\boldsymbol{r}_{il} - \boldsymbol{r}_{jl}\right|$ and $r_{ijl} = \left|\boldsymbol{r}_{il} - \boldsymbol{r}_{jl}\right|$ are a unit vector and length of a relative position vector $\boldsymbol{r}_{il} - \boldsymbol{r}_{jl}$, respectively. Fourth term $\boldsymbol{f}_{fil}$ is a thermal fluctuation force, which satisfies following fluctuation-dissipation relation,

$$\langle \boldsymbol{f}_{fil}(t_1) \cdot \boldsymbol{f}_{fjl}(t_2) \rangle = 2\eta k_B T \delta_{ij} \delta(t_1 - t_2) \tag{5}$$

Here, $\langle \ldots \rangle$ is an average about randomness of impurities. Fifth term $\boldsymbol{f}_{bil}$ is a bending force. This force comes from the vortex core energy. In an anisotropic superconductor with electron mass m_{ab} in ab-plane and M_c along c-axis, the vortex core energy is given as [8,9,10],

$$E_l(\theta) = E_0 \varepsilon_\theta \ln\left(\frac{\lambda}{\varepsilon_\theta \xi}\right) \tag{6}$$

Here, $\varepsilon_\theta = \varepsilon^2 \cos^2\theta + \sin^2\theta$. θ is the angle between the vortex and the ab-plane and $\varepsilon^2 = \frac{m_{ab}}{M_c}$ is a mass ratio. E_0 is a vortex core energy when the vortex is parallel to the c-axis, and $\xi = \xi_0 \left(1 - \frac{T}{T_c}\right)^{-1/2}$ is the coherence length. Phenomenologically, we can take $E_0 = \left(\frac{\Phi_0}{4\pi\lambda}\right)^2$, but microscopically it includes energy of bound states. So, in this study, we take it as a material depending parameter. If vortex is slanted from the c-axis, as show in figure 2, energy of vortex between l-th and $(l+1)$-th layers $E_b(l, l+1)$ is given as

$$E_{bi}(l, l+1) = L_i E_l(\theta) \tag{7}$$

Here, $L_i = |\boldsymbol{r}_{il+1} - \boldsymbol{r}_{il}|$ is a length of i-th vortex between l-th and $(l+1)$-th layers (see figure 2). In order to minimize this energy, there appears a force to minimize the length of the vortex. Then bending force on i-th vortex on l-th layer is given as

$$\boldsymbol{f}_{il}^b = -\frac{\partial}{\partial \boldsymbol{r}_{il}} \left(E_{bi}(l, l+1) + E_{bi}(l-1, l) \right) \tag{8}$$

We call this force a bending force. The last term $\boldsymbol{f}_{mil}$ is a magnetic energy. If an external field $\boldsymbol{H}_{ext}$ is applied to the superconductor, an interaction energy between external field and quantized fluxes is given as,

$$E_M = -\frac{1}{4\pi} \sum_i \boldsymbol{\Phi}_i \cdot \boldsymbol{H}_{ext} \tag{9}$$

Then the magnetic force is given by differentiate about the vortex position,

$$\boldsymbol{f}_{mil} = -\frac{\partial}{\partial \boldsymbol{r}_{il}} E_M \tag{10}$$

We solve these equations numerically, and obtain dynamics of vortices.

3. Numerical Results

In this study, we consider 20 layers and each layer is $10\lambda_0 \times 10\lambda_0$ square plate with periodic boundary condition. Distance between layers is set to 0.3 λ_0 We put 60 splayed columnar defects and their splay plane is parallel or perpendicular to the driving force, which are shown in figures 3(a) and 3(b), respectively. We set the radius of columnar defect $r_p = 0.4\lambda_0$, strength of pinning force $f_p = f_0$, and splay angle 0.1π. We set the external field $\boldsymbol{H}_{ext} = (0,0,0.5)\Phi_0/(4\pi\lambda_0^2)$ and $T/T_c = 0.7$, and we consider 20 vortices. We set random positions of vortices and start an initial simulation with $f_d = 0$ to obtain nearly thermal equilibrium state. Then, we apply the driving force.

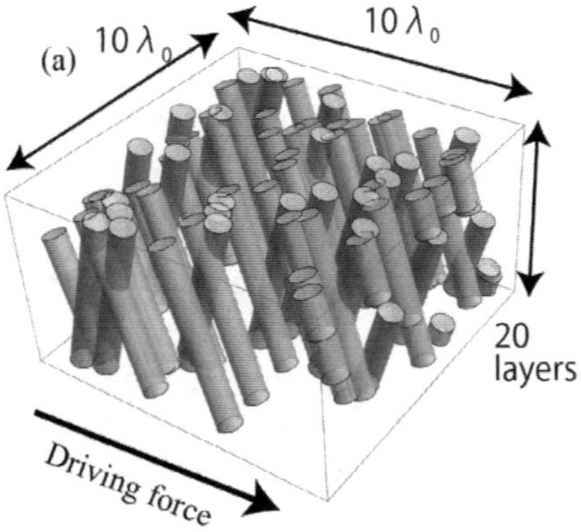

Figure 3. 60 columnar defects. (a) Splay plane is parallel to the driving force. (b) splay plane is perpendicular to the driving force.

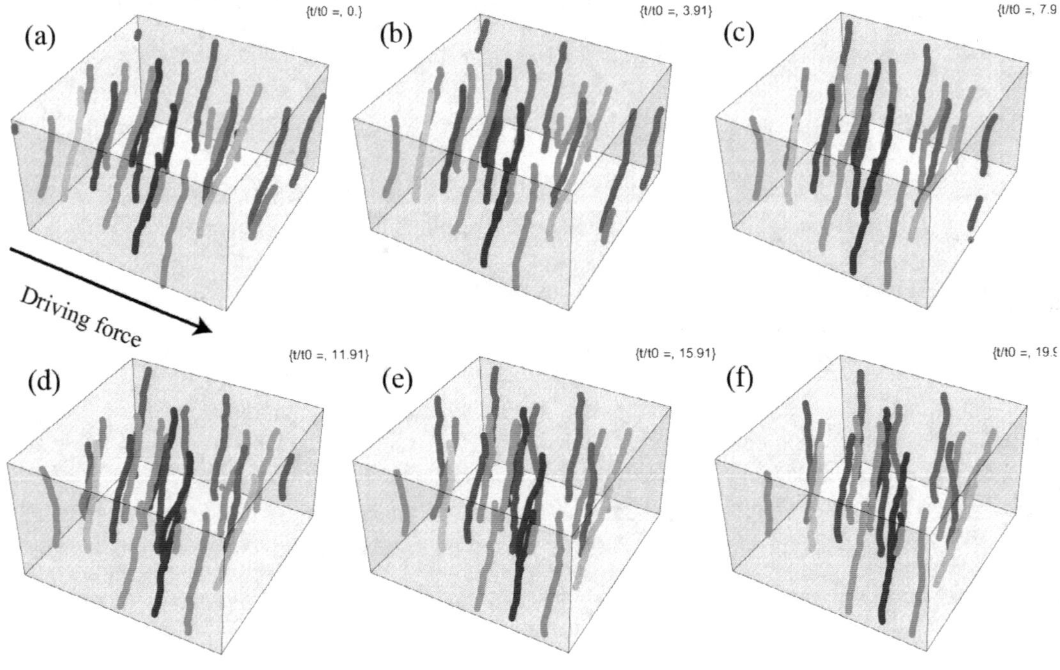

Figure 4. Configurations of 20 vortices at $t/t_0 = 0$(a), 3.91 (b), 7.91(c), 11.91 (d), 15.91 (e) and 19.91(f). The driving force is parallel to the splay plane of columnar defects.

In figure 4, we show time development of 20 vortices for driving force $f_{dil} = 0.1 f_0$ which is parallel to the splay plane of columnar defects. In this simulation, we set $E_0/(f_0\lambda_0) = 0.5$ and anisotropy $\varepsilon^2 = 0.2$. The figure shows vortex configurations at $t/t_0 = 0$(a), 3.91 (b), 7.91(c), 11.91 (d), 15.91

(e) and 19.91(f). Here, $t_0 = \eta\lambda_0 / f_0$ is a characteristic time of the equation of motions (equation (1)). Vortices are disconnected at boundaries because of the periodic boundary condition. All of vortices gradually move to the driving force direction. Structures of vortices are not straight, but wandering and slanted. This is because of pinning force form columnar defects as well as the thermal fluctuation force. However, the columnar defects cannot pin all of vortices.

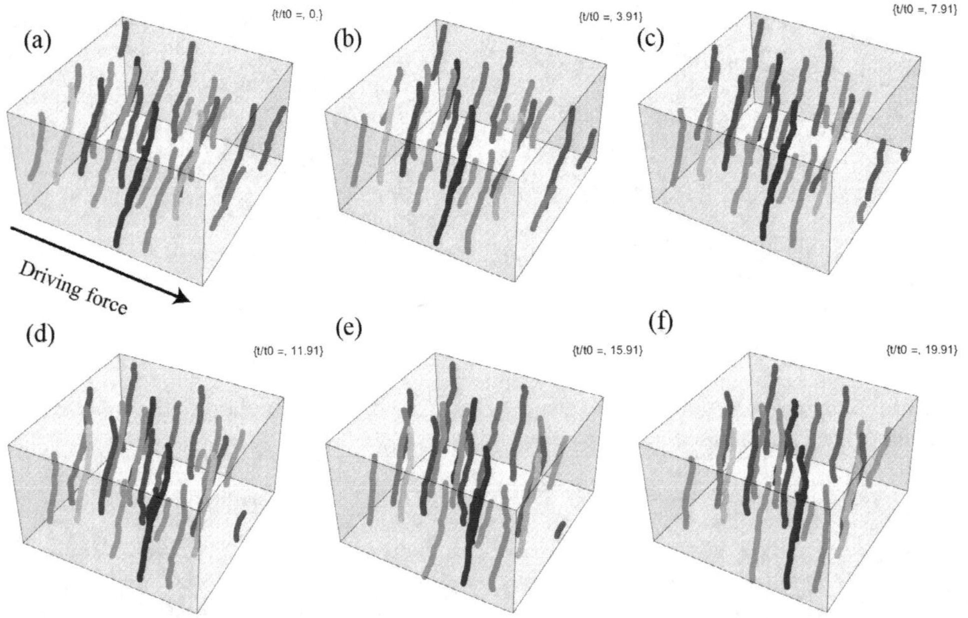

Figure 5. Configurations of 20 vortices at $t/t_0 = 0$(a), 3.91 (b), 7.91(c), 11.91 (d), 15.91 (e) and 19.91(f). The driving force is perpendicular to the splay plane of columnar defects.

Next, we show time development of 20 vortices for driving force $f_{dil} = 0.1 f_0$ which is perpendicular to the splay plane of columnar defects, in figure 5. In this simulation, we also set $E_0/(f_0\lambda_0) = 0.5$. The figure shows vortex configurations at $t/t_0 = 0$(a), 3.91 (b), 7.91(c), 11.91 (d), 15.91 (e) and 19.91 (f). Again, all of vortices gradually move to the driving force direction. Slanted angles of vortices are somewhat different from those in figure 4. However, we cannot find explicit difference between perpendicular and parallel splayed columnar defects in these simulations.

Next, we show effects of the core energy E_0. In figure 6, we show time development of 20 vortices for $E_0/(f_0\lambda_0) = 0.1$. The driving force is set to $f_{dil} = 0.1 f_0$ and perpendicular to the splay plane of columnar defects and anisotropy is set to $\varepsilon^2 = 0.2$. Figure 6 shows configurations of vortices at $t/t_0 = 0$(a), 3.91 (b), 7.91(c), 11.91 (d), 15.91 (e) and 19.91(f). Comparing with figure 5, vortices easily bend, and they entangle with each other. When the core energy is small, the bending force becomes small from equation (8), and then vortices become soft.

Finally, we investigate effects of anisotropy of superconductor. In figure 7, we show time development of 20 vortices for $\varepsilon^2 = 1.0$, i.e. isotropic case. The driving force is set to $f_{dil} = 0.1 f_0$ and perpendicular to the splay plane of columnar defects and the core energy is set to $E_0/(f_0\lambda_0) = 0.5$. Figure 7 shows configurations of vortices at $t/t_0 = 0$(a), 3.91 (b), 7.91(c), 11.91 (d), 15.91 (e) and 19.91(f). Wandering of vortices becomes small and vortices are almost parallel to the c-axis. Isotropic

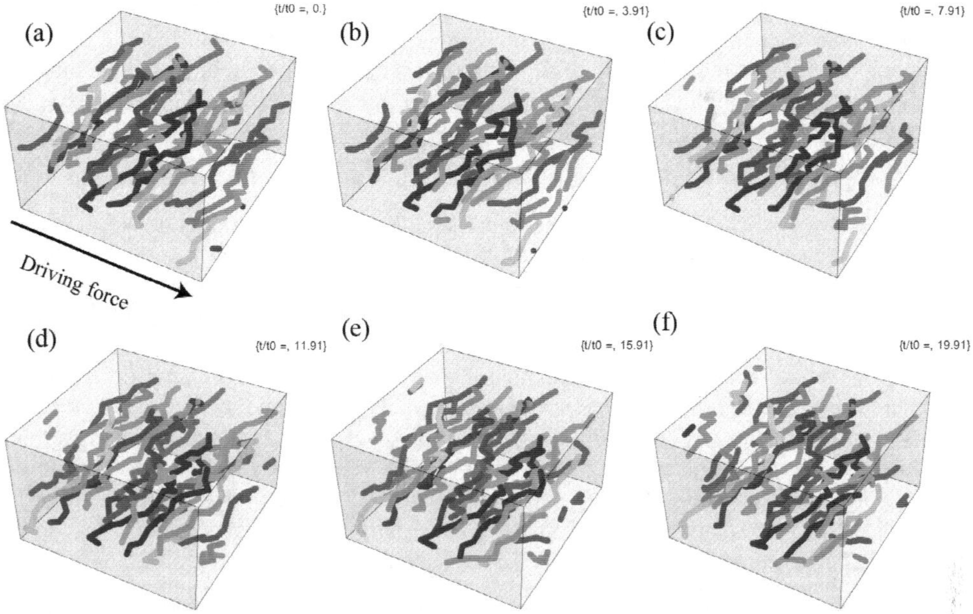

Figure 6. Configurations of 20 vortices at $t/t_0 = 0$(a), 3.91 (b), 7.91(c), 11.91 (d), 15.91 (e) and 19.91(f). The driving force is perpendicular to the splay plane of columnar defects and the core energy $\varepsilon_0/(f_0\lambda_0) = 0.1$.

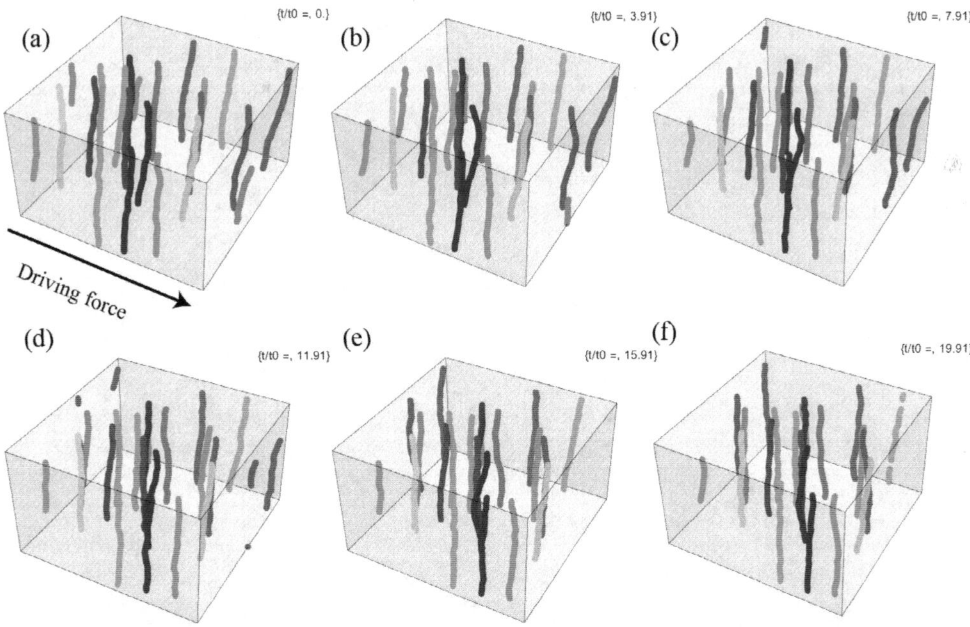

Figure 7. Configurations of 20 vortices at $t/t_0 = 0$(a), 3.91 (b), 7.91(c), 11.91 (d), 15.91 (e) and 19.91(f). The anisotropy is $\varepsilon^2 = 1.0$, the driving force is parallel to the splay plane of columnar defects and the $\varepsilon_0/(f_0\lambda_0) = 0.5$.

case, the core energy (equation (6)) do not depend on the direction of the vortex, and then the bending force is also independent from the direction of the vortex. However, in the anisotropic case, the core energy is small when the vortex is parallel to the ab-plane (equation 6) and then the bending force is small. So, vortices are hard in this isotropic case. We think effects of columnar defects become small in an isotropic superconductor.

4. Summary

In this study, we have developed three-dimensional molecular dynamics method for vortices, which is applicable to superconductors with splayed columnar defects. We have obtained time developments of vortices under external field and currents. In these simulations, vortices move slowly and their structure shows some wandering due to the pinning force from the defects and the thermal fluctuation force. For weak vortex core energy, vortices become soft or flexible. On the other hand, in an isotropic superconductor, vortices become hard and cannot be affected by splayed columnar defects. We think splayed columnar defects is more effective in anisotropic superconductors.

In this study, vortices cannot be pinned by splayed columnar defects. We think this is because we only considered two-dimensional pinning force. Also, in the MD method in this study, several points should be improved. Shapes of columnar defects should be elliptic. Vortex-vortex interaction in anisotropic superconductors should be incorporated. The difference between perpendicular and parallel columnar defects may come from these points, especially from three-dimensional pinning potential. Such extensions of the 3DMD are future problems.

References

[1] Reichhardt C, Olson C J, Groth J, Field S and Nori F 1996 *Phys. Rev.* B **53** R8898
[2] Misko V R and Peeters F M 2006 *Phys. Rev.* B **74** 174507
[3] Kusafuka T, Kato M and Tsuchiya Y 2020 *J. Phys.: Conf. Ser.* **1590** 012002
[4] Kusafuka T, Kato M and Sato O 2019 *J. Phys.: Conf. Ser.* **1293** 012021
[5] Park A, Pyon S, Ohara K Ito N and Tamegai T 2018 *Phys. Rev.* B **97** 064516
[6] van Otterlo A, Scalettar R T and Zimányi G T 1998 *Phys. Rev. Lett.* **81** 1497
[7] Olson C J, Scalettar R T, Zimányi G T and Grønbech-Jensen N 2000 *Phys. Rev.* B **62** R3612
[8] Blatter G, Feigel'man M V, Geshkenbein V B, Larkin A I and Vinokur V M 1994 *Rev. Mod. Phys.* **66** 1125
[9] Klemn R A and Clem J R 1980 *Phys. Rev.* B **21** 1868
[10] Barndt E H 1977 *J. Low Temp. Phys.* **26** 735

Journal of Physics: Conference Series **2323** (2022) 012015 doi:10.1088/1742-6596/2323/1/012015

Simulation of switching to voltage state in an intrinsic Josephson junction

Takamasa Wachi[1] and Masaru Kato[1,2]

[1]Department of Physics and Electronics, Osaka Prefecture University, 1-1 Gakuencho, Nakaku, Sakai, Osaka, 599-8531, Japan
[2]Department of Physics and Electronics, Osaka Metropolitan University, 1-1 Gakuencho, Nakaku, Sakai, Osaka, 599-8531, Japan

sbb01206@st.osakafu-u.ac.jp

Abstract. When a current applied to an intrinsic Josephson junction (IJJ) in a copper oxide high-temperature superconductor is increased, the IJJ enters a voltage state and then the voltage increases stepwise. After these voltage states, when the current is lowered, THz oscillation occurs. The voltage state is described as a running state of a particle in a washboard potential, using the RCSJ model of the Josephson junction (JJ). In this study, we investigate these running states in the IJJ, using a coupled layered JJ model with the finite element method. Especially, we focus on time evolution of phase differences in all JJs in the IJJ under weak external current.

1. Introduction

When two superconductors are connected by an insulator, electric current flows spontaneously. This is called the Josephson effect, and this junction is called the Josephson junction (JJ). There are two kind of Josephson effects. The first is the DC Josephson effect. When two superconductors have different phases, a direct current will flow even if there is no potential difference between superconductors. The second is the AC Josephson effect. When a DC voltage is applied to the junction, the phase difference increases with time and then an AC current flows. This is the AC Josephson effects. Because of this AC current, an electromagnetic wave is emitted from the junction. Conventionally, this junction has been fabricated, artificially.

In copper oxide high-temperature superconductors, there are intrinsic Josephson junctions (IJJ). They do not require complicated process and each junction is clean at atomic scale, because JJs come from unique structures of the crystals. Moreover, if oscillations of electromagnetic wave in all junctions are coherent, huge electromagnetic wave is expected, and frequency of the electromagnetic wave emission is within the THz region.

Another aspect of the IJJ is transition to a voltage state at low applied current. When an applied current to an IJJ in a copper oxide high-temperature superconductor is increased, the IJJ enters the voltage state and then the voltage increases stepwise [1]. After these voltage states, when the current is lowered, THz oscillation occurs. The voltage state is described as a running state of a particle in a washboard potential based on the RCSJ model of the JJ. In the RCSJ model, the phase difference (φ) of the JJ can be considered as a coordinate of a particle in a tilted washboard potential. When the external current becomes large, overall slope of the potential becomes large. When φ exceeds π, the AC current flows and the JJ becomes in the voltage state.

In this study, we investigate these running states in the IJJ, using coupled layered JJ model with the finite element method. Especially we focus on the time evolution of phase differences in JJs under weak external current.

2. Numerical Method

We consider an alternately stacked superconducting and insulating layers, as a model of the IJJ. In the superconducting layer, we consider electromagnetic fields $E_x(x,y,l,t)$,$E_y(x,y,l,t)$ and $B_z(x,y,l,t)$ where l is the layer number. In the insulating layer, we consider electromagnetic fields $E_z(x,y,l,t)$, $B_x(x,y,l,t)$, $B_y(x,y,l,t)$, and the phase difference of the order parameter between upper and lower superconducting layers $\varphi(x,y,l,t)$. We set thickness of superconducting layers s=3Å and thickness of insulating layers d=12Å. We call the bottom layer first layer.

In the numerical simulation, we use the finite element method and solve the following equations [2-4]. First equation is a generalized Josephson relation,

$$\frac{\partial}{\partial t}\varphi(x,y,l,t) = \frac{2ed}{\hbar}(1-\alpha\Delta_z^2)E_z(x,y,l,t) \tag{1}$$

Here Δ_z^2 is a second rank difference in z direction, which is defined for an arbitral function $f(l)$ as,

$$\Delta_z^2 f(l) = f(l+1) + f(l-1) - 2f(l) \tag{2}$$

α is a capacitive coupling constant $\alpha = \frac{\varepsilon\mu^2}{sd}$, and μ is the Debye length of the superconducting charge. Second equations are about spatial variations of phase differences under a magnetic field and given as,

$$\frac{\partial}{\partial x}\varphi(x,y,l,t) = \frac{2ed}{\hbar c}(1-\eta\Delta_z^2)B_y(x,y,l,t) \tag{3}$$

$$\frac{\partial}{\partial y}\varphi(x,y,l,t) = -\frac{2ed}{\hbar c}(1-\eta\Delta_z^2)B_x(x,y,l,t) \tag{4}$$

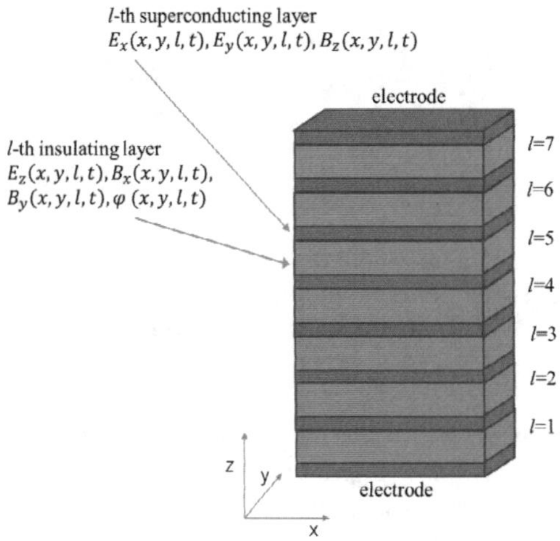

Figure 1. The numerical model of the intrinsic Josephson junction. Black layers are superconducting layers and blue layers are insulating layers.

where η is an inductive coupling constant $\eta = \frac{{\lambda_c}^2}{sd}$, and λ_c is the c-axis penetration depth. Third equation is the Maxwell's equation with a thermal fluctuation current J_f, which is given as,

$$\frac{1}{c}\frac{\partial}{\partial t}\varepsilon E_z(x,y,l,t) = \frac{\partial}{\partial x}B_y(x,y,l,t) - \frac{\partial}{\partial y}B_x(x,y,l,t)$$
$$-\frac{4\pi}{c}\{J_c \, sin\varphi(x,y,l,t) + \sigma E_z(x,y,l,t) + J_f(x,y,l,t)\} \tag{5}$$

where J_c is the Josephson critical current density, ε is a permittivity, and σ is a conductivity of insulating layers.

In order to solve these equations, we use the finite element method. In the finite element method, we divide each layer into triangular elements. We divide each layer into triangular elements as shown in figure 2. All physical quantities are expanded using area coordinates. Then equations (1), (3), (4) and (5) become as,

$$\sum_j I_{ij}{}^n \frac{\partial}{\partial t}\varphi_j{}^{nl} = \sum_j I_{ij}{}^n \{E_{zj}{}^{nl} - \alpha(E_{zj}{}^{nl+1} + E_{zj}{}^{nl-1} - 2E_{zj}{}^{nl})\} \tag{6}$$

$$\sum_j \varphi_j{}^{nl} J_{ji}{}^{nx} = \sum_j I_{ij}{}^n \{B_{yj}{}^{nl} - \eta(B_{yj}{}^{nl+1} + B_{yj}{}^{nl-1} - 2B_{yj}{}^{nl})\} \tag{7}$$

$$\sum_j \varphi_j{}^{nl} J_{ji}{}^{ny} = -\sum_j I_{ij}{}^n \{B_{xj}{}^{nl} - \eta(B_{xj}{}^{nl+1} + B_{xj}{}^{nl-1} - 2B_{xj}{}^{nl})\} \tag{8}$$

$$\sum_j \frac{\partial E_{zj}{}^{nl}}{\partial t} I_{ij}{}^n = \sum_j B_{yj}{}^{nl} J_{ji}{}^{nx} - \sum_j B_{xj}{}^{nl} J_{ji}{}^{ny} - \sum_j (S_{\varphi j}{}^{nl} + \beta E_{zj}{}^{nl} + J_{fj}{}^{nl}) I_{ij}{}^n \tag{9}$$

Here, $\varphi_j{}^{nl}$, $E_{zj}{}^{nl}$, $B_{yj}{}^{nl}$ and $B_{xj}{}^{nl}$ are values of φ, E_z, B_y, B_x at j-th node in n-th element on l-th layer, respectively, and we define I_{ij}^n, J_{ij}^{nx}, J_{ij}^{ny} and $S_{\varphi j}{}^{nl}$ as follow,

$$I_{ij}{}^n = \int \lambda_i{}^n(x,y)\lambda_j{}^n(x,y)dS \tag{10}$$

$$J_{ji}{}^{nx} = \int \{\nabla_x \lambda_j{}^n(x,y)\}\lambda_i{}^n(x,y)dS \tag{11}$$

$$J_{ji}{}^{ny} = \int \{\nabla_y \lambda_j{}^n(x,y)\}\lambda_i{}^n(x,y)dS \tag{12}$$

$$S_{\varphi j}{}^{nl} = sin\varphi_j{}^{nl} \tag{13}$$

Here $\lambda_j{}^n(x,y)$ is j-th area coordinate of n-th element. We solve these equations (6), (7), (8), and (9).

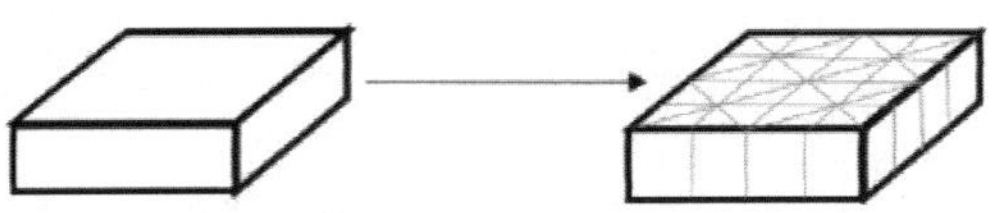

Figure 2. Finite elements for each layer

3. Results

In this study, we set $\sigma E_p / J_c$= 0.01, J_f / J_c= 0.001, α = 1.0 and η = 1.0, where $E_p = \hbar\omega_p / 2ed$, and ω_p is the plasma frequency. $\sigma E_p / J_c$ is a ratio of normal current to the Josephson current. If $\sigma E_p / J_c$ is small, portion of the Josephson current in a total current is relative large respectively.

Figure 3 shows the time evolution of the spatial average of phase differences $<\varphi>$ for j_{ext} / j_c = 1.0 (a) and 2.0 (b). Blue, orange, green and red lines show time evolution of $<\varphi>$ for l = 1, 2, 3 and 4, respectively. Figure 4 shows the time evolution of the spatial distribution of the phase differences $\varphi(t)$ at $\omega_p t$ = 0, 3.5, 7.0, in the first, second, third and fourth JJ layers, respectively, the color changes from light blue to red as the spatial average of phase differences increase. From figures 3(a) and 4(a), we can see when the applied voltage is j_{ext} / j_c = 1.0, all JJs do not become the voltage state. Also, from figures 3(b) and 4(b), we can see when the applied voltage is j_{ext} / j_c = 2.0, the first JJ becomes the voltage state first.

Second, we investigate switching of the second JJ. Figures 5(a) and 6(a) show that when the applied current is j_{ext} / j_c = 4.0, the second JJ does not switch to the voltage state. Figures 5(b) and 6(b) show that when the applied current is j_{ext} / j_c = 5.0, the second JJ switches to the voltage state.

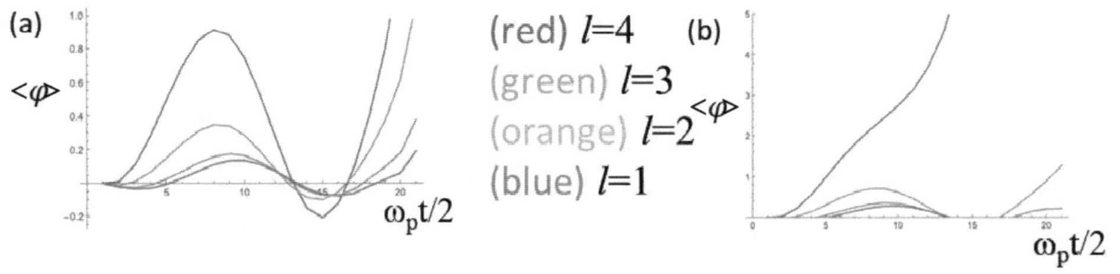

Figure 3. The time evolution of the spatial average of phase differences in the first, second third and fourth JJ layer at J_{ext} / J_c = 1.0 (a) and J_{ext} / J_c = 2.0 (b).

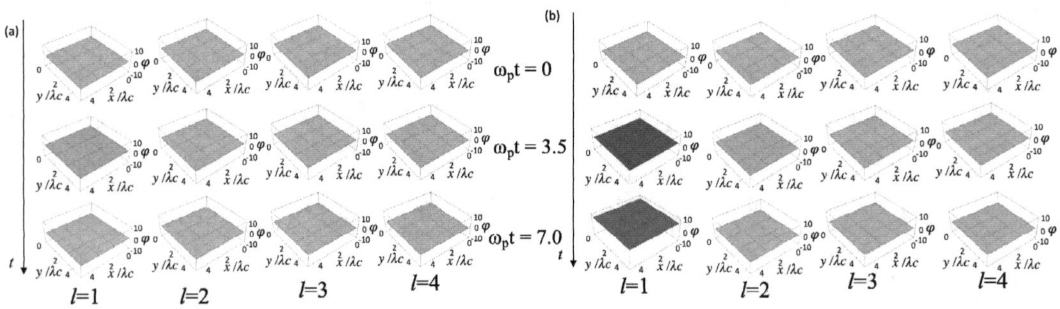

Figure 4. Time evolution of the spatial distribution of the phase differences in the first, second third and fourth JJ layer for $\omega_p t$ = 0, 3.5, 7.0 at J_{ext} / J_c = 1.0 (a) and J_{ext} / J_c = 2.0 (b), where λ_c is magnetic field penetration depth in the c-axis direction.

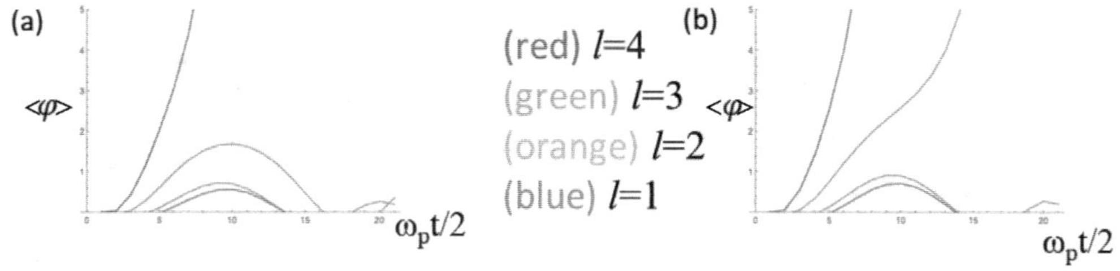

Figure 5. The time evolution of the spatial average of phase differences in the first, second third and fourth JJ layer at $j_{ext}/j_c = 4.0$ (a) and $j_{ext}/j_c = 5.0$ (b).

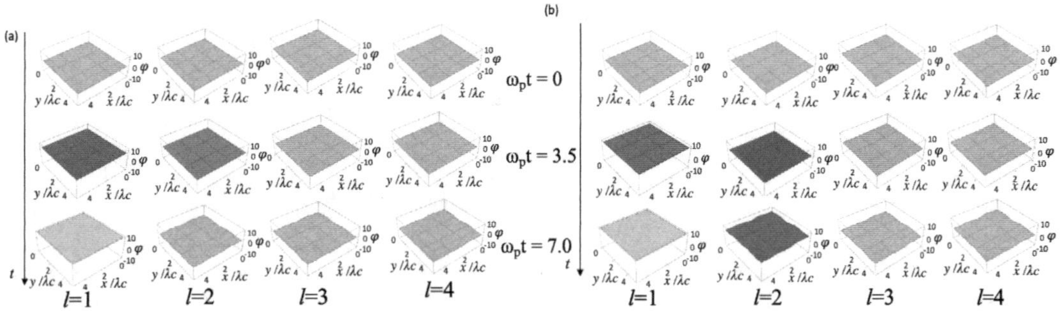

Figure 6. The time evolution of the spatial distribution of the phase differences in the first, second third and fourth JJ layer for $\omega_p t = 0, 3.5, 7.0$ at $j_{ext}/j_c = 4.0$ (a) and $j_{ext}/j_c = 5.0$ (b).

Next, we investigate switching of the third JJ. Figures 7(a) and 8(a) show that when the applied current is $j_{ext}/j_c = 9.0$, the third JJ does not switch to the voltage state. Figures 7(b) and 8(b) show that when the applied current is $j_{ext}/j_c = 10.0$, the third JJ switches to the voltage state.

Then, we investigate switching of the third JJ. Figures 9(a) and 10(a) show that when the applied current is $j_{ext}/j_c = 12.0$, the fourth JJ does not switch to the voltage state. Figures 9(b) and 10(b) show that when the applied current is $j_{ext}/j_c = 13.0$, the fourth JJ switches to the voltage state.

Next, we show the time evolution of the spatial distribution of the phase difference when the inductive coupling constant η is 10.0. Figures 11(a) and 12(a) show that when the applied voltage is $j_{ext}/j_c = 2.0$, first JJ does not become the voltage state. Figures 11(b) and 12(b) show that when the applied voltage is $j_{ext}/j_c = 7.0$, all JJs becomes the voltage state, so when the inductive coupling constant between layers is large, the inner JJs switch easily to the voltage state.

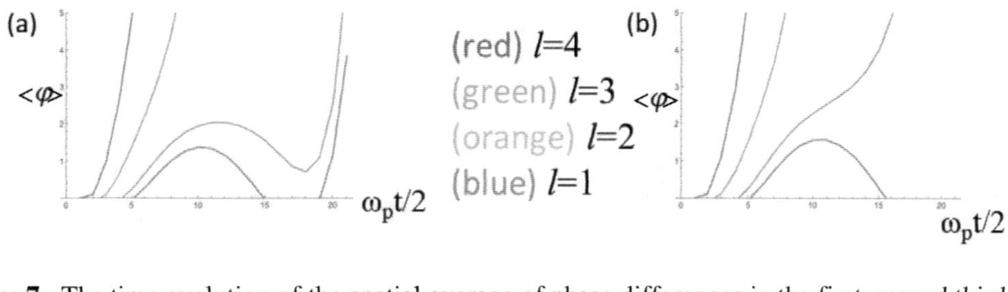

Figure 7. The time evolution of the spatial average of phase differences in the first, second third and fourth JJ layer at $j_{ext}/j_c = 9.0$ (a) and $j_{ext}/j_c = 10.0$ (b).

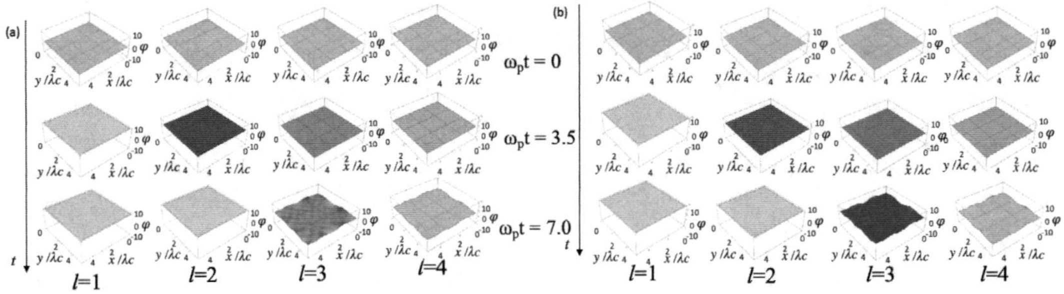

Figure 8. The time evolution of the spatial distribution of the phase differences in the first, second third and fourth JJ layer for $\omega_p t = 0, 3.5, 7.0$ at $j_{ext}/j_c = 9.0$ (a) and $j_{ext}/j_c = 10.0$ (b).

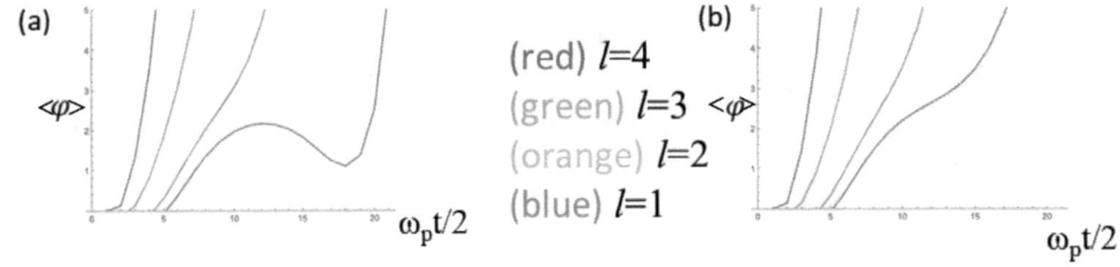

Figure 9. The time evolution of the spatial average of phase differences in the first, second third and fourth JJ layer at $j_{ext}/j_c = 12.0$ (a) and $j_{ext}/j_c = 13.0$ (b).

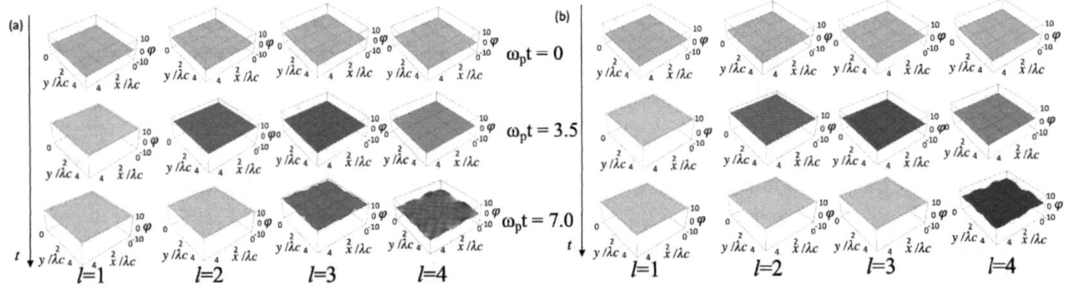

Figure 10. The time evolution of the spatial distribution of the phase differences in the first, second third and fourth JJ layer for $\omega_p t = 0, 3.5, 7.0$ at $j_{ext}/j_c = 12.0$ (a) and $j_{ext}/j_c = 13.0$ (b).

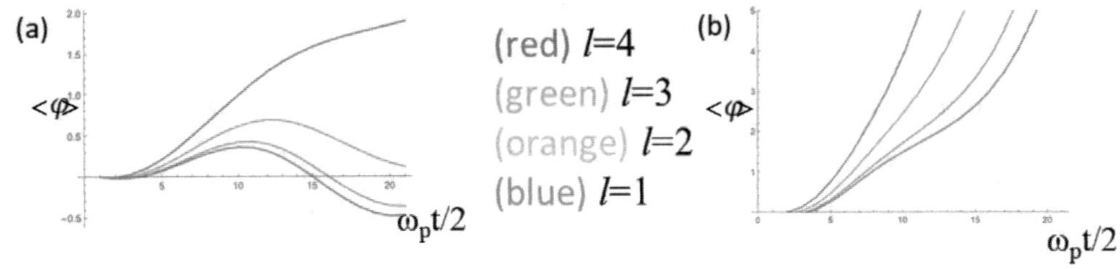

Figure 11. The time evolution of the spatial average of phase differences in the first, second third and fourth JJ layer at $j_{ext}/j_c = 2.0$ (a) and $j_{ext}/j_c = 7.0$ (b).

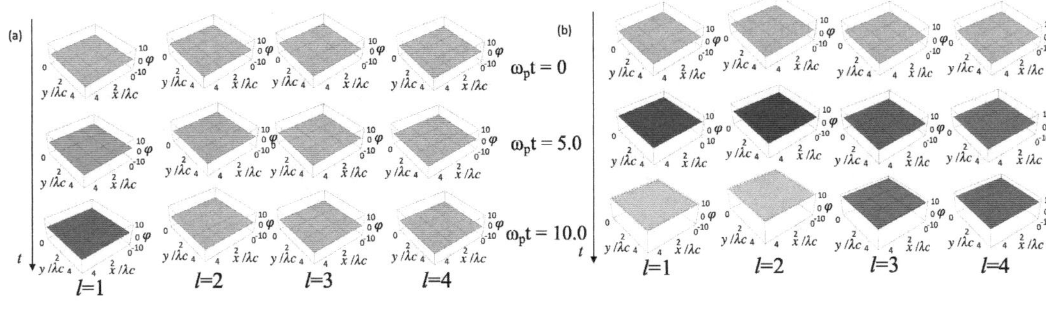

Figure 12. The time evolution of the spatial distribution of the phase differences in the first, second third and fourth JJ layer for $\omega_{pt} = 0, 5.0, 10.0$ at $j_{ext}/j_c = 2.0$ (a) and $j_{ext}/j_c = 7.0$ (b).

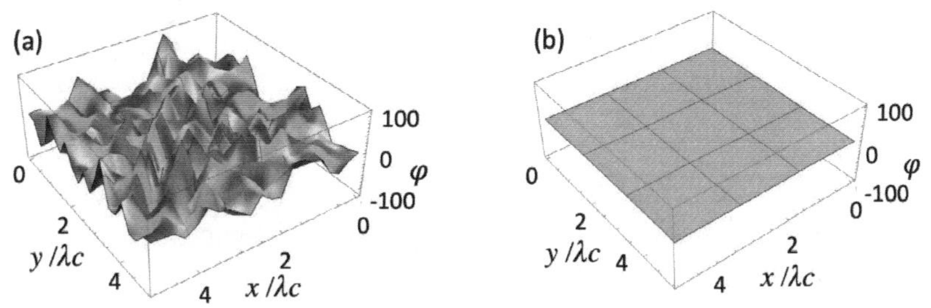

Figure 13. The time evolution of the spatial distribution of the phase differences for $j_{ext}/j_c = 1.0$, $\eta = 1.0$ (a) and $\eta = 10.0$ (b).

Finally, we discuss effects of the inductive coupling between layers on the time evolution of the spatial distribution of the phase difference. In figure 13, we show spatial distributions of ϕ at $t = 10.0/\omega_p$ for $\eta = 1.0$ (a) and 10.0 (b). These figures show that the stronger the inductive coupling constant, the smaller spatial variation of the phase difference due to the influence of the thermal fluctuation current.

4. Summary

In this study, we have investigated the switch to the voltage states of each JJ in the IJJ. We found that when the external current is increased first JJ become voltage state first. Then, second JJ becomes voltage state due to the coupling between first and second JJs. Also, we found that when the coupling between layers becomes stronger, inner JJs switch easily total voltage state and the effect of thermal fluctuation current becomes smaller. We assume that capacitive coupling and inductive coupling are $O(1)$. In real cuprate systems, these parameters are $\alpha \sim 0.01$ and $\eta \geq 1 \times 10^5$. Simulations on such system are future problems. Also, in future, we will investigate how coherence of THz oscillations in all JJ occurs, using present model.

References

[1] Kakeya I and Wang H 2016 *Supercond. Sci. Techmol.* **29** 073001
[2] Koyama T, Matsumoto H, Machida M, and Kadowaki K 2009 *Phys. Rev.* B **79** 104522
[3] Koyama T and Tachiki M 1996 *Phys. Rev.* B **54** 16183
[4] Kitano H, Takahashi Y, Kakehi D, Yamaguchi H, Koizumi S, Ayukawa S 2016 *J. Phys. Soc. Jpn.* **85** 054703

Tracking the Meissner Hole and Relaxation of Magnetization in (Ba,Rb)Fe_2As_2

T Ren, S Pyon and T Tamegai

Department of Applied Physics, The University of Tokyo, 7-3-1 Hongo, Bunkyo-ku, Tokyo 113-8656, Japan

E-mail: rentong@g.ecc.u-tokyo.ac.jp

Abstract. We investigated the unstable interface between vortices and anti-vortices in single crystals of iron-based superconductors ($Ba_{0.67}Rb_{0.33}$)Fe_2As_2 ($T_c \sim 37.6$ K). These interfaces, also known as the Meissner hole, were captured with magneto-optical imaging in both pristine crystal and one with artificial point defects. Local excess current were observed flowing along the interface. In addition, we confirmed that the creep mode of such an interface follows the prediction of the critical state model, accompanied with a finite relaxation rate of magnetization.

1. Introduction

When the applied field is reduced to zero from a sufficiently large value in superconductors with a small demagnetization factor (thin slab with parallel field), a full critical state is formed. A further reduction of the applied field toward negative values, called remagnetization, simply introduces antivortiecs with a well-defined boundary between pre-existing vortices. However, the situation becomes complicated in superconductors with a demagnetization factor close to unity (thin sample with perpendicular field). Near the boundary separating two regions with opposite polarities of vortices, vortices naturally form closed loops due to the presence of in-plane components of the field. Vortex loops have minimum diameters surrounding vortex free region, called "Meissner hole". Experiments by Vlasko-Vlasov showed the presence of such structures in $YBa_2Cu_3O_7$ [1]. What is more, at the skin layer of the Meissner hole, there flows a local excess current which is twice as large as the critical current in the same area. The Meissner hole is believed to affect the low-field vortex dynamics of superconductors since its unstable turbulent structure blurs the interface between vortices and anti-vortices.

In this work, we report high-resolution images of the Meissner holes captured by magneto-optical (MO) imaging in 122-type iron-based superconductor ($Ba_{0.67}Rb_{0.33}$)Fe_2As_2 single crystals. This is the third material in which the Meissner hole is confirmed experimentally after $YBa_2Cu_3O_7$ [1] and Ba($Fe_{1-x}Co_x$)$_2As_2$ [2]. To investigate how the artificial defects could influence the evolution of the Meissner hole, we analyzed and compared the vortex dynamics in both pristine and 3 MeV proton irradiated crystals. We also tracked the motion of the Meissner hole, and compared it with the relaxation of magnetization.

2. Experimental Methods

Single crystals of ($Ba_{0.67}Rb_{0.33}$)Fe_2As_2 were synthesized by FeAs self-flux method using a procedure similar to that described in [3]. 3 MeV proton irradiations were performed using

HIMAC at the National Institute of Radiological Sciences.

The magnetization of the pristine and irradiated crystals were tracked using a superconducting quantum interference device (SQUID) magnetometer (MPMS-5XL, Quantum Design). The relaxation of the magnetization in a single crystal under its critical state is discussed using the Anderson's model [4], where the normalized relaxation rate S is defined as

$$S \equiv \left| \frac{d \ln M}{d \ln t} \right|. \tag{1}$$

In the investigation of the Meissner hole, the distribution of the magnetic field at the crystal surface was taken using differential MO imaging, where integrated backgrounds were subtracted from the integrated target image [5, 6]. By doing so, it is possible to obtain high-resolution MO images with suppressed fluctuations caused by the environments.

3. Result and Discussion

3.1. Relaxation of Magnetization in Single Crystals

As it might be one's interest to relate the vortex dynamics in superconductor, especially at low fields, to the formation and dynamics of the Meissner hole, we first tracked the magnetization of both pristine and proton irradiated single crystals at various temperatures and fields. Figure 1 shows the field dependence of the normalized relaxation rate of magnetization in both pristine and proton irradiated crystals. Here t is the elapsed time after the critical state is created. At higher fields, the behaviour of $S(H)$ in Fig. 1(a) could be explained in the framework of the weak collective creep theory with a glassy exponent μ [7, 8]. The theory predicts the relation $S \propto \mu^{-1}$, so that $S(H)$ shows a characteristic peak when vortices creep individually and μ takes a relatively small value of 1/7, and that $S(H)$ becomes field independent at larger field where vortices creep in small bundles whith $\mu = 5/2$. However, the strong suppression of $S(H)$ at

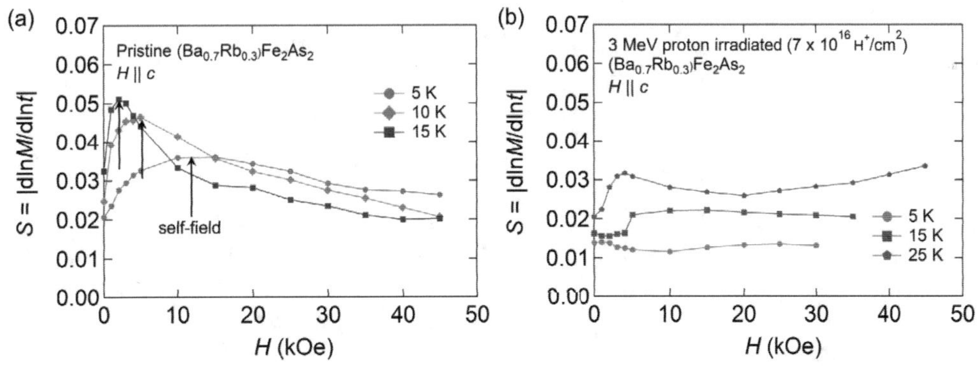

Figure 1. Magnetic field dependence of the normalized relaxation rate of magnetization $S(H)$ in (a) a pristine crystal and (b) a crystal that was irradiated with 3 MeV protons to a dose of $7{\times}10^{16}$ H^+/cm^2. The black arrows mark the location of the self-field.

fields below the self-field of the sample (marked with black arrows) is still under debate. In some cases, such suppression is attributed to the appearance of the Meissner hole [9].

After the introduction of point defects via proton irradiation, the extra strong pinning centers dramatically enhance the pinning force [10, 11], causing considerable variation in S behaviors. The low-field peak is almost entirely suppressed due to the large self-field, as shown in Fig. 1(b). The suppression of $S(H)$ at fields lower than the self-field, however, is still observable. These behaviors of S in the prstine and proton irradiated $(Ba_{0.67}Rb_{0.33})Fe_2As_2$ is very similar to those reported for $(Ba_{0.6}K_{0.4})Fe_2As_2$ [12].

3.2. Meissner Hole in Pristine Crystal

To distinguish the initial magnetization from the remagnetized state which we are actually interested in, we start from observing the virgin penetration process of the specimen. The crystal was firstly cooled to 15 K in zero-field. Then an external field was applied up to 600 Oe at a step of 100 Oe. Figure 2(a) shows the MO image of the crystal being penetrated from the perfect Meissner state. Weakly irregular penetration of vortices from some of edges are probably originated from the inhomogeneous distribution of pinning centers and/or microcracks in the crystal site distribution. The induction drops smoothly from the edge of the crystal as shown in Fig. 2(c).

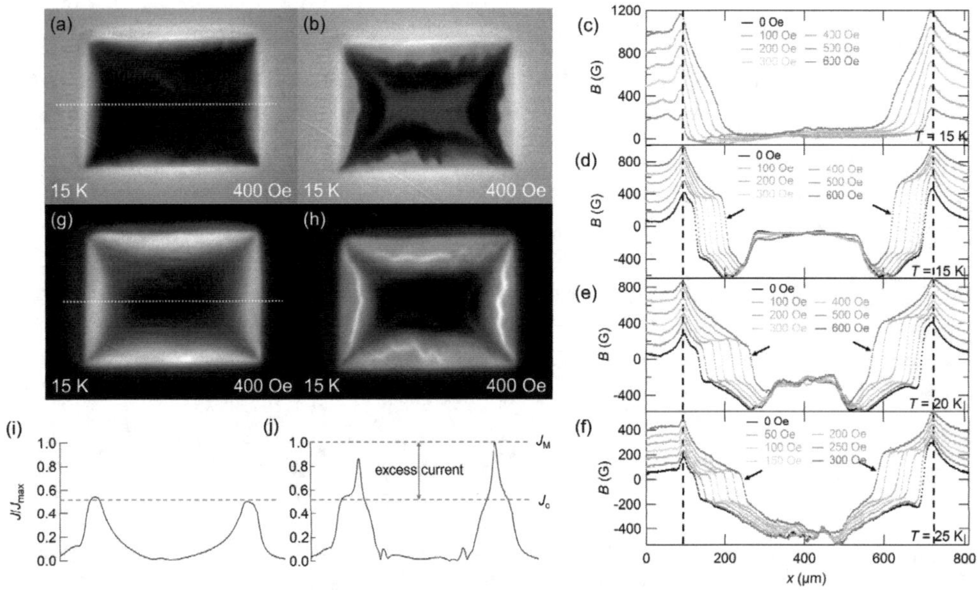

Figure 2. MO images of crystals after (a) virgin penetration and (b) remagnetization process. $B_z(x)$ profile taken along the white dotted line in (c) virgin penetration case and (d)-(f) remagnetized crystal at different temperatures. (g) and (h) are mappings of in-plane current density $|\boldsymbol{J}|$ converted from B mappings in (a) and (b), respectively, using inverse Biot-Savart law. $|\boldsymbol{J}|$ profiles normalized by $|J_{\max}|$ along the dotted line in (g) and (h) are shown in (i) and (j), respectively.

Next, we see the remagnetization process in the same crystal. The magnetization process is almost the same as that for the virgin penetration, except for that the crystal was initially penetrated by a negative field of -1,000 Oe before the positive field was applied. Figure 2(b) shows the crystal being remagnetized by the same positive field as that shown in Fig. 2(a). The remagnetization front between vortices and anti-vortices appears as the boundary between the bright and dark region. The front shows distinct behavior compared with that in virgin penetration. The boundaries are extremely curly and tend to penetrate more into the sample than the virgin case. The meandering remagnetization front progress towards interior of the crystal as the field increases while maintaining almost the same pattern, generating the $B_z(x)$ profile displayed in Fig. 2(d). Unlike the virgin penetration case, sharp drops (marked with black arrows) are spotted at the remagnetization front where $B_z = 0$. This local magnetic field distortion is interpreted to be induced by the integrated Meissner current, which is twice as large as the integrated J_c. We then further extend the observation to higher temperatures like 20 K and 25 K (see Figs. 2(e), (f)).

Since the evidence provided by the B_z profile points directly to the existence of a local excess current, we convert the B_z map into a $|\boldsymbol{J}|$ map using inverse Biot-Savart's Law [2]. The in-plane current density are mapped in Fig. 2(h). As a reference, $|\boldsymbol{J}|$ distribution in the virgin penetration is shown in Fig. 2(g). The brightness in the graph represents the strength of in-plane current density. It is clearly shown that very much alike the shape of the remagnetization front, conduction path of the excess current in Fig. 2(h) is curly and mostly continuous, while the current flows uniformly in Fig. 2(g). Figures 2(j) and (i) are corresponding normalized $|\boldsymbol{J}|$ profiles along the dotted lines in Fig. 2(h) and (g), respectively. The graphs are normalized by the maximum value in the profile, named J_M. In the remagnetized state, a spike-like structure is spotted at the tip of the peak in the profile. On the other hand, in the profile of virgin penetration, no significant structure is found at the same locations. Meanwhile, the magnitude of J_c is roughly half of J_M. Thus, we can safely conclude that in the low-field remagnetized state of the pristine $(Ba_{0.67}Rb_{0.33})Fe_2As_2$, there exist an unstable interface between vortices and antivortices associated with the Meissner hole, in the skin layer of which a doubled current is flowing.

3.3. Dynamics of the Meissner Hole

The motion of the Meissner hole could be analyzed from the aspect of magnetic relaxation, as shown below. In the critical state model [13, 14], a constant J_c is assumed to flow in the crystal. In the remagnetized state, when an external field H_{ext} is applied, the location of the Meissner hole x_M, where $|B_z| = 0$, can be expressed as

$$x_M = \frac{H_{\text{ext}}}{J_c}. \tag{2}$$

On the other hand, according to Anderson's model [4], J_c is relaxing proportional to $\ln t$. Such a relaxation in J_c causes the Meissner hole to move toward the center of the crystal. By defining a normalized creep rate of the Meissner hole, $S_{\text{MH}} = \left|x_M^{-1} d\ln x_M / d\ln t\right|$, one finds that it is exactly the same as the normalized magnetic relaxation rate S,

$$S_{\text{MH}} = \left|\frac{1}{x_M}\frac{dx_M}{d\ln t}\right| = \left|\frac{d\ln x_M}{d\ln t}\right| = \left|\frac{d\ln J_c}{d\ln t}\right| = \left|\frac{d\ln M}{d\ln t}\right| \equiv S. \tag{3}$$

To verify this assumption, we tracked the motion of the Meissner hole for a little bit over 1 min. For extra information on the interaction between defects and vortices in a single crystal, a crystal that was partly irradiated with protons was used for discussion. Figures 3(a) and (b) show the Meissner hole created in such a crystal, where the left part remains pristine, while the right part was irradiated with 3 MeV protons. In Figs. 3(c) and (d), we plot time evolution of the B profile across the red solid lines in Figs. 3(a) amd (b) for both regions. Significant motion of the front can be seen within $\sim$ 20 s from the spacing between profiles. Data near the induction steps are linearly fitted to obtain the location of the Meissner hole x_M where $B_z(x_M) = 0$, and the results are plotted in Figs. 3(e) and (f), respectively. The data points are fitted with a logarithmic relation $\ln x_M = S\ln t + C$. In this way, we estimated S_{MH} as 0.051 and 0.025 for the pristine and irradiated region, respectively. The creep rate of the Meissner hole in the pristine region is twice as fast as that in the irradiated region, indicating that the motion of the Meissner hole is affected by the extra point defects. Comparing S_{MH} with the value of S evaluated from the magnetic relaxation measurements shown in Fig. 1 (see table 1), we find that these results are nearly the same, which confirms our former assumption.

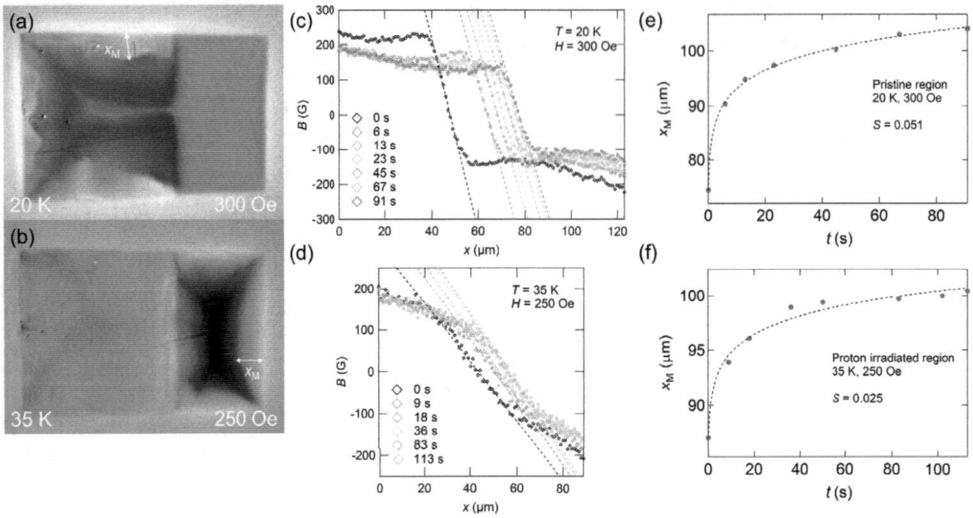

Figure 3. MO images taken on remagnetized $(Ba_{0.33}Rb_{0.67})Fe_2As_2$ single crystal showing Meissner hole appears in (a) the pristine region and (b) the proton irradiated region. $B_z(x)$ profiles along the red lines in (a) and (b) are shown in (c) and (d), respectively, which are considered as the march route of the vortex-antivortex front. The result obtained from (c) and (d) are depicted in (e) and (f), respectively, and fitted to the dashed curve to extract S.

Table 1. The normalized relaxation rate S obtained from two different methods for the pristine and irradiated regions.

	Pristine	Proton irradiated
S_{MH} (from MO)	0.051 (300 Oe, 20 K)	0.025 (250 Oe, 35 K)
S (from SQUID)	0.046 (300 Oe, 20 K)	0.025 (250 Oe, 35 K)

4. Summary

As a microscopic exploration for vortex dynamics in iron-based superconductors, we investigated the remagnetization process in both pristine and 3 MeV proton irradiated $(Ba_{0.67}Rb_{0.33})Fe_2As_2$. Novel behavior of remagnetization fronts was captured in high-resolution MO images at various temperatures. In the profile of magnetic induction, a step near the interface between vortices and anti-vortices clearly indicates the existence of a local excess current, known as the Meissner current. Using inverse Biot-Savart's law, such current was observed clearly in images converted from B mapping. To investigate the dynamics of the Meissner hole, we tracked its location x_M for ~1.5 min. By doing so, the normalized relaxation rate of magnetization S is extracted, which has nearly the same value as that obtained from magnetic relaxation measurements. The motion of the Meissner hole, indicated by S, is apparently affected by the extra point defects created by proton irradiation.

Acknowledgments

This work is partly supported by a Grant in Aid for Scientific Research (A) (17H01141) from the Japan Society of Science (JSPS).

Reference

[1] Vlasko-Vlasov V, Welp U, Crabtree G, Gunter D, Kabanov V and Nikitenko V 1997 *Phys. Rev. B* **56** 5622
[2] Mohan S, Tsuchiya Y, Nakajima Y and Tamegai T 2011 *Phys. Rev. B* **84** 180504
[3] Ren T, Pyon S and Tamegai T 2021 *Journal of Phys.: Conf. Series* **1975** 012013
[4] Anderson P W 1962 *Phys. Rev. Lett.* **9** 309
[5] Soibel A, Zeldov E, Rappaport M, Myasoedov Y, Tamegai T, Ooi S, Konczykowski M and Geshkenbein V B 2000 *Nature* **406** 282
[6] Yasugaki M, Itaka K, Tokunaga M, Kameda N and Tamegai T 2002 *Phys. Rev. B* **65**(21) 212502
[7] Feigel'Man M, Geshkenbein V, Larkin A and Vinokur V 1989 *Phys. Rev. Lett.* **63** 2303
[8] Feigel'Man M, Geshkenbein V and Larkin A 1990 *Physica C Supercond.* **167** 177
[9] Tamegai T, Taen T, Yagyuda H, Tsuchiya Y, Mohan S, Taniguchi T, Nakajima Y, Okayasu S, Sasase M, Kitamura H, Murakami T, Kambara T and Kanai Y 2012 *Supercond. Sci. Technol.* **25** 084008
[10] Blatter G, Feigel'man M V, Geshkenbein V B, Larkin A I and Vinokur V M 1994 *Rev. Mod. Phys.* **66** 1125
[11] Willa R, Koshelev A E, Sadovskyy I A and Glatz A 2017 *Supercond. Sci. Technol.* **31** 014001
[12] Taen T, Ohtake F, Pyon S, Tamegai T and Kitamura H 2015 *Supercond. Sci. Technol.* **28** 085003
[13] Bean C P 1962 *Phys. Rev. Lett.* **8** 250
[14] Bean C P 1964 *Rev. Mod. Phys.* **36** 31

Peak Effects in 2*H*-$NbSe_2$ Single Crystals Introduced with Artificial Defects by Particle Irradiation

W J Li[1], S Pyon[1], S Okayasu[2] and T Tamegai[1]

[1]Department of Applied Physics, The University of Tokyo, 7-3-1 Hongo, Bunkyo-ku, Tokyo 113-8656, Japan
[2]Advanced Science Research Center, Japan Atomic Energy Agency, Tokai, Ibaraki 319-1195, Japan

E-mail: liwenjie@g.ecc.u-tokyo.ac.jp

Abstract. Effects of splayed columnar defects in $NbSe_2$ single crystals introduced by 800 MeV Xe and 320 MeV Au irradiations were studied. Peak effects have been observed for both samples. For 320 MeV Au irradiated samples, the peak field is found to increase with increasing the splay angle, while in samples irradiated by 800 MeV Xe the peak field shows an opposite dependence. The temperature dependence of the peak field for these two samples are also different. Based on the analysis of pinning force density, this different temperature dependence of the peak field may originate from the different pinning mechanisms.

1. Introduction

Particle irradiation is known as one of the typical ways to introduce artificial defects in superconductors [1]. Artificial defects can strongly enhance critical current density (J_c), which has been achieved in both conventional [2] and unconventional superconductors [1,3]. Other than J_c enhancement, these artificial defects can also act as pair breakers and suppress the critical temperature (T_c) [4]. In addition, in superconductors with competing ground state such as charge-density wave (CDW), the particle irradiation can be a good way to study the relationship between superconductivity and another ground state. For example, the competition between superconductivity and CDW is reported in $NbSe_2$ single crystals after 2.5 MeV electron irradiation [5]. Competition between CDW and superconductivity has also been studied in $Lu_5Ir_4Si_{10}$ single crystal after 4 MeV proton irradiation, where monotonic suppression of T_{CDW} and steady enhancement of T_c with increasing dose were observed [6].

The particle irradiation not only affects the J_c, T_c, and T_{CDW}, it can also induce some other phenomena, such as the appearance of an anomalous peak effect which is reported in unconventional superconductors introduced with columnar defects by heavy-ion irradiations [1,7]. The anomalous peak effect is usually observed in magnetization measurements and the peak field locates approximately equals to 1/3 B_Φ (B_Φ: dose equivalent matching field) [1,7]. Recently, a similar peak effect located at ~1/5 B_Φ was reported in $NbSe_2$ introduced with tilted columnar defects by 1.4 GeV Pb irradiation [8]. On the other hand, other peak effects which are located at around H_{c2} are frequently observed in conventional superconductors [9,10]. Pippard explained it based on the idea that the rigidity of the flux lattice falls to zero rapidly close to H_{c2} than the order parameter, which results in an enhancement of J_c in a narrow region close to H_{c2} [9]. In recent years, the mechanism of this peak effect is reinterpreted to come from the phase transition from the ordered Bragg glass phase at low fields to the disordered state at high fields, resulting in a reentrant disorder-driven transition [11]. Actually, by introducing point defects by chemical doping of Co into 2*H*-$NbSe_2$, this peak effect is found to shift from high fields to low fields [12], which is one of the characteristics of the order-disorder transition. So it is necessary to study how the density of columnar defects affects the peak effect.

In this study, we irradiated 2*H*-$NbSe_2$ single crystals by 800 MeV Xe and 320 MeV Au to introduce splayed columnar defects. How the peak effect is affected by the density of columnar defects and splay angle θ_{CD} are studied.

2. Experimental details

Single crystals of $NbSe_2$ were prepared by iodine vapor transport method [13]. The details on the sample synthesis have been described in Ref. [2]. 800 MeV Xe irradiation was conducted at NIRS-HIMAC in Chiba. 320 MeV Au irradiation was conducted using the tandem accelerator at JAEA in Tokai. By controlling the angle between the beam and the crystal *c*-axis, splayed columnar defects were introduced. The magnetization measurements were conducted by a commercial SQUID magnetometer (MPMS-XL5, Quantum Design). J_c was estimated from the magnetic hysteresis loops (MHLs) by using the extended Bean model [14]. In this model, J_c [A/cm^2] is given by

$$J_c = \frac{20\Delta M}{a(1-a/3b)} \quad (b > a), \tag{1}$$

where ΔM [emu/cm^3] is the difference of magnetization when the external field is swept down and up. a [cm] and b [cm] are sample width and length, respectively.

3. Results and discussion

Figures 1 (a) and (b) show the external field dependence of J_c for $NbSe_2$ introduced with splayed columnar defects (B_Φ = 2 T + 2 T, θ_{CD} = ±5°) by 800 MeV Xe and 320 MeV Au irradiations, respectively. J_c is strongly enhanced for both samples compared to that in a pristine sample [2], and the peak effects are observed in both samples. The J_c value at the peak field (H_p) in the two samples decreases monotonically with increasing temperature, while H_p follows different temperature dependence as shown in Figs. 1 (c) and (d). Our preliminary transmission electron microscope observations show that the 320 MeV Au irradiation creates clear columnar defects in $NbSe_2$, while the 800 MeV Xe irradiation creates only faint traces of columnar defects. Such a difference in the morphology of columnar defects is caused by the difference in the energy deposition rates in the two ion beams (S_e (320 MeV Au) = 3.27 keV/Å and S_e (800 MeV Xe) = 2.74 keV/Å).

Figure 2 shows B_Φ dependence of H_p for samples irradiated by 320 MeV Au. H_p shows a linear increase as a function of B_Φ, which is approximated by the following formula; $H_p = CB_\Phi$. The parameter C depends on the temperature. The increasing trend of H_p with B_Φ is opposite to the peak effect induced by the order-disorder transition [12]. However, in $NbSe_2$ with splayed columnar defects created by 800 MeV Xe, H_p increases with increasing θ_{CD} as shown in Fig. 3 (b). Thus behavior is different from $NbSe_2$ introduced with splayed columnar defects created by 320 MeV Au [15]. Instead, this behavior shares a similarity with the peak effect induced by the order-disorder transition, where with increasing point defects, peak field shifts from high fields to low fields. One possible explanation of this similarity is a crossover of pinning from columnar pinning to point-like pinning. At a relatively small θ_{CD}, the vortex tries to occupy all parts of columnar defects, while with increasing θ_{CD} only a small part of columnar defects are occupied by the vortex as shown in the inset of Fig. 3 (b). The effective length of columnar defects occupied by a vortex is ~ $2\xi/\sin\theta_{CD}$, where ξ is the coherence length of the superconductors. It means that with increasing θ_{CD} the occupied part is decreased, and the pinning of vortices by columnar defects become more like point pinning. This explanation applies only to the case where the distance between neighboring vortices is not too large, otherwise the vortices always try to follow the direction of columnar defects.

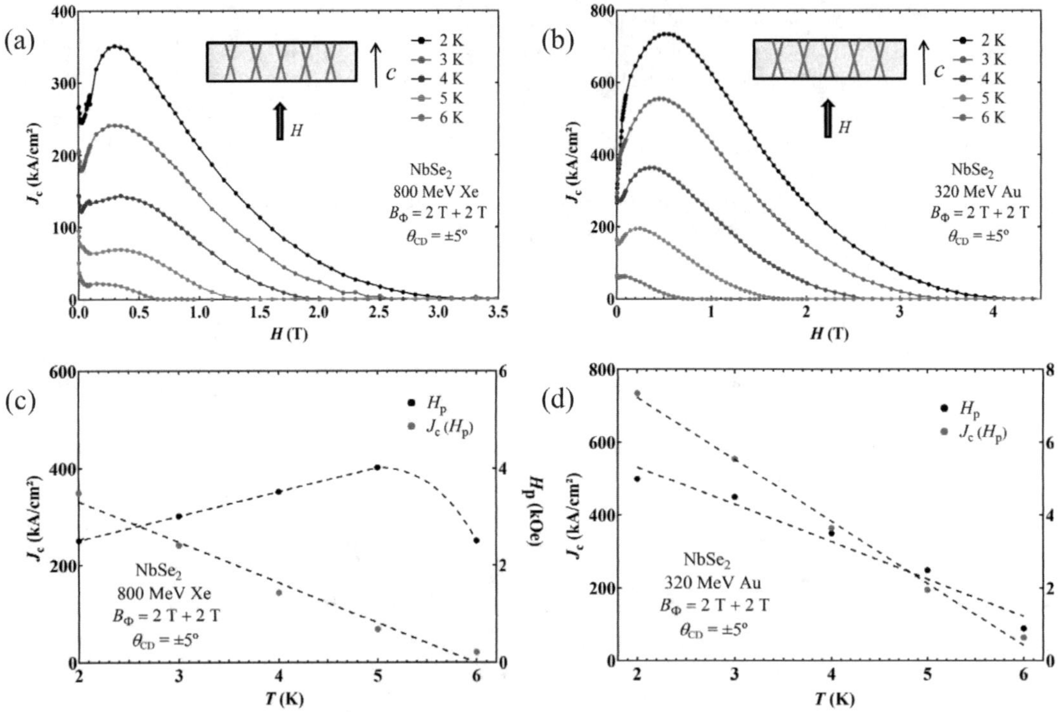

Figure 1. Magnetic field dependence of J_c for (a) 800 MeV Xe and (b) 320 MeV Au irradiated $NbSe_2$ single crystals with splayed columnar defects with $\theta_{CD} = \pm 15°$. Temperature dependence of J_c at H_p ($J_c(H_p)$) and H_p for (c) 800 MeV Xe and (d) 320 MeV Au irradiated $NbSe_2$ single crystals with splayed columnar defects with $\theta_{CD} = \pm 15°$.

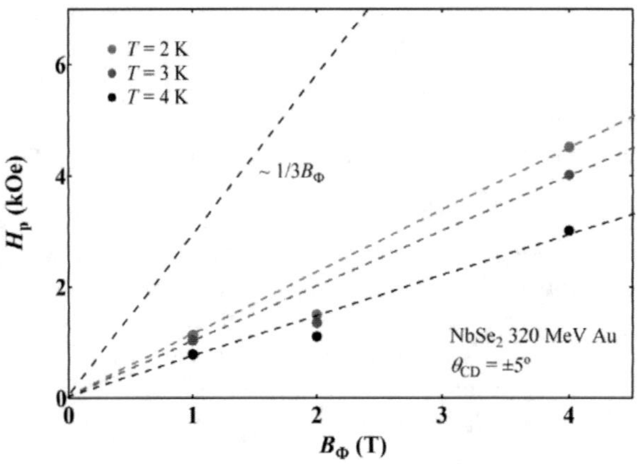

Figure 2. The B_Φ dependence of H_p for $NbSe_2$ with splayed columnar defects ($\theta_{CD} = \pm 5°$) created by 320 MeV Au at $T = 2$, 3, and 4 K.

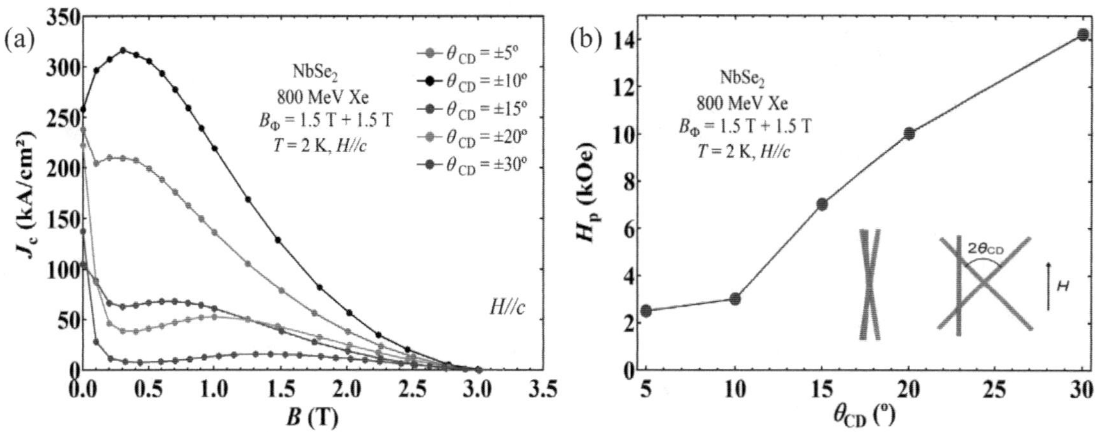

Figure 3. (a) Magnetic field dependence of J_c for $NbSe_2$ with splayed columnar defects (1.5 T + 1.5 T) created by 800 MeV Xe at $T = 2$ K for different splay angles. (b) The θ_{CD} dependence of H_p for $NbSe_2$ with splayed columnar defects (1.5 T + 1.5 T) created by 800 MeV Xe at $T = 2$ K. The inset shows schematic figures of vortex pinning crossover from columnar pinning to point-like pinning. Blue and red solid lines represent columnar defects and vortices, respectively.

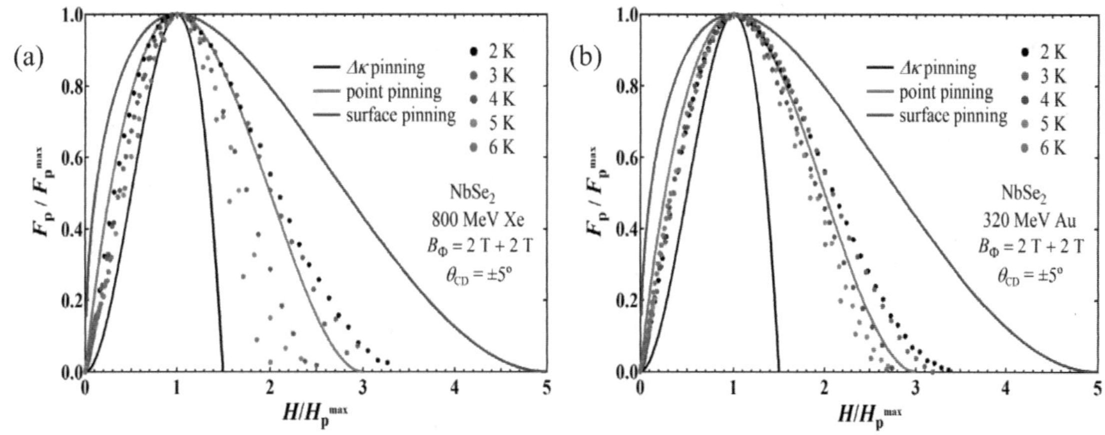

Figure 4. The magnetic field dependence of the normalized pinning force using the extended Dew-Hughes model for $NbSe_2$ single crystals with splayed columnar defects irradiated by 800 MeV Xe (a) and 320 MeV Au (b). The magnetic is normalized by the field H_p^{max}, where the pinning force is maximum.

The pinning mechanisms of vortices in superconductors have been categorized into several models by Dew-hughes [16], such as point pinning, surface pinning, and $\Delta\kappa$ pinning. In the original Dew-hughes model, the external field was normalized by the upper critical field (H_{c2}), while it was replaced by the irreversibility field (H_{irr}) in superconductors with large thermal fluctuations such as cuprates [17]. However, it is not easy to estimate these field scales, when they are extremely large. A better way is to replace H_{c2} or H_{irr} by H_p^{max} [18], which is the field where the maximum pinning force is achieved. Using this extended method, we normalized the pinning force of $NbSe_2$ with splayed columnar defects created by 800 MeV Xe and 320 MeV Au irradiations as shown in Figs. 4 (a) and (b), respectively. In sample irradiated with 800 MeV Xe, the F_p/F_p^{max}-H/H_p^{max} curve at different temperatures cannot be scaled to one curve. It means that the pinning mechanisms at different temperatures are different. In this sample, the normalized curves at $T = 2$ and 3 K are close to the curve for 'point pinning', while they approach to the curve for '$\Delta\kappa$ pinning' as the temperature is increased. The nonmonotonic temperature dependence of H_p in the 800 MeV Xe irradiated sample could be related to this crossover of the pinning mechanisms with temperature. On the other hand, in the sample irradiated by 320 MeV Au, the normalized pinning force curves are well reproduced by 'point pinning' in the whole temperature range. However, it should be noted that the reason why the normalized pinning curve for a superconductor with columnar defects can be fitted with the 'point pinning' is not clear at the present stage.

4. Summary

Effects of splayed columnar defects created by 800 MeV Xe and 320 MeV Au irradiations in $NbSe_2$ single crystals have been studied. In both samples, J_c is enhanced after the irradiation, and it decreases monotonically with temperature. In addition, both samples show similar peak effects with broad peaks below 5 kOe. However, temperature dependences of H_p are different; H_p shows nonmonotonic temperature dependence in the 800 MeV Xe-irradiated samples. While H_p follows a linear dependence with increasing B_Φ in samples irradiated by 320 MeV Au. The scaled pinning force depends on temperature in the 800 MeV Xe irradiated sample, which may be related to the nonmonotonic temperature dependence of H_p.

Acknowledgements

This work is partly supported by a Grant in Aid for Scientific Research (A) (17H01141) from the Japan Society for the Promotion of Science (JSPS).

References

[1] Civale L, Marwick A D, Worthington T K, Kirk M A, Thompson J R, Krusin-Elbaum L, Sun Y, Clem J R and Holtzberg F 1991 *Phys. Rev. Lett.* **67** 648
[2] Li W, Pyon S, Takahashi A, Miyawaki D, Kobayashi Y and Tamegai T 2020 *J. Phys.: Conf. Ser.* **1590** 012003
[3] Tamegai T, Taen T, Yagyuda H, Tsuchiya Y, Mohan S, Taniguchi T, Nakajima Y, Okayasu S, Sasase M, Kitamura H, Murakami T, Kambara T and Kanai Y 2012 *Supercond. Sci. Technol.* **25** 084008
[4] Nakajima Y, Taen T, Tsuchiya Y, Tamegai T, Kitamura H and Murakami T, 2010 *Phys. Rev. B* **82** 220504(R)
[5] Cho K, Konczykowski M, Teknowijoyo S, Tanatar M A, Guss J, Gartin P, Wilde J M, Kreyssig A, McQueeney R, Goldman A I, Mishra V, Hirschfeld P J and Prozorov R 2018 *Nat. Commun.* **9** 2796
[6] Leroux M, Mishra V, Opagiste C, Rodière P, Kayani A, Kwok W and Welp U 2020 *Phys. Rev. B* **102** 094519
[7] Park A, Pyon S, Ohara K, Ito N, Tamegai T, Kambara T, Yoshida A and Ichinose A 2018 *Phys. Rev. B* **97** 064516
[8] Eley S, Khilstrom K, Fotovat R, Xiao Z L, Chen A, Chen D, Leroux M, Welp U, Kwok W K and Civale L 2018 Sci. Rep. **8** 13162
[9] Pippard A B 1969 Philos. Mag. **19** 217
[10] Campell A M, Evetts J 1972 *J. Adv. Phys.* **72** 199
[11] Paltiel Y, Zeldov E, Myasoedov Y, Rappaport M L, Jung G, Bhattacharya S, Higgins M J, Xiao Z L, Andrei E Y, Gammel P L and Bishop D J 2000 *Phys. Rev. Lett.* **85** 3712
[12] Iavarone M, Capua R Di, Karapetrov G, Koshelev A E, Rosenmann D, Claus H, Malliakas C D, Kanatzidis M G, Nishizaki T and Kobayashi N 2008 *Phys. Rev. B* **78** 174518
[13] Naito M and Tanaka S 1982 *J. Phys. Soc. Jpn.* **51** 219
[14] Bean C P 1964 *Rev. Mod. Phys.* **36** 31
[15] Li W, Pyon S, Okayasu S and Tamegai T 2021 *J. Phys.: Conf. Ser.* **1975** 012003
[16] Dew-Hughes D 1974 *Philos. Mag.* **30** 293
[17] Koblischka M R, van Dalen A J J, Higuchi T, Yoo S I and Murakami M 1998 *Phys. Rev. B* **58** 2863
[18] Higuchi T, Yoo S I and Murakami M 1999 *Phys. Rev. B* **59** 1514

Vortex Dynamics and Critical Current Densities in 90 K-phase and 60 K-phase $YBa_2Cu_3O_{7-\delta}$ with Splayed Columnar Defects

J Wang[1], N Ito[1], S Pyon[1], T Nishizaki[2] and T Tamegai[1]
[1] Department of Applied Physics, The University of Tokyo, 7-3-1 Hongo, Bunkyo-ku, Tokyo 113-8656, Japan
[2] Department of Electrical Engineering, Kyushu Sangyo University, 2-3-1 Matsukadai, Higashi-ku, Fukuoka 813-8503, Japan

E-mail: tamegai@ap.t.u-tokyo.ac.jp

Abstract. Enhancement of critical current density (J_c) was observed in $YBa_2Cu_3O_{7-\delta}$ with splayed columnar defects (CDs). In the case of 90 K-phase $YBa_2Cu_3O_{7-\delta}$ with splayed CDs, the enhanced J_c drops suddenly at the depinning temperature of vortices from CDs (T_{dp}), which is interpreted to be caused by the excitation of double kinks traversing neighboring CDs. In addition to the optimal 90 K-phase, where the superconductivity dominates, there is another stable 60 K-phase and other phases with different T_c. This study reports temperature dependence of J_c and the step-like decrease of J_c in $YBa_2Cu_3O_{7-\delta}$ with CDs and with different T_c. Results show that the T_{dp} is strongly related to the T_c. However, additional entropic reduction of T_{dp} may depend on parameters other than T_c.

1. Introduction

Columnar defects (CDs) created by heavy-ion irradiation were proved to enhance the critical current density (J_c) in cuprate superconductors [1]. Further enhancement of J_c was also demonstrated by splaying the direction of CDs [2]. Similar enhancements of J_c by the introduction of CDs were also reported in iron-based superconductors (IBSs) [3, 4] and $NbSe_2$ [5, 6]. For the temperature dependence of enhanced J_c, a sudden drop was observed at around 40 K below the dose-equivalent matching field B_Φ in the case of optimally-doped 90 K-phase $YBa_2Cu_3O_{7-\delta}$ [2]. The sudden drop of J_c is considered to occur above the depinning temperature (T_{dp}), above which vortices become mobile due to excitation of kinks traversing neighbouring CDs [7]. The word *depinning* here stands for the depinning of vortices from CDs. However, similar drops in J_c have not been observed in IBSs with CDs [3]. In addition to the optimally-doped phase with $T_c = 90$ K, $YBa_2Cu_3O_{7-\delta}$ has more phases depending on the oxygen content δ. There is another stable phase with $T_c \sim 60$ K, where charge-density wave competes with the superconductivity. The two phases are named 90 K- and 60 K-phase, respectively. $YBa_2Cu_3O_{7-\delta}$ with different oxygen contents can be obtained by oxygen annealing under different pressures and temperatures [8].

In this study, we synthesized $YBa_2Cu_3O_{7-\delta}$ single crystals and used the oxygen annealing to obtain three different kinds of crystals with $T_c = 92$ K, 67 K, and 46 K to confirm whether the sudden drop of J_c will appear in other phases of $YBa_2Cu_3O_{7-\delta}$. 90 K-phase crystals were irradiated by 2.6 GeV U ions and 800 MeV Xe ions from two directions with a total dose-equivalent field of $B_\Phi = 4$ T, each with $B_\Phi = 2$ T. Crystals with $T_c = 67$ K and 46 K were irradiated by 800 MeV Xe ions with the same conditions. We controlled splay angles (θ_{CD}), which are angles between the incident beam direction and the c-axis, by tilting the ab-plane of samples during the irradiation. The temperature and θ_{CD} dependences of J_c were studied by evaluating J_c from the magnetization of the sample by using the Bean model.

2. Experimental Methods

90 K-phase $YBa_2Cu_3O_{7-\delta}$ single crystals were grown by a self-flux method using an yttria crucible. The synthesis method was described in a previous study [9]. The typical transition widths of 90 K-phase $YBa_2Cu_3O_{7-\delta}$ single crystals is 2.5 K. After the synthesis, crystals were annealed with 1 bar oxygen flow and 450°C for 7 days, resulting in $T_c = 92$ K. Other crystals were grown using a gold crucible method described in a previous study [10]. Crystals after the synthesis were annealed with 1 bar oxygen for 7 days under 650°C and 850°C, with the final $T_c = 67$ K and 46 K, respectively. The typical transition widths of 67 K and 46 K $YBa_2Cu_3O_{7-\delta}$ single crystals are 6 K and 3 K, respectively. The 2.6 GeV U irradiation was performed using the RIKEN Ring Cyclotron for the 90 K-phase crystals, and the total dose for each sample was $B_\Phi = 4$ T. The 800 MeV Xe irradiation was performed for all the crystals using Heavy Ion Medical Accelerator at NIRS and the total dose for each sample was $B_\Phi = 4$ T.

Magnetization of 2.6 GeV U irradiated 90 K-phase crystals was measured by the superconducting quantum interference device (SQUID) magnetometer (MPMS-XL5, Quantum Design), while other crystals were measured by another SQUID magnetometer (MPMS3, Quantum Design).

3. Results and Discussion

The J_c [A/cm^2] values were evaluated from the magnetization data using the Bean model with the following formula:

$$J_c = \frac{20\Delta M}{a\left(1 - \frac{a}{3b}\right)}, \tag{1}$$

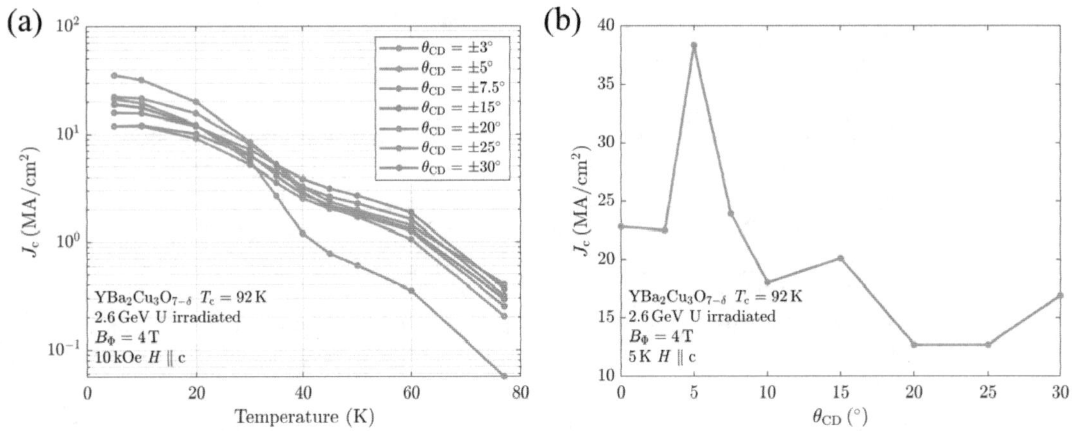

Figure 1. (a) The temperature dependence of J_c at 10 kOe for 2.6 GeV U irradiated 90 K-phase $YBa_2Cu_3O_{7-\delta}$ with different θ_{CD}. (b) The θ_{CD} dependence of J_c under the self-field at 5 K of 2.6 GeV U irradiated 90 K-phase $YBa_2Cu_3O_{7-\delta}$.

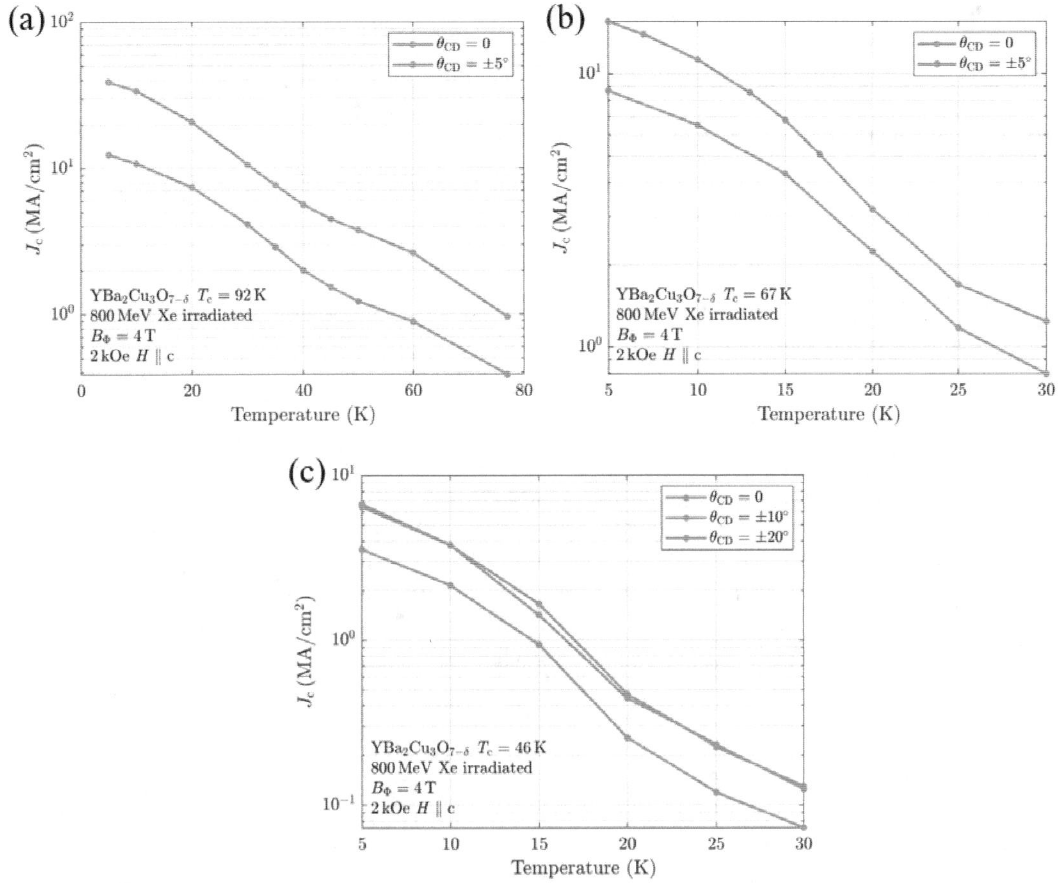

Figure 2. The temperature dependence of J_c at 2 kOe with different θ_{CD} for 800 MeV Xe irradiated (a) 90 K-phase $YBa_2Cu_3O_{7-\delta}$, (b) 60 K-phase $YBa_2Cu_3O_{7-\delta}$, and (c) $YBa_2Cu_3O_{7-\delta}$ with $T_c \sim 46$ K.

where ΔM [emu/cm^3] is the difference between magnetizations in the field-decreasing and field-increasing processes, while a [cm] and b [cm] ($a < b$) are the width and the length of the sample, respectively [11, 12].

For the evaluation of temperature dependence of J_c, we used data at 10 kOe and 2 kOe for 2.6 GeV U irradiated samples and 800 MeV Xe irradiated samples, respectively. The reason for this is that sudden decrease of J_c only occurs below the B_Φ [2]. In the case of 800 MeV Xe irradiation, the real dose-equivalent matching field could be lower than the estimated one because the intensity of the beam has a Gaussian distribution, and the value of it could be over-estimated during the irradiation in the real situation. The temperature dependence of J_c of 90 K-phase $YBa_2Cu_3O_{7-\delta}$ with different θ_{CD} are shown in Figs. 1(a) and 2(a) for 2.6 GeV U and 800 MeV Xe irradiated samples, respectively. For both cases, a sudden drop of J_c can be observed at around 35 K under each condition, which agrees with the previous study [2]. It is clear that θ_{CD} up to 30° does not change the temperature dependence of J_c too much. In addition, the θ_{CD} dependence of J_c under the self-field at 5 K of 2.6 GeV U irradiated 90 K-phase $YBa_2Cu_3O_{7-\delta}$ is shown in Fig. 1(b). A sharp increase of J_c at $\theta_{CD} = \pm 5°$ is reproduced similar to that reported in Ref. [2]. The temperature dependence of J_c with different θ_{CD} for

$YBa_2Cu_3O_{7-\delta}$ with $T_c = 67$ K and 46 K are shown in Figs. 2(b) and 2(c), respectively. Similar to the case of 90 K-phase $YBa_2Cu_3O_{7-\delta}$, sudden drops in J_c are observed at ~19 K and ~17 K for 60 K-phase $YBa_2Cu_3O_{7-\delta}$ and $YBa_2Cu_3O_{7-\delta}$ with $T_c = 46$ K. According to the model proposed in the previous study, the sudden decrease of J_c can be explained by the result of double-kink excitations at T_{dp}, which can be estimated by the following equation:

$$T_{dp} \approx T_c\left(\frac{\nu}{1+\nu}\right), \quad \nu = \frac{r_0}{4\xi_{ab}(0)}\frac{1}{\sqrt{G_i}}. \tag{2}$$

Here r_0 is the radius of CDs and, $\xi_{ab}(0)$ is the in-plane coherence length, and the G_i is the Ginzburg number which reflects the strength of fluctuations of the superconductor given by the following formula:

$$G_i \propto \frac{\gamma^2 T_c^2 \lambda^4}{\xi_{ab}^2(0)}, \tag{3}$$

where γ is the anisotropy parameter and λ is the penetration depth [13]. All parameters necessary to calculate G_i and estimated and measured T_{dp} are summarized in Table 3. In the case of 90 K-phase $YBa_2Cu_3O_{7-\delta}$, G_i is estimated as $\sim 1\times10^{-2}$, resulting T_{dp} being about $0.9T_c$ [7]. In the case of 60 K-phase $YBa_2Cu_3O_{7-\delta}$, the estimated T_{dp} is about $0.8T_c$, according to the measured values of $\xi_{ab}(0)$, γ, and λ [14, 15, 16]. However, even for the 90 K-phase $YBa_2Cu_3O_{7-\delta}$, the measured T_{dp} value of 40 K does not agree with the estimated one, which should be ~ 80 K. One of possible reasons for the discrepancy is that the real T_{dp} is less than the estimated one due to the nonideal pinning efficiency of the real CDs allowing easier wandering of vortices. In Ref. [7], this entropic reduction phenomenon was described to reduce the value of T_{dp} in the real material to roughly half of its estimated value predicted by Eq. (2). Here we define the reduction parameter given by the ratio of measured T_{dp} and the estimated one. The reduction parameter for 90-K phase $YBa_2Cu_3O_{7-\delta}$ is 0.44. If we assume the same entropic reduction of T_{dp} occurs irrespective of the value of T_c, the sharp drop of J_c in 60 K-phase $YBa_2Cu_3O_{7-\delta}$ should appear at around $0.35T_c$, which is about 23 K. This value agrees with our experimental result that the sudden drop of J_c happened at around 19 K with the reduction value being 0.38. However, the estimated value of T_{dp} in $YBa_2Cu_3O_{7-\delta}$ with $T_c = 46K$ is about $0.37T_c$, which is exactly what we measured experimentally. The absence of entropic reduction could be related to the sufficiently distorted crystal structure of underdoped $YBa_2Cu_3O_{7-\delta}$ with plenty of oxygen defects in CuO chains. However, the relevance of structural defects to the vortex dynamics in $YBa_2Cu_3O_{7-\delta}$ with CDs needs to be confirmed by additional experiments in crystals with CDs and controlled amount of point defects introduced by proton irradiation.

Finally, we compare the temperature dependence of J_c in cuprate and IBSs with CDs. In the case of $Ba_{0.6}K_{0.4}Fe_2As_2$, anisotropy is only about 2.5. Together with other experimental data [13], we estimate T_{dp} for $Ba_{0.6}K_{0.4}Fe_2As_2$ as $0.92T_c$. This value of T_{dp} is similar to that for 90 K-phase $YBa_2Cu_3O_{7-\delta}$. However, there is no evidence that similar sudden drops in the temperature dependence of J_c show up IBSs, which implies that entropic reductions of T_{dp} are different due to the difference of detailed structure of CDs in these two classes of superconductors.

Our results show that the sudden drop of J_c occurs not only in 90 K-phase $YBa_2Cu_3O_{7-\delta}$, but also $YBa_2Cu_3O_{7-\delta}$ with $T_c = 67$ K and 46 K. These results support the theoretical assumption about the relationship between J_c and kink motions of vortices for $YBa_2Cu_3O_{7-\delta}$ with CDs in the previous study [7], and also gives a strong evidence that the T_{dp} and related physical parameters will influence the excitation and pinning proprieties of votrices in $YBa_2Cu_3O_{7-\delta}$ with CDs.

4. Summary

We prepared $YBa_2Cu_3O_{7-\delta}$ crystals with three different T_c by the oxygen annealing under different conditions. Splayed CDs with splay angles up to 30° are introduced into 92 K

Material	r_0 (Å)	$\xi_{ab}(0)$ (Å)	λ^{-2} (μm^{-2})	γ	G_i	ν	$T_{\rm dp}$ (K)	$T_{\rm dp}$ Measured (K)	Reduction Parameter
90 K-phase $YBa_2Cu_3O_{7-\delta}$	~ 40	12	69	7	$\sim 10^{-2}$	8.3	$\sim 0.9T_c$	$\sim 0.4T_c$	0.44
60 K-phase $YBa_2Cu_3O_{7-\delta}$	~ 40	24	58	17	$\sim 10^{-2}$	4	$\sim 0.8T_c$	$\sim 0.3T_c$	0.38
46 K $YBa_2Cu_3O_{7-\delta}$	~ 40	29	9	26	~ 0.33	0.59	$\sim 0.37T_c$	$\sim 0.37T_c$	1
$Ba_{0.6}K_{0.4}Fe_2As_2$	~ 25	19	27	2.5	$\sim 7.7 \times 10^{-4}$	12	$\sim 0.92T_c$	$\sim T_c$	~ 1.1

Table 1. Comparison of physical parameters with $T_{\rm dp}$ and their reduction parameters in different superconductors.

$YBa_2Cu_3O_{7-\delta}$ crystals with by 2.6 GeV U irradiation and 800 MeV Xe irradiation, while the 67 K and 46 K crystals were irradiated only by 800 MeV Xe irradiation. In the case of 90 K-phase $YBa_2Cu_3O_{7-\delta}$ with CDs, we confirm that J_c drops suddenly $\sim$ 35 K irrespective of the value of $\theta_{\rm CD}$, and J_c takes the maximum at $\theta_{\rm CD} = \pm 5°$. We also confirm that similar drops of J_c occur at $\sim$ 19 K and 17 K in $YBa_2Cu_3O_{7-\delta}$ with T_c = 67 K and 46 K, respectively. These results imply that the critical temperature and related physical parameters influence the excitation of double kinks and the pinning proprieties of vortices of $YBa_2Cu_3O_{7-\delta}$.

Acknowledgement
This work was supported by a Grant-in-Aid for Scientific Research (A) (17H01141) from the Japan Society for the Promotion of Science (JSPS).

Reference

[1] Civale L, Marwick A, Worthington T, Kirk M, Thompson J, Krusin-Elbaum L, Sun Y, Clem J and Holtzberg F 1991 *Phys. Rev. Lett.* **67** 648
[2] Krusin-Elbaum L, Marwick A, Wheeler R, Feild C, Vinokur V, Leaf G and Palumbo M 1996 *Phys. Rev. Lett.* **76** 2563
[3] Nakajima Y, Tsuchiya Y, Taen T, Tamegai T, Okayasu S and Sasase M 2009 *Phys. Rev. B* **80** 012510
[4] Tamegai T, Taen T, Yagyuda H, Tsuchiya Y, Mohan S, Taniguchi T, Nakajima Y, Okayasu S, Sasase M, Kitamura H *et al.* 2012 *Supercond. Sci. Technol* **25** 084008
[5] Eley S, Khilstrom K, Fotovat R, Xiao Z L, Chen A, Chen D, Leroux M, Welp U, Kwok W K and Civale L 2018 *Sci. Rep.* **8** 1
[6] Li W, Pyon S, Okayasu S and Tamegai T 2021 *J. Phys.: Conf. Ser.* **1975** 012003
[7] Krusin-Elbaum L, Civale L, Thompson J and Feild C 1996 *Phys. Rev. B* **53** 11744
[8] Lindemer T B, Hunley J F, Gates J E, Sutton Jr A L, Brynestad J, Hubbard C R and Gallagher P K 1989 *J. Am. Ceram. Soc.* **72** 1775
[9] Naito T, Nishizaki T, Watanabe Y and Kobayashi N 1997 *Advances in Superconductivity IX (Springer Verlag, Tokyo) 601*
[10] Holtzberg F and Feild C 1990 *Eur. J. Solid State Inorg. Chem.* **27** 107
[11] Bean C P 1962 *Phys. Rev. Lett.* **8** 250
[12] Bean C P 1964 *Rev. Mod. Phys.* **36** 31
[13] Eley S, Miura M, Maiorov B and Civale L 2017 *Nat. Mater.* **16** 409
[14] Ando Y and Segawa K 2002 *Phys. Rev. lett.* **88** 167005
[15] Naito T, Iwasaki H, Nishizaki T, Haraguchi S, Kawabata Y, Shibata K and Kobayashi N 2003 *Phys. Rev. B* **68** 224516
[16] Baglo J C 2014 *A detailed doping survey of the low-energy electrodynamics of* $YBa_2Cu_3O_{6+x}$ Ph.D. thesis University of British Columbia

Structure-property-process parameter correlation in the Advanced MOCVD growth of 4+ μm REBCO thick films over 10+ m

M Paidpilli[1,2,3], K Boyina[1], V Shyam[1], E Galstyan[1,2,3], G Majkic[1,2,3], and V Selvamanickam[1,2,3]

[1]Department of Mechanical Engineering, University of Houston, Houston, USA
[2]Advanced Manufacturing Institute, University of Houston, Houston, USA
[3]Texas Center for Superconductivity, University of Houston, Houston, USA

Email: maheshpaidpilli@gmail.com

Abstract. REBCO ($REBa_2Cu_3O_{7-\delta}$) coated conductor with J_e > 1000 A/mm^2 at operating condition is being utilized in fusion and high-field-magnet applications. We have demonstrated short REBCO samples with an engineering current density (J_e) ~ 4,600 A/mm^2 (based on 0.1 mm thick tape) at 4.2 K, 20 T using Advanced Metal-Organic Chemical Vapor Deposition (A-MOCVD). In order to achieve such high performance uniformly over long tapes, defective regions in thick films need to be avoided. In this work, we investigated parameters that cause dropout or variation in the critical current along the length of long tapes made by A-MOCVD. By addressing the engineering challenges of growing 4+ μm thick REBCO films in a single pass, 10+ m long tapes with high critical current that varies < 5% along the length have been fabricated.

1. Introduction

In recent years, demand for high-field superconducting magnets has increased in many scientific disciplines, including biology, chemistry, geology, and materials science [1]. The most extensive application of superconducting magnets is the ITER project for magnetic confinement fusion [2]. Generally, particle accelerators and detectors are built using superconducting coils due to their power efficiency. Future technological opportunities in fusion, high energy physics and other magnet-based applications lie in replacing traditional low-temperature superconductor (LTS), which presently produce up to 23.5 T, with high-temperature superconductor (HTS) [3]. These applications are the driver for developing long-length HTS tapes capable of generating ultrahigh fields.

REBCO can operate at higher temperatures, which will be very helpful in segmented coils and fusion systems where removal of Joule heating is required. In REBCO coated conductor, the superconducting ceramic (1-5μm thick) is a small fraction of the total cross-section, while the Hastelloy substrate comprises most of cross-sectional (30-100 μm thick), leading to yield stress> 600 MPa [4]. REBCO films are usually deposited on a biaxially-textured buffer layer either by chemical routes including metal-organic chemical vapor deposition (MOCVD) [5], metal-organic deposition (MOD) [6], or by physical vapor deposition routes including pulsed laser deposition (PLD) [7,8] and reactive co-evaporation (RCE) [9]. University of Houston developed Advanced MOCVD to grow REBCO layer with high critical current density (J_c) and film thickness as high as 5 μm in a single pass [10,11]. A critical current (I_c) of

1836 A has been achieved at 4.2 K, 20 T which corresponds to an engineering current density of 4,600 A/cm^2 based on 0.1 mm thick tape [12].

In this paper, we report initial work to optimize the R&D A-MOCVD tool process recipe to produce epitaxial 4+ μm thick film REBCO tapes in lengths over 10 meters in a single pass. Principal component analysis (PCA) is employed to correlate processing parameters with the non-uniformity of the critical current of tape in the R&D A-MOCVD tool. This work resulted in fabrication of 10 m long REBCO tape with a uniform critical current and high infield performance at 4.2 K, 13 T.

2. Materials and Methods

2.1. Experimental

The A-MOCVD reactor with laminar flow of precursor vapor over the substrate described in [13,14] was used to deposit REBCO film on a 12 mm-width substrate with configuration of $LaMnO_3$ (50 nm) / MgO (50 nm) / Y_2O_3 (7 nm) / Al_2O_3 (80 nm) / C-276 Hastelloy (50 μm) over 10 m length in a single pass. The substrate was directly ohmic heated through a sliding electrical contact. The temperature of the moving substrate tape was measured using optical probes at three tape locations. A 2 μm thick silver shunt layer was deposited over 10 m REBCO tapes using DC sputtering and then the tapes were oxygenated at 500 °C for two hours. The precursor composition used for all 20 tapes was $Gd_{0.65}Y_{0.65}Ba_xCu_{2.6}$ with barium content in the range of x = 1.94 to 1.95 and 5 vol.% Zr dopant. 2D scanning hall probe microscopy (SHPM) [15] at a speed of 36 cm/min was used to characterize the self-field I_c of the tapes. Defective sections of the tapes were analyzed using General Area Detector Diffraction System (GADDS) and Scanning Electron Microscope (SEM). A sample was cut from the end section of a uniform critical current tape to measure in-field properties using vibrating sample magnetometer (VSM) in the direction $B\perp tape$, and 4.2 K to 77 K. A 12 mm wide REBCO tape was laser slit to 4 mm wide tape and plated on both sides with 20 μm thick copper to measure transport I_c using the four-probe method at 77 K. The transport I_c was determined using a voltage criterion of 1 μV/cm.

2.2. Principal component analysis

Algorithm for PCA analysis [16,17]:

Assume we have a dataset $[x_1, x_2, x_3, \dots\dots, x_N]$

1. Calculate the means vector of all data : $\bar{x} = \frac{1}{N}\sum_{i=1}^{N} x_i$
2. Compute the covariance matrix: $Cov(x) = \frac{1}{N}\sum_{i=1}^{N}(x_i - \bar{x}_i)(x_i - \bar{x}_i)^T$
3. Compute the eigenvector and their corresponding eigenvalue of the covariance matrix, then sort the eigenvector in the decreasing order of the eigenvalues.
4. For projecting raw data into a K-dimensional subspace: choose K eigenvectors corresponding to the largest eigenvalues of the covariance matrix to construct U_K matrix with columns forming an orthogonal system. Then, take dot product of eigenvector and standardized columns to extract principal component features.

3. Results and Discussion

We performed a root cause analysis of drop out in the I_c of 10 m tapes produced in A-MOCVD to identify process parameters that could be a reason for drop-in I_c.

3.1. Data Analysis

This section presents the description of preliminary data processing of the dataset and principal component analysis results.

3.1.1. Preliminary data processing:

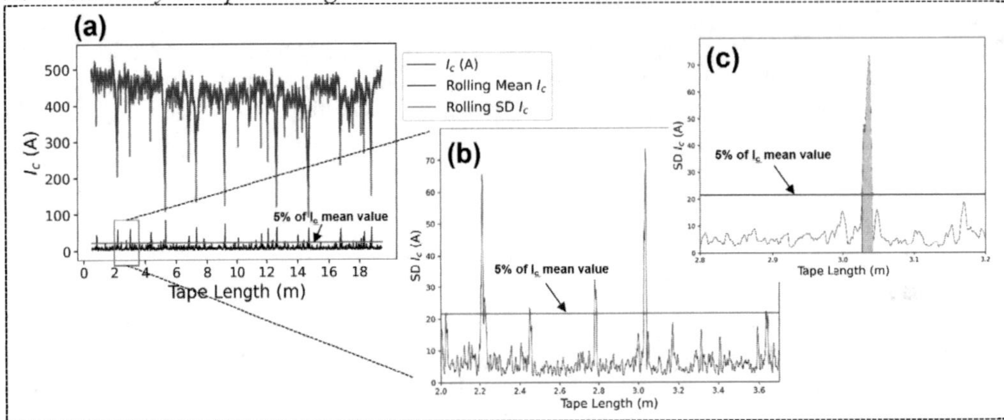

Figure 1. (a) Critical current of 19 m tape with moving mean and moving standard deviation (MSD) plot, (b) magnified plot of MSD shown as % of I_c mean value, and (c) magnified plot to illustrate the area calculated under the MSD curve between the points recorded in (b).

The data set consists of the process variables and I_c value of the tape. Nine process variables recorded during deposition were considered for data analysis including tape entry side temperature, tape middle location temperature, tape exit side temperature, applied current, voltage, mixer gas flow rate, precursor liquid flow rate, mixer gas injector opening, and precursor liquid injector opening. The I_c value of the tape was measured using 2D scanning hall probe microscopy.

The mean value and rolling or moving standard deviation (MSD) of I_c curve was determined [17]. The I_c of 19 m tape and its MSD curve are shown in Fig. 1(a). Anomaly or defect regions of the tape were defined where the MSD value is greater than the 5% of mean I_c value, as shown in Fig. 1(b). The area under MSD of I_c curve was calculated at all these defects using the trapezoidal rule [18,19]. Simultaneously, MSD curves for nine process variables of the tape were plotted, and the area under these MSD curves in the identified defect locations were calculated. A similar analysis was carried out for 20 long tapes made by A-MOCVD. Fig. 2 summarizes the comparison of calculated areas under the MSD curve of I_c and process parameters at the defective regions in these 20 tapes. In Fig. 2, the x-axis labeled as SD_Area_Ic represents the value of area under the MSD curve of I_c plot calculated at the defective regions. Whereas, y axis labeled as SD Area_ (Process Parameter Name) represents the value of area calculated under the MSD curve of individual processing or monitoring parameters recorded at same defective regions noted above.

3.1.2. Principal Component analysis:

Principal component analysis (PCA) was employed to analyze the observed process variables associated with non-uniformity of the critical current of the tapes. PCA allows transformation of a high-dimensional dataset to a lower-dimensional dataset by converting several correlated variables into several linearly uncorrelated variables, named principal components. The dataset for this analysis is the areas under the MSD curves for various process parameters and critical current at the defective location of 20 tapes as described in section 3.1.1.

PCA method builds new uncorrelated variables that successively maximize the variance. The criteria Kaiser's eigenvalue greater than 1.0 rule (Kl) [20] sometimes over or underestimated the number of components to retain for showing a correlation in 2D or 3D space. Hence, it is suggested that multiple approaches be used to decide the number of principal components (PCs) to include. Another criterion say the total variance explained by all retained PCs should be between 70% to 80% variance [21]. In this case, we consider two criteria to determine the number of PCs: Fig. 3 shows that the first two PCs proportionate more than 75% of variance and eigenvalue greater than 1.0. It is reasonable to retain only stronger PCs (have greater variance) than an individual input variable. Hence, first two PCs are enough to explain the entire dataset.

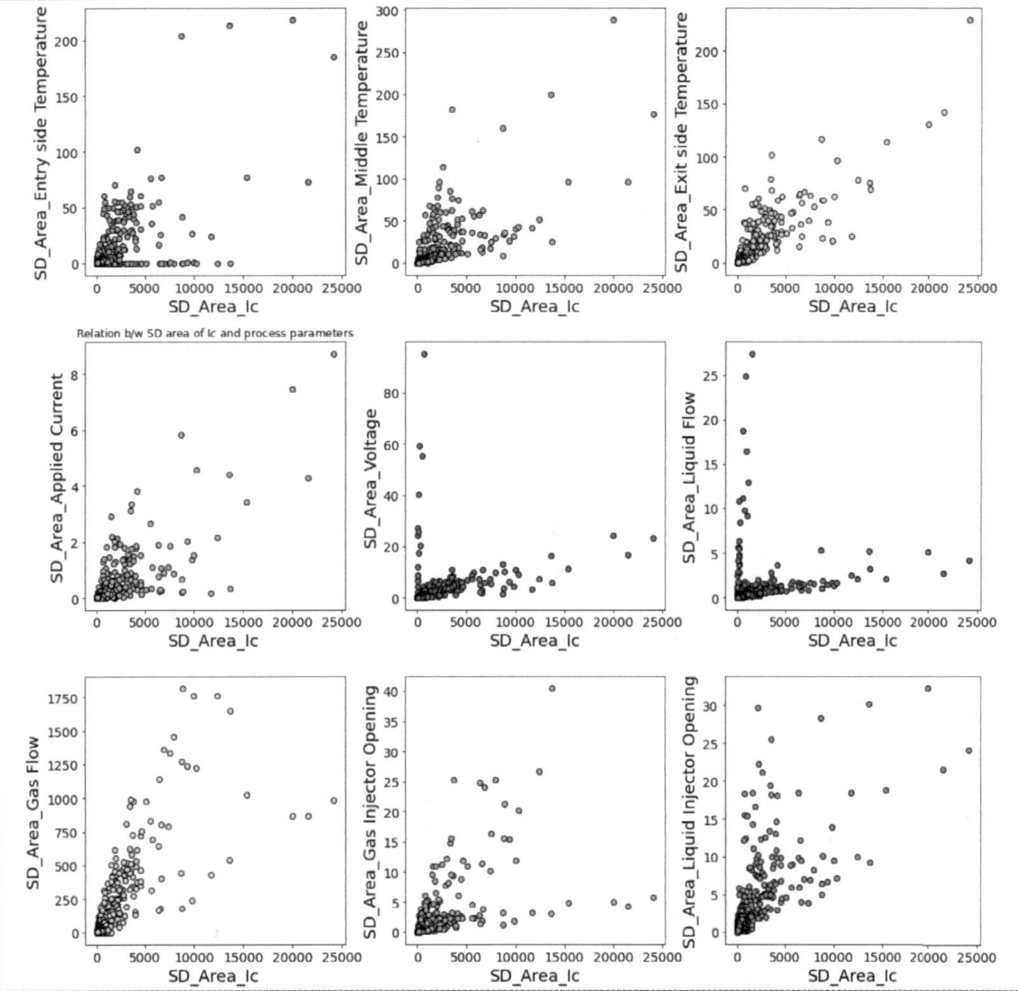

Figure 2. Relation between MSD area of I_c and process parameters at defective regions of 20 long tapes made by A-MOCVD.

Fig. 3(b) shows the correlation between all data variables in 2-dimensional space. The loading plot graphs the coefficients of each variable for the first component versus the coefficients for the second component in 2-dimensional space, showing the correlation between process variables. Positive loadings indicate a variable and a principal component are positively correlated, while negative loadings indicate a negative correlation between a variable and a principal component. The angles between the vectors also hint at how variables correlate with one another: a small angle indicates a positive correlation, a 90° angle implies no correlation, and a large angle (close to 180°) suggests a negative correlation.

The relation between all variables is established through PCA-enabled data processing, and results indicate reactor exit side temperature variable correlates well with the drop of I_c values. This interpretation was made because both exit side temperature and I_c variable are in the positive coordinate of PC1 and PC2 components. Additionally, the vector of exit side temperature and I_c are very close to each other. The coefficients of the first three principal components are documented in Table 1. On further investigation, we found that the tape at the exit side of the reactor was sliding towards the exhaust direction of the precursor, and this movement in the tape likely generated defects in the tape. To resolve this issue, we installed a guide to restrict the shift in tape alignment, and the result of that installation is presented in section 3.3. Hence, PCA-enabled data processing was an effective strategy to determine the process variable that most influenced the I_c uniformity of long tapes made by Advanced MOCVD.

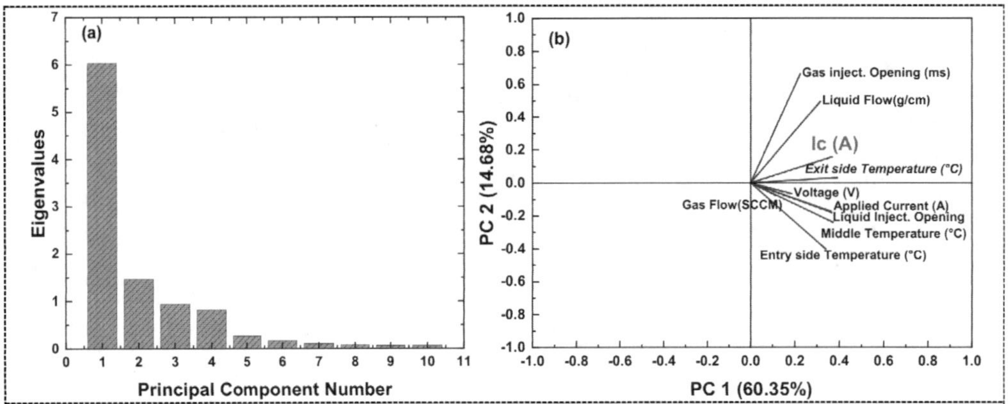

Figure 3. (a) Eigenvalues for principal component and (b) Principal component analysis (PCA) of all the measured process variables and I_c for 20 tapes.

Table 1. A Coefficients of first three principal components of eigenvectors.

	Coefficients of PC1	**Coefficients of PC2**	**Coefficients of PC3**
I_c (A)	**0.37**	**0.16**	0.01
Entry side Temperature (°C)	0.34	-0.40	-0.05
Middle Temperature (°C)	0.37	-0.24	-0.06
Exit side Temperature (°C)	**0.39**	**0.03**	-0.05
Applied Current (A)	0.37	-0.18	-0.05
Voltage (V)	0.19	-0.07	-0.08
Liquid Flow (g/cm)	0.31	0.49	-0.01
Gas Flow (SCCM)	0.11	-0.07	0.99
Gas inject. Opening (ms)	0.22	0.66	0.03

3.2. Structural analysis of defective regions:

Fig 4 (left) shows the SHPM longitudinal I_c profile of a 6 m tape. The figure on the left also shows SEM micrographs of selected regions (b, c, and d). Regions b & d reveal *a*-axis grains, and the region c shows no *a*-axis grains. This observation is consistent with the I_c values of tape. Fig. 4 (right) shows GADDS X-ray Diffraction data obtained from 2theta scans in the range of 7°–55° from the selected defective regions (a & e). The data was obtained at every 1 cm distance in the selected region. The diffraction patterns reveal that the intensity of (003) peak is proportional to the I_c value of the tape. The full-width-at-half-maximum (FWHM) values of these regions were determined by XRD ω–scan of the (Gd,Y)BCO (005) peak. The results indicate out-of-plane texture values above 1.6° near the defective region and below 1.2° at the uniform I_c region.

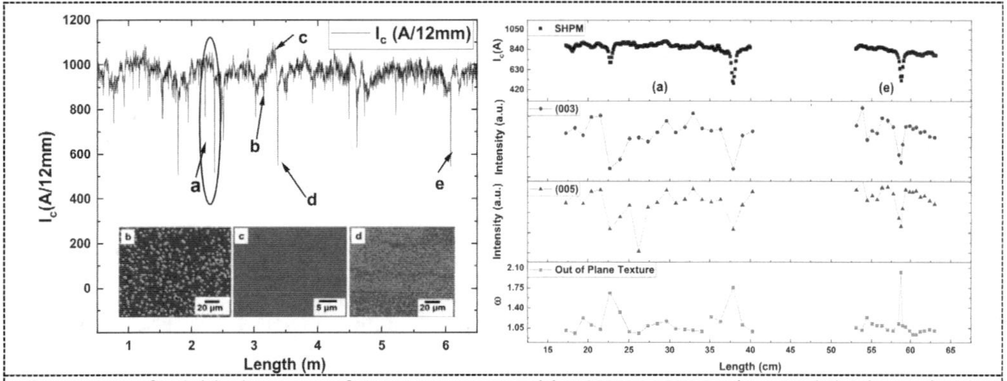

Figure 4. (Left) Critical current of 6 m tape measured by SHPM. SEM micrograph is shown for the selected regions (b, c, and d), (Right) X-ray diffraction peak intensity of (003) and (005) reflections from selected regions (a & e) and Out-of-plane texture values, as a function of distance.

3.3. 10 m Length Uniform REBCO Tape

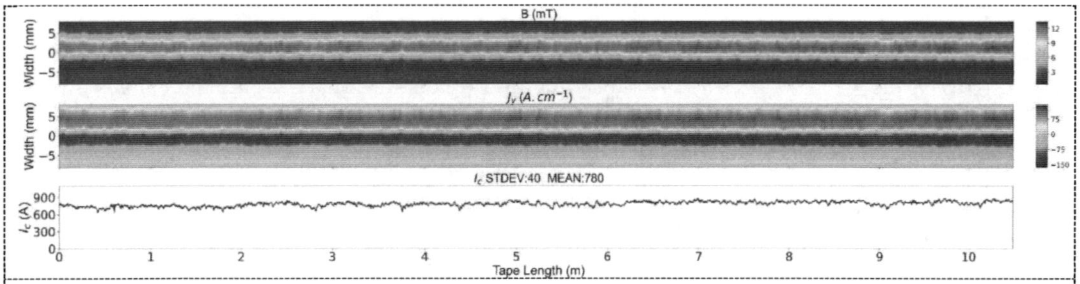

Figure 5. Critical current and magnetic field map measured by SHPM on 11 m tape.

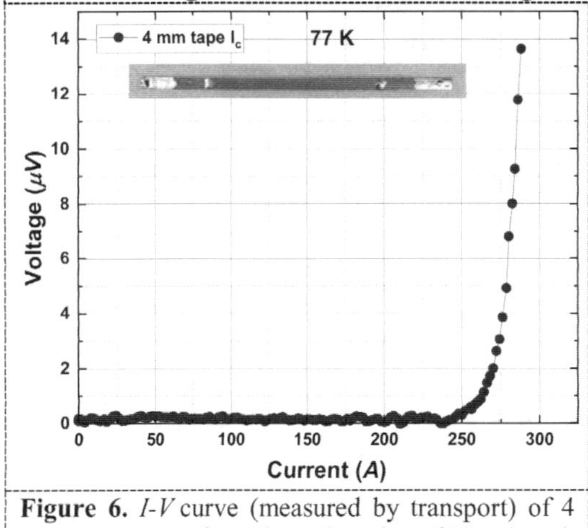

Figure 6. *I-V* curve (measured by transport) of 4 mm segment cut from the end section of 10 m length tape.

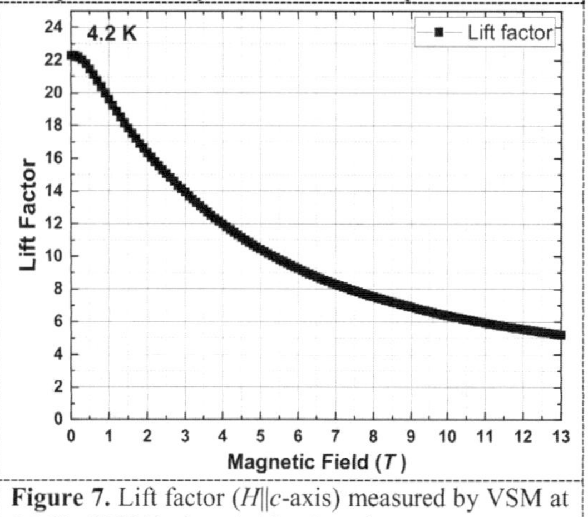

Figure 7. Lift factor ($H \| c$-axis) measured by VSM at 4.2 K of 5% Zr-doped, 10 m length tape.

Fig. 5 shows the SHPM, longitudinal I_c profile, and B map of a 10 m tape with 5% Zr addition. The standard deviation in I_c over the tape length is about 5 % of the mean value. The transport I_c of 10 m length tape was measured at one end of the tape over a 4 mm wide section. The I-V curve is illustrated in Fig. 6. A critical current of 275 A/4 mm was obtained. Hence, the I_c of the 12 mm wide tape is expected to be 825 A. The lift factor is the ratio of I_c (4.2 K, $B\perp tape$) to the I_c (77 K, 0 T). Fig. 7 shows the dependence of applied magnetic field up to 13 T along the *c-axis* on the lift factor of the tape at 4.2 K. A lift factor of 5.22 was measured at 13 T. So, we expect the I_c of the 10 m tape is expected to be 4300 A/12 mm at 4.2 K, 13 T. However, it is worth noting that estimating the exact I_c value at 4.2 K, 13 T may be difficult due to the difference in the criterion of transport I_c and VSM I_c. Since I_c criterion used in VSM is much more stringent than the 1 μV/cm used in transport measurements [22].

4. Conclusion

PCA has been employed to discriminate and quantify the effect of various process variables on the self-field I_c of long tapes made by A-MOCVD. PCA results showed the variation in right optical probe temperature near the exit side of the reactor is well correlated with the SHPM I_c dropout. Based on this understanding, we modified the MOCVD reactor at the exit side. This modification resulted in self-field I_c uniformity over 10 m length. The defective regions of the tapes were poorly textured, with a preponderance of *a-axis* grains. Excellent I_c uniformity has been achieved in 4+ μm thick Zr-added (Gd,Y)BCO tape over 10 m length produced by A-MOCVD reactor in a single pass with precursor-to-film-conversion efficiency up to 45%. Based on the lift factor measured in VSM and transport I_c of 4 mm wide tape, the expected I_c at 4.2 K, 13 T is around 4300 A/12mm over 10 m. We are now scaling up the tapes to over 50 m length in our pilot manufacturing A-MOCVD tool.

Acknowledgments

This work is supported by the U.S. Department of Energy (DOE) Advanced Manufacturing Office award DE-EE0007869, the U.S. DOE Office of High Energy Physics award DE-SC0016220, and the U.S. DOE SBIR award DE-SC0020717.

References

[1] Budinger T F, Bird M D, Frydman L, Long J R, Mareci T H, Rooney W D, Rosen B, Schenck J F, Schepkin V D, Sherry A D and Sodickson D K 2016 *MAGMA.* **29** 617-39
[2] Barabash V, Team T I, Peacock A, Fabritsiev S, Kalinin G, Zinkle S, Rowcliffe A, Rensman J W, Tavassoli A A, Marmy P and Karditsas P J 2007 *J. Nucl. Mater.* **367** 21-32
[3] Gupta R, Anerella M, Ghosh A, Joshi P, Kirk H, Lalitha S L, Palmer R, Sampson W, Wanderer P, Witte H and Shiroyanagi Y 2013 *IEEE Trans. Appl. Supercond.* **24** 1-5
[4] Gorospe A, Nisay A, Shin H S 2014 *Phys. C Supercond. its Appl.* **504** 47-52
[5] Selvamanickam V, Galinski G B, Carota G, DeFrank J, Trautwein C, Haldar P, Balachandran U, Chudzik M, Coulter J Y, Arendt P N and Groves J R 2000 *Phys. C Supercond.* **333** 155-62
[6] Smith J A, Cima M J and Sonnenberg N 1999 *IEEE Trans. Appl. Supercond.* **9** 1531-4
[7] Watanabe T, Kuriki R, Iwai H, Muroga T, Miyata S, Ibi A, Yamada Y and Shiohara Y 2005 *IEEE Trans. Appl. Supercond.* **15** 2566-9
[8] Lee S, Petrykin V, Molodyk A, Samoilenkov S, Kaul A, Vavilov A, Vysotsky V, Fetisov S 2014 *Supercond. Sci. Technol.* **27** 044022
[9] Matias V, Rowley E J, Coulter Y, Maiorov B, Holesinger T, Yung C, Glyantsev V and Moeckly B 2009 *Supercond. Sci. Technol.* **23** 014018
[10] Paidpilli M, Pratap R, Kochat M, Galstyan E, Goel C, Majkic G and Selvamanickam V 2021 *IEEE Trans. Appl. Supercond.* **31** 1-5
[11] Galstyan E, Pratap R, Paidpilli M, Majkic G, Selvamanickam V 2021 *IEEE Trans. Appl. Supercond.* **31** 1-5
[12] Majkic G, Pratap R, Paidpilli M, Galstyan E, Kochat M, Goel C, Kar S, Jaroszynski J, Abraimov D and Selvamanickam V 2020 *Supercond. Sci. Technol.* **33** 07
[13] Majkic G, Pratap R, Galstyan E, Xu A, Zhang Y and Selvamanickam V 2016 *IEEE Trans. Appl. Supercond.* **27** 1-5
[14] Paidpilli M, Boyina K, Galstyan E, Majkic G and Selvamanickam V 2021 *MRS Adv.* **6** 718–722
[15] Chen S, Li X F, Luo W and Selvamanickam V 2019 *IEEE Trans. Appl. Supercond.* **29** 1-4

[16] H. Hotelling, 1933 J. Educ. Psychol. 24 417–441
[17] Montgomery D C 2020 *Introduction to statistical quality control* (John Wiley & Sons)
[18] Yeh S T 2002 *Proc. of the 27th Annual SAS® User Group Int.* (SUGI'02) 1-5
[19] Tallarida R J and Murray R B 1987 Area under a Curve: Trapezoidal and Simpson's Rules. In: Manual of Pharmacologic Calculations. Springer, New York, NY 77-81
[20] Kaiser H F 1960 *Educ. Psychol. Meas.* **20** 141-151
[21] Sidou L F and Borges E M 2020 *J. Chem. Educ.* **97** 1666-76
[22] Majkic G, Pratap R, Galstyan E and Selvamanickam V 2019 *IEEE Trans. Appl. Supercond.* **29** 1-5

Fabrication of superconducting coils and (Ba,*A*)Fe_2As_2 (*A*:Na, K) long round wires with large superconducting cores

S Pyon[1], T Ito[1], T Tamegai[1], H Kajitani[2], N Koizumi[2], S Awaji[3], H Kito[4], S Ishida[4] and Y Yoshida[4]

[1]Department of Applied Physics, The University of Tokyo, Hongo, Bunkyo-ku, Tokyo 113-8656, Japan
[2]Naka Fusion Institute, National Institutes for Quantum Science and Technology (QST), 801-1 Mukoyama, Naka-shi, Ibaraki 311-0193, Japan
[3]High Field Laboratory for Superconducting Materials, Institute for Materials Research, Tohoku University, Sendai 980-8577, Japan
[4]National Institute of Advanced Industrial Science and Technology (AIST), 1-1-1 Umezono, Tsukuba, Ibaraki 305-8568, Japan
E-mail: pyon@ap.t.u-tokyo.ac.jp

Abstract. We fabricated superconducting coils using 10-20 m-class round wires of 122-type iron-based superconductors (IBSs). Round wires and coils are fabricated by powder-in-tube method and hot-isostatic-press technique. Transport critical current (I_c) of the whole (Ba,K)Fe_2As_2 coil is 46 A under the self-field at 4.2 K, and magnetic field at the center of the coil reaches 0.3 T. Although the edge of the long wire in the coil is damaged, the rest of the part is relatively homogeneous. Furthermore, the largest transport critical current density (J_c) and I_c in (Ba,K)Fe_2As_2 wires picked up from the coil reach 49 kAcm^{-2} and 44 A at 4.2 K under a magnetic field of 10 T, respectively. This value exceeds the previous highest transport J_c of (Ba,K)Fe_2As_2 round wires. We also fabricated a (Ba,Na)Fe_2As_2 coil using long round wire with large superconducting core by react and wind method. Transport I_c of the coil is significantly low due to cracks perpendicular to electric current flow direction, although magnetic J_c in the round wire picked up from the coil reaches 40 kAcm^{-2} at 4.2 K under 4 T.

1. Introduction

After the discovery of iron-based superconductors (IBSs) in 2008, they have attracted much attention due to their high critical temperature, T_c, high upper critical field, H_{c2}, and high critical current density, J_c, and are considered to be promising materials for future high-field applications [1-8]. Among IBSs, 122-type compounds of (Ba,*A*)Fe_2As_2 (*A*: Na, K) are the most actively studied materials as candidates of superconducting wire/tape for practical use. The transport J_c of (Ba,K)Fe_2As_2 powder-in-tube (PIT) round wire processed by hot-isostatic pressing (HIP) has reached practical level of 100 kAcm^{-2} at 4.2 K under the self-field for the first time [9]. The J_c of round wires and tapes have been further improved by reduction of impurities, improvement of texturing, and increase of the density of the superconducting core, using pressing technique. In the case of textured tapes, texturing of grains in the superconducting core and high densification by uniaxial pressing are effective to increase J_c [10]. The transport J_c of (Ba,K)Fe_2As_2 tapes at a high field of 10 T at 4.2 K has reached 150 kAcm^{-2}, which exceeds the practical level of 100 kAcm^{-2} [11]. In the case of round wires, which are suitable for winding coils, texturing of the grains is difficult compared with that in tapes. However, it has been demonstrated that the degree of texturing of the grains can be enhanced to a certain degree by changing the deformation process, leading to larger J_c [12]. The transport J_c under a high field of 10 T at 4.2 K for wires of (Ba,K)Fe_2As_2 and (Ba,Na)Fe_2As_2 have reached 47 and 54 kAcm^{-2}, respectively [13,14]. These promising wires and tapes have been applied to superconducting coils, dozens or hundreds of meters of wires or tapes. Fabrications of 100 m-class (Ba,K)Fe_2As_2 or (Sr,K)Fe_2As_2 tapes for coils are reported [15-18]. Recently, we fabricated 10 m-class (Ba,Na)Fe_2As_2 and (Ba,K)Fe_2As_2 round wires and coils with these wires by wind and react method [14,19]. In the case of (Ba,Na)Fe_2As_2 coils with the round wire, critical current I_c of the whole coil reaches ~60 A under the self-field at 4.2 K, and magnetic field at the center of the coil reaches 0.26 T [14]. Furthermore, the largest transport J_c and I_c in (Ba,Na)Fe_2As_2 wires picked up from the coil reach 54 kA cm^{-2} and 51.8 A at 4.2 K under a magnetic field of 10 T, respectively [14]. To increase the value of generated magnetic field of the coil with round wires, longer wire length and higher I_c of the wire are demanded.

In this study, we fabricate two kinds of superconducting coils using (Ba,*A*)Fe_2As_2 round wires, to improve properties of superconducting coils compared with our previous coils [14]. First, a superconducting coil using ~19 m length of (Ba,K)Fe_2As_2 round wire is fabricated by wind and react method. Second, a superconducting coil using ~13 m length of (Ba,Na)Fe_2As_2 round wire, in which the core size is larger than that in the previous report, is fabricated by react and wind method. Larger superconducting core is advantageous to increase I_c and to enhance generated magnetic field in the coil. Short pieces of round wires with large superconducting core are also fabricated. React and wind method is more convenient for the fabrication of coils compared with wind and react method, although bending strain and micro cracks in the superconducting core are easily introduced. Generated magnetic field of the coils and I_c and J_c in coils and wires are evaluated.

2. Experiments

Superconducting wires of (Ba,*A*)Fe_2As_2 were fabricated by *ex situ* PIT method and HIP process. Polycrystalline powders of (Ba,*A*)Fe_2As_2 are prepared by solid state reaction after mixing the raw materials of elements by ball milling method, as described in Ref. 12 and 14. To enlarge area of superconducting core to increase I_c of the wire, we used Ag and Cu tubes with different outer and inner diameters (OD and ID). In the case of the (Ba,K)Fe_2As_2 wire, OD and ID of Cu and Ag tubes are 3.0 mm and 2.0 mm, and 4.5 mm and 3.0 mm, respectively. Expected size of superconducting core is the same to that in the previous report [14]. On the other hand, for (Ba,Na)Fe_2As_2 wire with larger superconducting core, OD and ID of Cu and Ag tubes are 4.0 mm and 3.0 mm, and 6.3 mm and 5.3 mm, respectively. Expected ratio of the diameter of the wire and the core is schematically shown in Figs. 1 (a) and (b). Ag tubes filled with the polycrystalline powders were cold-drawn using dies with circular holes. The wires were formed into a round shape with a diameter of the same sizes of ID of Cu tubes. After cutting them into short pieces, they were inserted into Cu tube and cold-drawn using dies with circular holes. The wires were formed into a round shape with a diameter of ~2.3 mm. Then, the tubes were swaged using a rotary swaging machine, and formed into round shapes with diameters of ~1.0 mm. The other shorter pieces of (Ba,Na)Fe_2As_2 were inserted into Cu tube and redrawn into square shapes with a groove roller down to diagonal dimensions of ~1.1 mm. The total lengths of the long wires reach ~19 m and ~13 m for (Ba,K)Fe_2As_2 and (Ba,Na)Fe_2As_2, respectively. Cross sections of these wires are shown in Figs. 1 (c) and (d). They indicate that sizes of core are successfully controlled by selecting proper sizes of metal sheathes. Obtained long wire of (Ba,K)Fe_2As_2 was covered by glass sleeve for insulation, and the total diameter of the wire was increased to ~1.3 mm. Then, it was wound around stainless steel bobbin. After sealing the edges of the wire in the (Ba,K)Fe_2As_2 coil, it was heated for 4 hours at 700°C under the pressure of 175 MPa at National Institutes for Quantum and Radiological Science and Technology. We confirmed that a glass sleeve was not damaged after the HIP process at 700°C. On the other hand, obtained long wire of (Ba,Na)Fe_2As_2 was wound around a metal tube with a diameter of 6 cm, then heated at high pressure in the same manner. It should be noted that the internal volume of the HIP furnace is large enough to heat the coil homogeneously. So the nonuniform heating of the wire used for the coil can be ignored. The long (Ba,Na)Fe_2As_2 wire was made straight and covered by glass sleeve for insulation, then rewound around a metal tube with a diameter of 4 cm and its current–voltage (*I–V*) characteristics in liquid helium were measured in order to evaluate the transport I_c and J_c. After that was made straight again, it was rewound around stainless steel bobbin. All coils were not impregnated with epoxy resin. Optical photos of the coils are shown in Figs. 1(e) and (f). Dimensions of coils are summarized in Table 1. The short wires of (Ba,Na)Fe_2As_2 were heated for 4 hours at 700°C under the pressure of 120 MPa at National Institute of Advanced Industrial Science and Technology. The I_c measurements in high magnetic fields up to 14 T for shorter wires were carried out by using the 15T-SM at the High Field Laboratory for Superconducting Materials, IMR, Tohoku University. The I_c under the self-field and generated magnetic field by the superconducting coil were measured at The University of Tokyo. Generated field at the center of the coil was measured using a Hall probe. Voltage taps for the evaluation of *I-V* characteristics of the coil were set at the ends of the wires. The bulk magnetization of a short piece of the wire was measured to characterize magnetic J_c by a superconducting quantum interference device magnetometer (MPMS-5XL, Quantum Design).

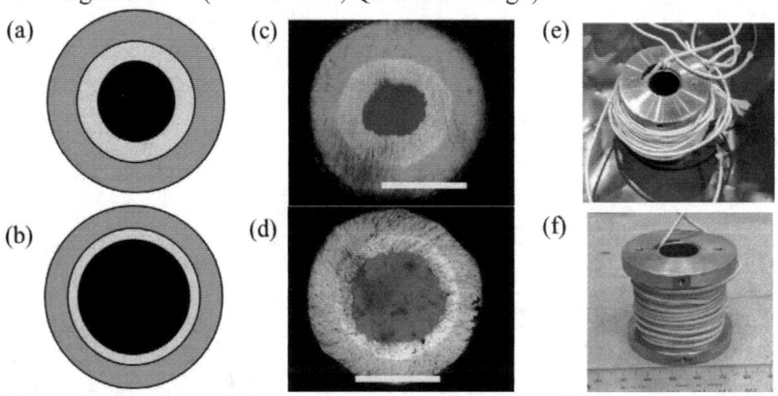

Figure 1. (a) Schematic drawings of the cross sections of superconducting wire of (a) (Ba,K)Fe_2As_2 and (b) (Ba,Na)Fe_2As_2, expected from the used metal sheathes. Optical micrographs of the cross sections of superconducting wires of (c) (Ba,K)Fe_2As_2 and (d) (Ba,Na)Fe_2As_2. Yellow bars in (c) and (d) indicate 0.5 mm. Optical photos of the superconducting coils of (e) (Ba,K)Fe_2As_2 and (f) (Ba,Na)Fe_2As_2.

Table 1. Specifications of (Ba,*A*)Fe_2As_2 (*A*: Na, K) coils.

Structure	Parameters	(Ba,K)Fe_2As_2	(Ba,Na)Fe_2As_2
Wires	Diameter	1.0 mm	1.0 mm
	Length	~19 m	~13 m
	Filament number	Mono core	Mono core
Coils	Inner diameter ($2R_1$)	19 mm	19 mm
	Outer diameter ($2R_2$)	47 mm	39 mm
	Height ($2l$)	25 mm	25 mm
	Numbers of turns (N)	189	140
	Coil constant (calculated)	0.006 T/A	0.0048 T /A

3. Results and discussion

The performance of the superconducting coil of (Ba,K)Fe_2As_2 is evaluated from the *I-V* characteristics as shown in Fig. 2(a). When electric current is lower than ~50 A, linear increase of *V* is observed, which corresponds to the resistance of ~4 mΩ in the coil. Origin of this finite resistance is discussed later. When we ignore the resistance of this part, I_c of the coil can be estimated as ~46 A at 4.2 K under the self-field. Electric current dependence of generated magnetic field is also evaluated using a Hall probe. At I_c = 46 A, generated magnetic field reaches 0.3 T. The ratio of the field and the current is ~0.0065 T/A. The magnetic field at the center of a coil with finite dimensions can be calculated using the following formula [20];

$$\mu_0 H = \frac{\mu_0 NI}{2(R_2 - R_1)} \ln \frac{R_2 + \sqrt{R_2^2 + l^2}}{R_1 + \sqrt{R_1^2 + l^2}},$$

where μ_0 is vacuum permeability, *N* is the numbers of turns, $2l$ is the height of the coil along its axis, $2R_1$ and $2R_2$ are ID and OD of the coil, respectively. Values of these parameters are also summarized in Table 1. The calculated coil constants $\mu_0 H/I$ of 0.006 T/A are close to the measured coil constants of 0.0065 T/A. After the characterization of generated magnetic field, the wire of the coil was unwound and short segments of the wire were picked up and evaluated. Fig. 2 (b) exhibits I_c of these short segments from (Ba,K)Fe_2As_2 coils. As shown in Fig. 2(b), I_c varies considerably along the length of the (Ba,K)Fe_2As_2 wire. The maximum values of I_c of (Ba,K)Fe_2As_2 wires reach 135, 150, 88, and 49 A at 0, 0.1, 1, and 10 T, respectively, for a segment around 8.5 m from the end of the wire. Central part of the wire is relatively homogeneous, although some degree of distribution of I_c value exists. However, I_c in both edge parts of the wire is significantly reduced. It may be caused mainly by inadequate handling. Furthermore, we found that ~1 m length of the wire around 1 m from the end do not contain both superconducting core and Ag. This may indicate that Ag tube was separated in Cu sheath during the drawing and/or swaging processes. The absence of superconducting core in a part of the wire is the origin of the finite resistance as discussed above. The reason for this accidental separation of the core is not clear. However, it is clear that the wire drawing process should be improved.

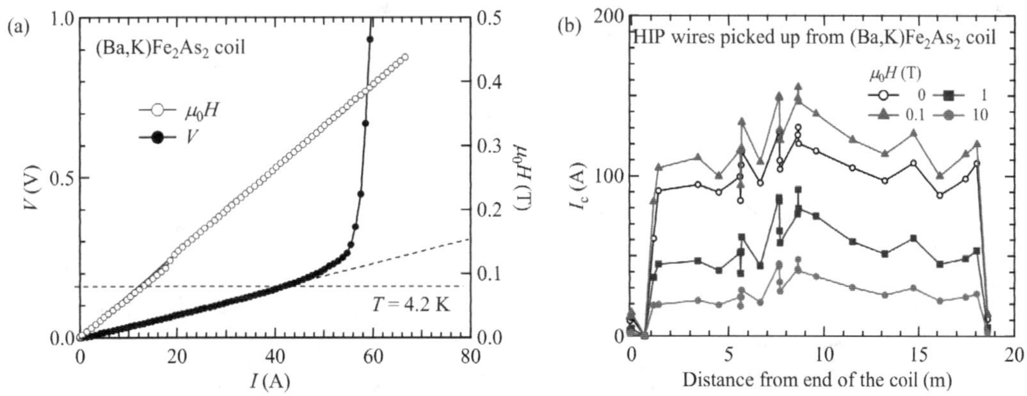

Figure 2. (a) *I–V* characteristics and electric current dependence of generated magnetic field of the (Ba,K)Fe_2As_2 coil. (b) Transport I_c of short segments of the wire picked up from (Ba,K)Fe_2As_2 coil. Measurements were performed at 4.2 K under several magnetic fields.

After the evaluation of the area of superconducting core of short segments of the wire from (Ba,K)Fe_2As_2, it is found that the value of transport J_c at high magnetic field of 10 T exceeds the record value of (Ba,K)Fe_2As_2 round wires. The magnetic field dependence of the transport J_c at 4.2 K for the wires picked up from the (Ba,K)Fe_2As_2 coils are shown in Fig. 3. Reported highest transport J_c of the round wire of both (Ba,K)Fe_2As_2 and (Ba,Na)Fe_2As_2 are also plotted [13,14]. As shown in Fig.

3, transport J_c under the magnetic field of 10 T of the wires picked up from the (Ba,K)Fe_2As_2 coil at ~8.5 m reaches 49 $kAcm^{-2}$ at 4.2 K. This value is slightly larger than the previous largest value of J_c of 47 $kAcm^{-2}$ of the (Ba,K)Fe_2As_2 round wire [13], although it is smaller than that of 54 $kAcm^{-2}$ of the (Ba,Na)Fe_2As_2 round wire [14]. It is demanded that homogeneous long wire with high J_c in the whole length of the wire should be fabricated.

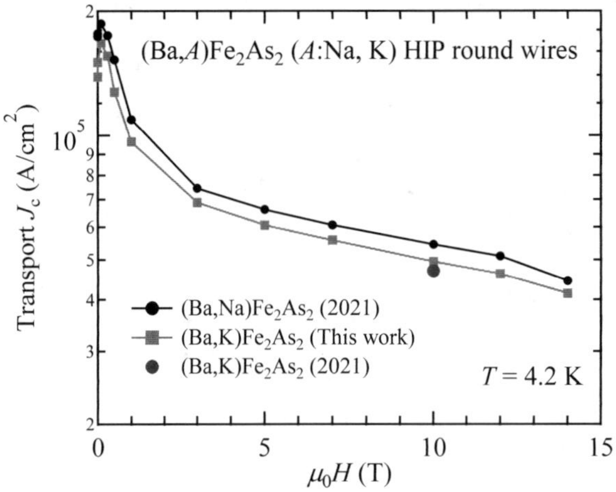

Figure 3. Magnetic field dependence of transport J_c at 4.2 K for the short segment picked up from the (Ba,K)Fe_2As_2 coil. Transport J_c of the (Ba,Na)Fe_2As_2 and (Ba,K)Fe_2As_2 HIP wires from the previous studies are also plotted for comparison [13,14].

Next, I–V characteristics of the (Ba,Na)Fe_2As_2 coil and long wire wound around 4 cm metal tube at 4.2 K are evaluated as shown in Fig. 4(a). Transport I_c for the (Ba,Na)Fe_2As_2 coil and long wire, which were evaluated by adopting the 1 μV cm^{-1} criterion, are ~ 2 and ~6 A, respectively. Obviously, rewinding of the wire to the coil adversely affected the superconducting properties. It is also expected that the HIP processed wire was also damaged after rewinding around 4 cm metal tube. After the characterization of the whole coil, the wire of the coil was unwound and short segments of the wire were picked up and evaluated. Fig. 4(b) exhibit I_c of these short segments of the wire from (Ba,Na)Fe_2As_2 coils. Local I_c is also low. After polishing the wire, it is clarified that low I_c is caused by many cracks in the whole length of the wire in the coil, as demonstrated by an optical photo in the inset of Fig. 4(b). These results suggest that multiple rewinding processes after the HIP process add significantly strong bending distortion to the superconducting core. The maximum bending strain of the (Ba,Na)Fe_2As_2 wire in the coil is calculated as 2.6% using the minimum bending diameter (19 mm) and diameter (0.5 mm) of superconducting core estimated from the optical photo in Fig. 1(d). This value is too large for a mono-core round wire. Even the I_c of a multi-core wire of MgB_2, in which the effect of bending strain on I_c is weaker than that in a mono-core wire, is reduced to less than half at the same bending strain [21]. For react and wind method, multi-core round wire should be developed and the relation between the core area and coil diameter should be considered.

It is noteworthy that relatively high J_c can be observed in the superconducting core without cracks. Fig. 5 show magnetic field dependence of magnetic J_c at 4.2 K for the short segments picked up from the (Ba,Na)Fe_2As_2 coil. Magnetic J_c of the (Ba,Na)Fe_2As_2 HIP wires from previous studies are also plotted for comparison [14,22]. The magnetic J_c is evaluated from the irreversible magnetization using the extended Bean model [23]. Measured segment of the wire, applied magnetic field, and J_c in the critical state are schematically described in the inset of Fig. 5. In this case, it is expected that cracks perpendicular to electric current flow direction do not affect J_c much. Magnetic field dependence of magnetic J_c of short segments picked up from the (Ba,Na)Fe_2As_2 coil is similar to that in the other (Ba,Na)Fe_2As_2 HIP wire, which is sintered at 175 MPa and with transport J_c at 4.2 K and at 10 T being relatively high value of 40 kAcm^{-2} [22]. Although the value of magnetic J_c is smaller than that of another HIP wire, which is sintered at 200 MPa and with transport J_c at 4.2 K and at 10 T being the highest value of 54 kAcm^{-2} [14], it is still large value. It should also be noted that we failed to fabricate the (Ba,Na)Fe_2As_2 short wires by rolling method because of thin Ag sheathes. During the deformation process using groove rolling, a part of Ag tube was broken and the superconducting core touched Cu sheath. These results suggest that react and wind method using thin metal sheath to enlarge superconducting core area requires a lot of improvements. Small core is more advantageous to reduce bending distortion. So the fabrication of multicore wire is one of the possible solution.

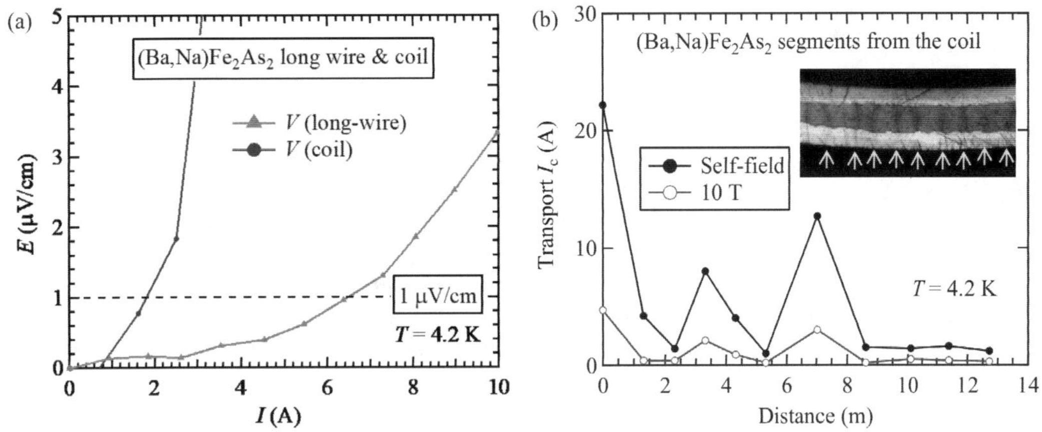

Figure 4. (a) *I*–*V* characteristics of the (Ba,Na)Fe_2As_2 coil and long wire wound around 4 cm metal tubes, respectively. (b) Transport I_c of short segments of wires picked up from (Ba,Na)Fe_2As_2 coil. Measurements were performed at 4.2 K under several magnetic fields. The inset photo shows the transverse cross section of a short segment of the wire picked from (Ba,Na)Fe_2As_2 coil.

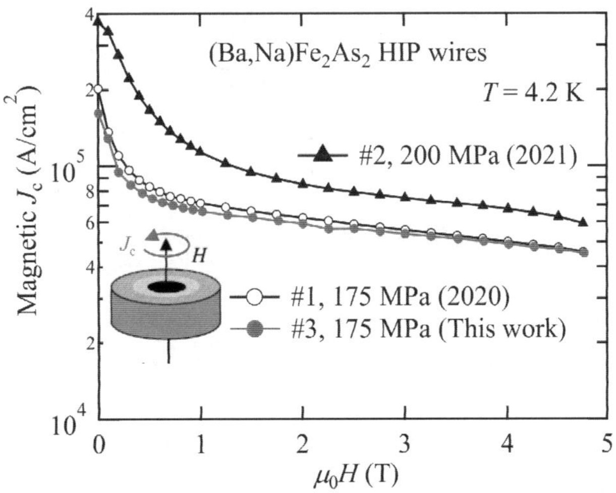

Figure 5. Magnetic field dependence of magnetic J_c at 4.2 K for the short segments picked up from the (Ba,Na)Fe_2As_2 coil. Magnetic J_c of the (Ba,Na)Fe_2As_2 HIP wires from previous studies are also plotted for comparison [14,22]. The inset shows the relation between sample and directions of the applied field and current.

4. Summary

Superconducting coils using 10-20 m class (Ba,*A*)Fe_2As_2 round wires are fabricated. Round wires and coils are prepared by PIT and HIP method. Transport I_c of the whole (Ba,K)Fe_2As_2 coil, which is fabricated by wind-and-react method, is 46 A under the self-field at 4.2 K, and magnetic field at the center of the coil reaches 0.3 T. This value is larger than the previous value of 0.26 T. The edge part of the long wire used for the (Ba,K)Fe_2As_2 coil is damaged, and accidental separation of Ag tube was observed. Further improvements of drawing process are demanded. On the other hand, other parts of the wire are relatively homogeneous. The largest transport J_c and I_c in (Ba,K)Fe_2As_2 wires picked up from the coil reach 49 kAcm^{-2} and 44 A at 4.2 K under a magnetic field of 10 T, respectively. This value exceeds the previous highest transport J_c of (Ba,K)Fe_2As_2 round wires. (Ba,Na)Fe_2As_2 coil using long round wire with large superconducting core is also fabricated by react-and-wind method. Significantly low transport I_c of wound long wire, the coil, and short segments from the coil are obtained. This is caused by cracks perpendicular to the electric current flow direction in the superconducting core. Potentially high J_c is demonstrated from magnetic J_c of 40 kAcm^{-2} at 4.2 K under 4 T. However, react-and-wind method and enlarging the superconducting core area strategy need much improvements in future.

Acknowledgments

A part of this study was performed at the High Field Laboratory for Superconducting Materials, Institute for Materials Research, Tohoku University (Project No. 202012-HMKGE-0025). This work was partially supported by a Grant-in-Aid for Scientific Research (A) (17H01141) from the Japan Society for the Promotion of Science (JPS).

References

[1] Hosono H, Yamamoto A, Hiramatsu H and Ma Y 2018 *Mater. Today* **21** 278
[2] Ma Y 2012 *Supercond. Sci. Technol.* **25** 113001
[3] Shimoyama J 2014 *Supercond. Sci. Technol.* **27** 044002
[4] Ma Y 2015 *Physica C* **516** 17
[5] Pallecchi I, Eisterer M, Malagoli A and Putti M 2015 *Supercond. Sci. Technol.* **28** 114005
[6] Hosono H, Tanabe K, Takayam-Muromachi E, Kageyama H, Yamanaka S, Kumakura H, Nohara M, Hiramatsu H and Fujitsu S 2015 *Sci. Technol. Adv. Mater.* **16** 033503
[7] Yao C and Ma Y 2019 *Supercond. Sci. Technol.* **32** 023002
[8] Pyon S, Tamegai T, Takano K, Kajitani H, Koizumi N and Awaji S 2019 *J. Phys.: Conf. Ser.* **1293** 012042
[9] Weiss J D, Tarantini C, Jiang J, Kametani F, Polyanskii A A, Larbalestier D C and Hellstrom E E 2012 *Nat. Mater.* **11** 682
[10] Gao Z, Togano K, Matsumoto A and Kumakura H 2014 *Sci. Rep.* **4** 4065
[11] Huang H, Yao C, Dong C, Zhang X, Wang D, Cheng Z, Li J, Awaji S, Wen H and Ma Y 2018 *Supercond. Sci. Technol.* **31** 015017
[12] Pyon S, Suwa T, Tamegai T, Takano K, Kajitani H, Koizumi N, Awaji S, Zhou N and Shi Z 2018 *Supercond. Sci. Technol.* **31** 055016
[13] Guo W, Yao C, Huang H, Dong C, Liu S, Wang C and Ma Y 2021 *Supercond. Sci. Technol.* **34** 094001
[14] Pyon S, Mori H, Tamegai T, Awaji S, Kito H, Ishida S, Yoshida Y, Kajitani H, Koizumi N, 2021 *Supercond. Sci. Technol.* **34** 105008
[15] Zhang X, Oguro H, Yao C, Dong C, Xu Z, Wang D, Awaji S, Watanabe K, and Ma Y 2017 *IEEE Trans. Appl. Supercond.* **27** 7300705
[16] Wang D, Zhang Z, Zhang X, Jiang D, Dong C, Huang H, Chen W, Xu Q and Ma Yanwei 2019 *Supercond. Sci. Technol.* **32** 04LT01
[17] Zhang Z, Wang D, Liu F. Jiange D, Wei S, Wang Y, Gong L, Zhang X, Zhang Z, Liu H, Tian

Chao, Ma Y and Xu Qingjin 2020 *IEEE Trans. Appl. Supercond.* **30** 4602404

[18] Zhang Z, Wang D, Wei S, Wang Y, Wang C, Zhang Z, Yao H, Zhang X, Liu F, Liu H, Ma Y and Xu Q 2021 *Supercond. Sci. Technol.* **34** 035021

[19] Pyon S, Mori H, Tamegai T, Awaji S, Kito H, Ishida S, Yoshida Y, Kajitani H, Koizumi N, 2021 *J. Phys.: Conf. Ser.* **1975** 012020

[20] Montgomery D B, 1969 *Solenoid Magnet Design*, Wiley-Interscience

[21] Kovác P, Hušek I, Melišek T, Kulich M and Kopera L 2016 *Supercond. Sci. Technol.* **29**, 045002

[22] Pyon S, Miyawaki D, Tamegai T, Awaji S, Kito H, Ishida S and Yoshida Y 2020 *Supercond. Sci. Technol.* **33** 065001

[23] Bean C P 1964 *Rev. Mod. Phys.* **36**, 31

Evaluation of the joint of superconducting tapes for railway feeder cable

T Akasaka[1,2], A Ishihara[1], T Onji[1], H Yamamoto[1], M Tomita[1], Y Zhong[2] and E S Otabe[2]

[1] Applied Superconductivity Laboratory, Maglev Systems Technology Division, Railway Technical Research Institute, 2-8-38 Hikari-cho, Kokubunji-shi, Tokyo 185-8540, Japan
[2] Department of Computer Science and Electronics, Kyushu Institute of Technology, 680-4 Kawazu, Iizuka, Fukuoka 820-8502, Japan

Abstract. Developing superconducting cable for railway power transmission should lead to increased regeneration efficiency, reduced power loss, equalization of load between substations. In order to realize a kilometer-class superconducting feeder cable, on-site joint technique between superconducting feeder cables is required. For joining superconducting cables at railway construction sites, we conducted solder joining of superconducting tapes and its evaluation in the laboratory. We performed electromagnetic analysis of the joints by the finite element method in order to obtain a prospect of the optimum conditions of the solder joining. The target value of the joint resistance to 10^{-7} Ω class will be achieved by the solder joining regardless of the solder thickness and the joint length. Referring to the analysis results, mechanical and electrical properties measurement of the joined superconducting tape was conducted. From the results of the tensile and bending test, it is found that the solder joined superconducting tape with the lap length of more than 20 mm achieve the mechanical and electrical targets. These results are being utilized for cable joining at railway construction sites.

1. Introduction

Railways characterized by economic efficiency and environmental harmony have increasing importance moreover as thorough energy efficiency is required [1, 2]. From now on, it will be important to find measures to further save energy over the entire railway transportation system [3]. Superconducting materials are expected to have a wide range of applications such as zero-resistance power transmission and strong magnets. Cables for railway transmission lines using superconductors enable improvement of regeneration efficiency, reduction of power loss, load levelling between substations, and consolidation of substations by suppressing voltage drop. Railway Technical Research Institute is developing superconducting feeder cables for railways with the aim of rationalization and saving energy in the electric power system of DC electric railways. We began research and development of superconducting cables for DC railway electrification system since 2007[4]. The train running test was conducted with 31 m superconducting cable in the test track for the first time in 2013. After laying method of a cable was considered, the train running test was also conducted with 310 m superconducting cable [5]. The first test of a commercial train using superconducting feeder cables was conducted in 2015 [6, 7]. The outcome which has been obtained so far is receiving attention from the viewpoint of energy-saving in the railway field [8]. One of the issues for practical use is the joining technology of superconducting cables. When transporting a superconducting cable by land, its length stored in a heat insulating pipe is limited to about 500 m due to the general vehicle and road size restrictions. When considering the introduction of superconducting cables between substations at intervals of several kilometers, it is necessary to join the superconducting cables. Therefore, it is indispensable to construct a simple joining technology that does not interfere with commercial operation and can be implemented at railway construction sites. For cables joined at the railway construction sites, only the electrical and mechanical properties at the liquid nitrogen temperature are important. However, for cables joined outside the railway construction sites, the mechanical properties at room temperature are also important because they must withstand the winding tension applied when the cable is fabricated and the bending stress applied when the cable is wound around a transport drum. As a first step for joining superconducting cables at railway construction sites, we conducted solder joining of superconducting tapes and its evaluation in the laboratory. Before conducting the experiment, we performed electromagnetic analysis of the joints by the finite element method in order to obtain a prospect of the optimum conditions of the solder joining [9]. It is desirable that the joints have no deterioration of I_c and have the same mechanical properties as before the joining. We set the target mechanical properties to 300 N for permissible tensile tension and 150 mm for permissible double bending diameters, and the target joining resistance to 10^{-7} Ω class for railway application. In this paper, we report the evaluation results of mechanical and electrical properties in joining superconducting tapes for the realization of long superconducting feeder cables.

2. Experimental work

2.1. Analysis method of the jointed superconducting tapes

In order to obtain the optimum solder joining conditions for the two sheets of superconducting tape, electromagnetic analysis was performed by the finite element method. In this analysis, JMAG-Designer manufactured by JSOL was used to simulate the flow of current through the joint model, and the electrical characteristics of the joint were calculated. Figure 1 shows the joint model. Two sheets of superconducting tape are soldered together. The conditions of the simulation are as follows. The cross-section, the length, the critical current, the initial conductivity, and the maximum conductivity of the superconducting tape were set to 4.2 mm × 0.23 mm, 50 mm, 180 A at 77 K, 1.0×10^{13} S/m, 1.0×10^{15} S/m, respectively, assuming the Bi-based superconducting tape to be actually used. The thickness and the resistivity of the solder were set to 0 to 0.2 mm and 1.4×10^{-7} Ω m, respectively, assuming the eutectic solder. The joint length of two sheets of the superconducting tape was set to 3 to 80 mm. The current increased step by step up to 200 A. In the n-value model, J_C-B characteristics data at 77 K were used.

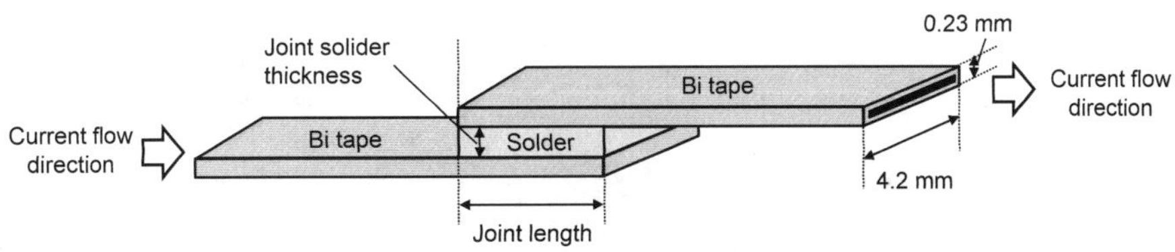

Figure 1. Model of joined superconducting tape

2.2. Experimental method using the jointed superconducting tapes

Referring to the analysis results, mechanical and electrical properties measurement of a single superconducting tape and joined superconducting tape was conducted. Bi-based superconducting tape (Sumitomo Electric Industries, Type HT-CA) was joined together with eutectic solder. Table 1 shows the joining conditions for the superconducting tape.

First, the electrical properties of a single and joined superconducting tape under tensile stress were measured and their critical current was evaluated. The lap lengths of the joined superconducting tape are 10 mm, 20 mm, and 30 mm. Figure 2 shows the appearance of the developed tensile testing equipment in liquid nitrogen. This equipment consists of commercial room temperature tensile testing equipment and a liquid nitrogen container, and it can measure tensile stress of the superconducting tape in liquid nitrogen. Energization tests were conducted under a maximum tensile stress of 440 N.

Second, the electrical properties of a single and joined superconducting tape under bending stress were measured and their critical current was evaluated. The lap of the joined superconducting tape are 20 mm length and the entire surface. Figure 3 shows the appearance of bend testing equipment in liquid nitrogen. In this equipment, both ends of the superconducting tape are fixed with electrodes, then both electrodes are rotated by gears, and as a result, a constant curvature is applied to the superconducting tape according to the rotation angle of the gears. Bending stress was applied continuously from a straight line. The critical current I_c was measured by the four-probe method according to JIS H 7305 using this equipment while tensile or bending stress was applied. The resistance was calculated from current-voltage slope.

Table 1 Condition of jointing for superconducting tape

Pressing force	400 N
Heating temperature	210 °C
Heat holding time	60 sec
Cooling time	10 min

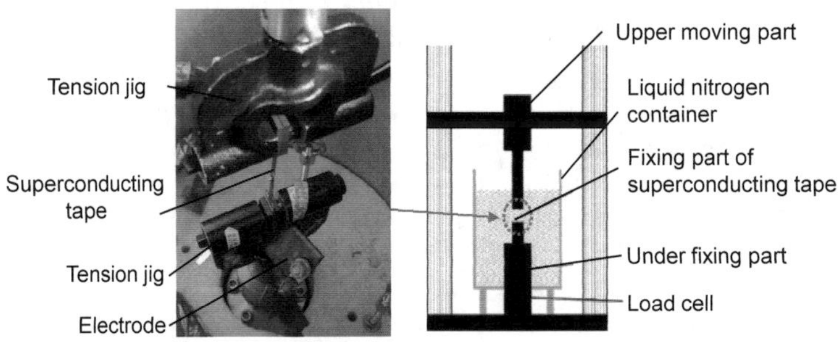

Figure2. Appearance of tensile testing equipment under liquid nitrogen for superconducting tape

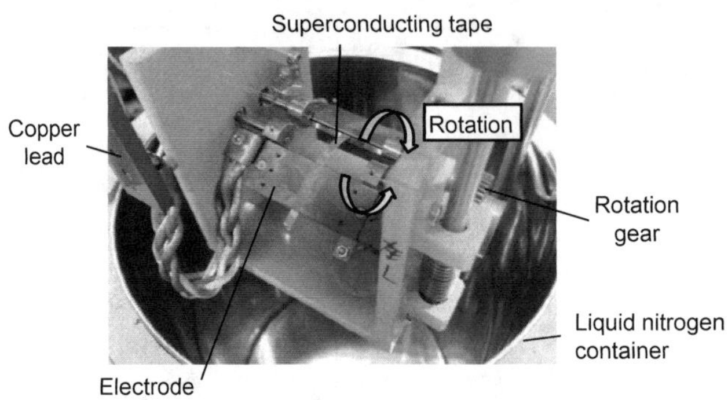

Figure 3. Appearance of bend testing equipment under liquid nitrogen for superconducting tape

3. Results and discussion

3.1. Analysis results of the jointed superconducting tapes

Figure 4 shows the calculation results of critical current for joining superconducting tape with the various solider thickness. The joint resistance was 10^{-7} Ω class even at the thickness of 0.2 mm, which is expected to be the maximum thickness in the experiment. Figure 5 shows the calculation results of critical current for joining superconducting tape with the various joint length. The joint resistance was 10^{-7} Ω class at all the joint lengths set this simulation. In conclusion, the target value of the joint resistance will be achieved by the solder joining regardless of the solder thickness and the joint length. However, if the solder thickness or joint length is small, it is expected that the mechanical strength of the joint will decrease.

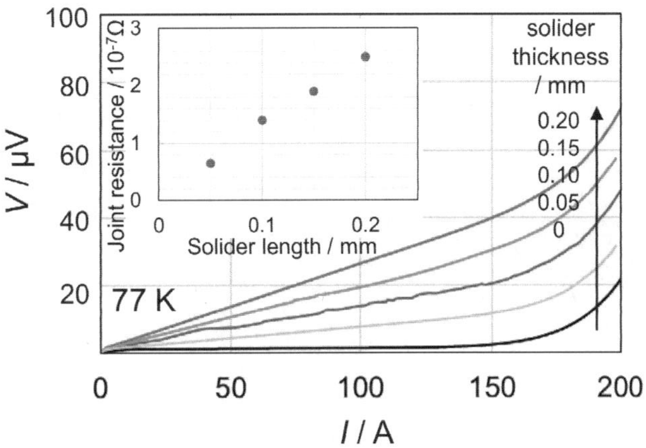

Figure 4. The calculation results of critical current for joining superconducting tape with the various solider thickness

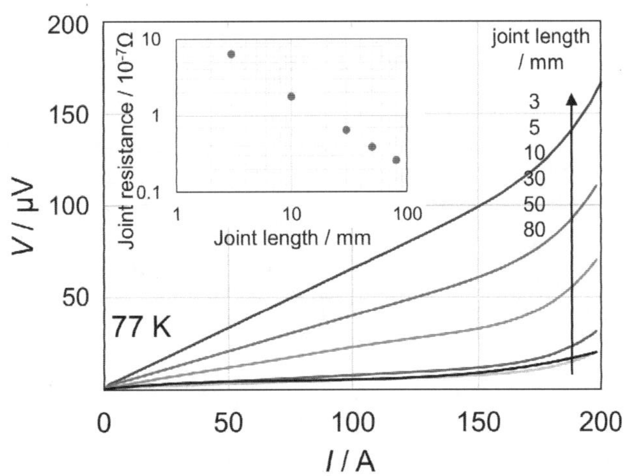

Figure 5. The calculation results of critical current for joining superconducting tape with the various joint length

3.2. Experimental results using the jointed superconducting tapes

Next, the superconducting tapes were joined actually for evaluation of mechanical property. Figure 6 shows the picture of jointed superconducting tapes using solider. Figure 7 shows the current-voltage characteristics of the superconducting tape under tensile stress. It was confirmed that the joining resistance was 10^{-7} Ω class for the joined tape with a lap length of 10 mm and 20 mm and did not depend on the magnitude of the tensile stress. Namely, it is expected that the solder alloy layer at the joint does not deteriorate if tensile stress of up to 400 N is applied. Figure 8 shows the tensile stress dependence of the critical current I_c. Unlike the joining resistance, I_c decreased systematically where the tensile stress exceeded a critical value. Defining the deterioration of I_c as its more than 5% decrease, I_c did not deteriorate by the tensile stress of up to 410 N for a sheet of tape, 210 N for the joined tape with a lap length of 10 mm, 350 N for the joined tape with a lap length of 20 mm, and 430 N for the joined tape with a lap length of 30 mm. It is found that if the lap length is more than 20 mm, the target of the allowable tensile stress of 300 N and the joining resistance of 10^{-7} Ω class is achieved.

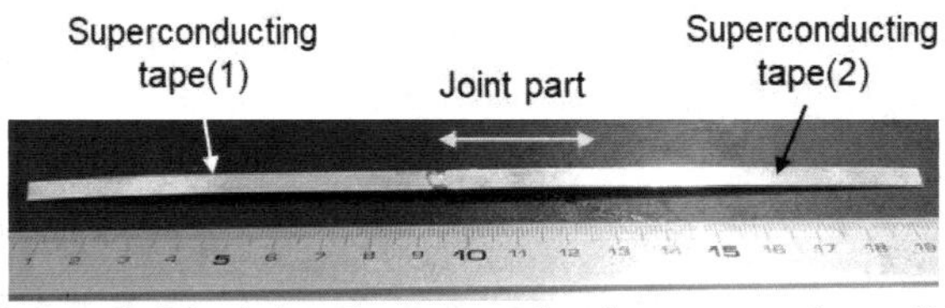

Figure 6. The jointed superconducting tapes using solider

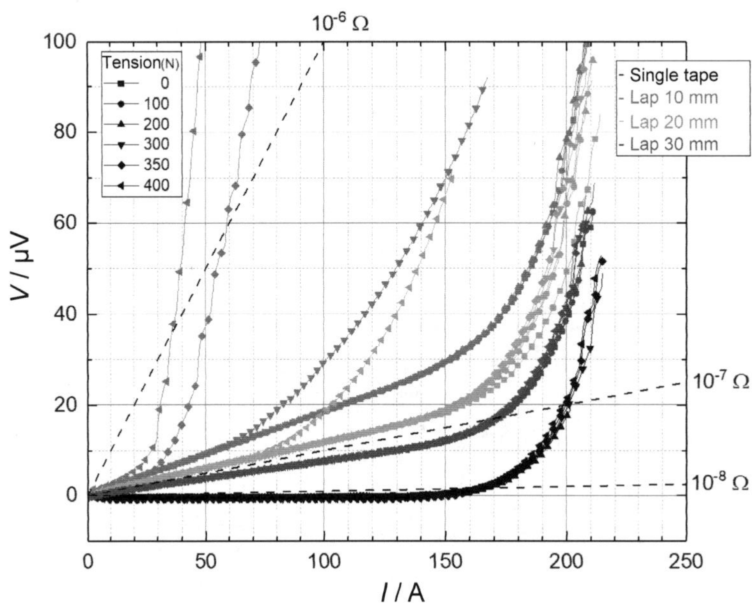

Figure 7. Current-voltage characteristics of the superconducting tape under tensile stress in liquid nitrogen

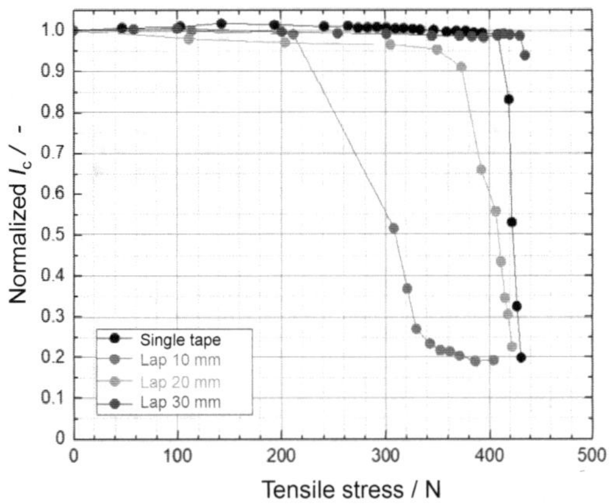

Figure 8. Tensile stress dependence of I_c

Figure 9 shows the current-voltage characteristics of the superconducting tape under bending stress. We refer to the bending diameter of 75 mm to "D75". Note that I_c of the samples with the entire lap length is twice that of a sheet of the superconducting tape because it is just two sheets of the superconducting tape stuck each other. From the results of the lap length of 20 mm, the joining resistance does not depend on the bending diameter. Namely, it is expected that the solder alloy layer at the joint does not deteriorate if bending stress was applied to it as well as tensile stress. Figure 10 shows the bending diameter dependence of I_c. Unlike the joining resistance, I_c decreased systematically where the bending diameter exceeded a critical value. Defining the deterioration of I_c as its more than 5% decrease, I_c did not deteriorate with the bending diameter up to D40 mm for the sheet of tape, D75 mm for the joined tape with a lap length of 20 mm, D120 mm for the joined tape with the entire surface. It is found that the target of the allowable bending diameter of 150 mm is achieved regardless of the lap length. From the results of the tensile test and the bending test, it is found that the solder joined superconducting tape with the lap length of more than 20 mm achieve the mechanical and electrical targets.

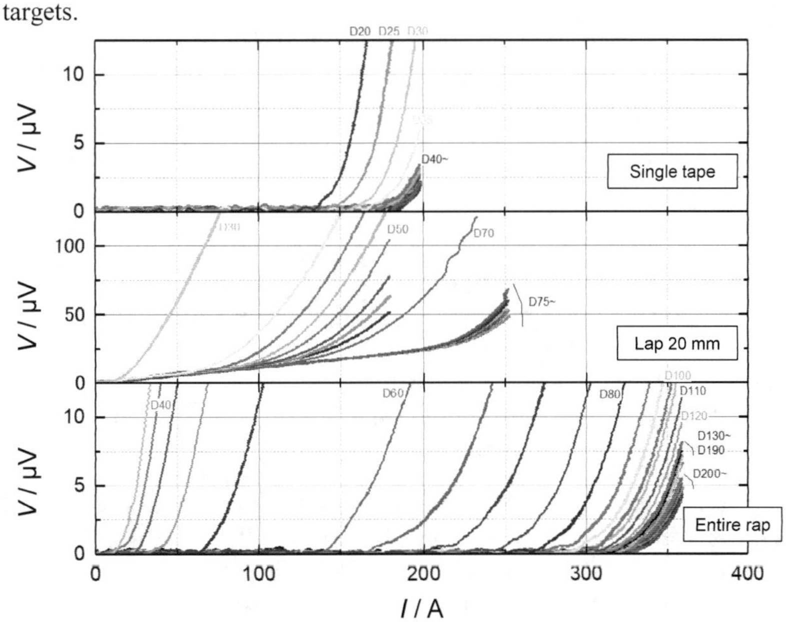

Figure 9. Current-voltage characteristics of the superconducting tape under bending stress in liquid nitrogen

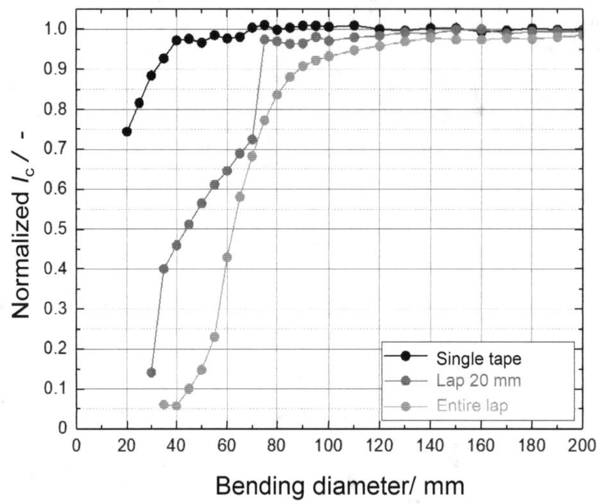

Figure 10. Bending diameter dependence of I_c

4. Conclusion

In this study, we conducted solder joining of superconducting tapes and its evaluation in the laboratory for joining superconducting cables at railway construction sites. At first, we performed electromagnetic analysis of the joints by the finite element method in order to obtain a prospect of the optimum conditions of the solder joining. As the results, the joint resistance to 10^{-7} Ω class will be achieved by the solder joining regardless of the solder thickness and the joint length. Referring to the analysis results, mechanical and electrical properties measurement of the joined superconducting tape was conducted. From the results of the tensile and bending test, it is found that the solder joined superconducting tape with the lap length of more than 20 mm achieve the target mechanical properties to 300 N for permissible tensile tension and 150 mm for permissible double bending diameters, without deterioration of I_c. Based on these results, we will test the cable joining at actual railway sites next. Then, we aim to introduce the long superconducting feeder cable into commercial railway lines.

Acknowledgments
This work was supported in part by the JST-Mirai Program Grant Number JPMJMI17A2 and JSPS KAKENHI Grant Number 19H00771.

References

[1] Smith R. 2001 *Jap Railway Trans Rev* **27** 16-24.
[2] Pietzcker RC, Longden T, Chen W, Fu S, Kriegler E, Kyle P and Luderer G. 2014 *Energy* **64** 95-108.
[3] Gunselmann W 2005 *Power electronics and applications. 2005 European conference on* p.10
[4] Tomita M, Fukumoto Y, Suzuki K and Muralidhar M 2010 *Physica C* **470** S.1007-1008
[5] Tomita M, Fukumoto Y, Ishihara A, Suzuki K, Caron H, Akasaka T, Kobayashi Y and Onji T 2020 *IEEE Transactions on Applied Superconductivity* **30** 5400107
[6] Tomita M, Suzuki K, Fukumoto Y, Ishihara A, Akasaka T and Kobayashi Y 2017 *Energy* **122** 579-587

[7] Tomita M, Suzuki K, Fukumoto Y, Ishihara A, Torii T, Akasaka T and Kobayashi Y 2021 *IEEE Transactions on Applied Superconductivity* **31** 4800104
[8] *Nature,* Research Highlights, This week 2017 **542** 275

[9] Zhang R, Kinoshita Y, Otabe E S, Akasaka T, Ishihara A and Tomita M 2020 *Physica C* **577** 1353733

Sharp edged Gd_2BaCuO_5 particles addition effect on growth and properties of top-seeded melt-grown $GdBa_2Cu_3O_{7-\delta}$ bulk superconductors

Shinnosuke Ishibashi [1], Rikako Hagiwara [1], Sugali Pavan Kumar Naik [1,2,*], Hiraku Ogino [2], Paweł Pęczkowski [3], Shigeyuki Ishida [2], Hiroshi Eisaki [2], Naomichi Sakai [4], Tetsuo Oka [4] and Taichiro Nishio [1, #]

[1] Department of Physics, Tokyo University of Science, 1 Chome-3 Kagurazaka, Shinjuku City, Tokyo, 162-8601, Japan
[2] Research Institute for Advanced Electronics and Photonics, National Institute of Advanced Industrial Science and Technology (AIST), 1-1-1 Central 2 Umezono Tsukuba, Ibaraki, 305-8568, Japan
[3] Institute of Physical Sciences, Faculty of Mathematics and Natural Sciences, School of Exact Sciences, Cardinal Stefan Wyszyński University, K. Wóycickiego 1/3 Street, 01-938, Warsaw, Poland
[4] Shibaura Institute of Technology, 3-7-5 Toyosu, Koto-ku, Tokyo 135-8548, Japan.
*spavankumarnaik@yahoo.in and #nishio@rs.tus.ac.jp

Abstract. We report on top-seeded melt growth of $GdBa_2Cu_3O_{7-\delta}$ (GdBCO) bulks added by sharp edged Gd_2BaCuO_5 nanoparticles ($Gd211_U$) refined by ultrasonication. GdBCO bulks mixed with 30 mol.% of Gd211U were examined to be grown isothermally at different temperatures in a range of 1015 °C−1035 °C using a top-seeded melt growth technique. Magnetic susceptibility measurements show that all samples become superconducting and especially samples grown at 1020 °C and 1025 °C have a maximum superconducting transition temperature (T_c) of 94.5 K. The critical current density (J_c) obtained from magnetic hysteresis loops for magnetic fields parallel to the *c*-axis is up to 65100 A/cm^2 (the self-field at 77 K) for a sample grown at 1025 °C, while J_c of a GdBCO bulk with conventional Gd211 particles grown at 1025 °C for comparison is 29000 A/cm^2 (the self-field at 77 K).

1. Introduction

$REBa_2Cu_3O_{7-\delta}$ (REBCO, RE123, RE – rare earth elements such as Nd, Sm, Gd, etc) - based material is known as one of the copper oxide high-temperature superconductors with a critical temperature (T_c) above the boiling point of the liquid nitrogen. REBCO compounds are ceramic in nature and the intrinsic grain boundaries limit their critical current densities (J_c), which is a negative factor for developing practical applications. Due to their reversible peritectic decomposition and recombination reactions (equations are given below), REBCO samples can be grown in large size bulk single grains which are able to trap the large amount of magnetic fields. The REBCO bulk is useful for constructing quasi-permanent magnets for magnetic resonance imaging, nuclear magnetic resonance, friction-free flywheels for energy storage, motors, etc [1-3]. Among REBCO superconductors, GdBCO is interesting due to producing fine sized 211 particles which support large J_c across the bulk samples [4-5]. In fact, the GdBCO single grain superconductors produced trapping of a world record magnetic field of 17.6 T at 26 K [3]. Also, the formation of lower-T_c based solid solutions of $RE_{1+x}Ba_{2-x}Cu_3O_{7-\delta}$ are minimal in the case of GdBCO compared to other REBCO candidates even when fabricated in air [6-7].

$$\text{Peritectic decomposition: } GdBa_2Cu_3O_{7-\delta} \rightarrow \text{Heating} \rightarrow Gd_2BaCuO_5 + L(Ba_3Cu_5O_8) + x\ O_2$$

$$\text{Peritectic reaction: } Gd_2BaCuO_5 + L(Ba_3Cu_5O_8) \rightarrow \text{Cooling} \rightarrow GdBa_2Cu_3O_{7-\delta}$$

GdBCO bulks can be prepared via several methods mainly, sintering, melt growth (MG) and infiltration growth (IG). Among, by MG and IG methods one can fabricate large size single grain bulks which are essential for practical applications [8-9]. However, MG process is rather simple compared to IG method for producing bulk single grains of REBCO, while IG process involves complexity and optimization of several parameters [9]. For improving J_c of the REBCO, several methods have developed, mainly by the addition of dopants, refining the secondary phases such as RE_2BaCuO_5 (RE211/211), etc [10-11]. Among these, 211 phase particles size, homogeneous dispersion and their content in the final bulk samples are crucial for bulk samples growth [12]. For this, addition of grain refining agents such as Pt, CeO_2, ball milling, etc methods were employed [13-14]. However, these methods involve difficulties such as chemical contamination.

Recently, for the first time, S. Pavan Kumar Naik et al., [15-16] developed an efficient and simple technique for producing nanometre sized 211 phase particles via employing high-energy ultrasonic irradiation. In general, the 211 phase produced by the conventional sintering method will be of the order of few tens of microns due to agglomeration. The ultrasonically pre-treated 211 particles will be no longer of typical spherical or elongated in shape, while applying the high-energy, the edges of the 211 particles will be damaged producing sharp edged and few tens of nanometre sized individual particles. The YBCO superconductors fabricated via top-seeded IG process and ultrasonically pre-treated Y211 powders exhibit superior superconducting properties. After this, ultrasonication method is also successfully applied to other materials and improved their superconducting properties [17]. However, it is very important to study the effect of these pre-treated 211 particles on the growth and superconducting properties of REBCO compounds fabricated via MG process too. For fabricating larger sized REBCO bulks, complete knowledge of the growth temperature window is necessary, which can be accomplished by isothermal under-cooling experiments. With this aim, in the present work, we studied the growth of GdBCO bulk superconductors with the addition of 30 mol% of ultrasonically pre-treated Gd211 phase particles via top-seeded MG (TSMG) process. Growth studies were conducted by isothermal experiments keeping the sample assemblies at particular growth temperatures and superconducting properties were investigated.

2. Experimental details

Gd_2O_3, $BaCO_3$, and CuO powders were thoroughly mixed in the molar ratio of 1:2:3 and 2:1:1 to obtain the respective 123 and 211 phase powders. These mixtures were then sintered at 860 °C-880 °C-900 °C for 12 h at each temperature along with intermediate mixing for obtaining pure phases. The Gd211 powder was mixed in ethanol solution and irradiated with 450 W ultrasonic radiation for 30 min employing MITSUI, UX-600 model. The sonicated Gd211 powder along with the ethanol was kept at 110 °C for overnight to remove the ethanol completely. Further, to remove the carbon related compounds if any present, dried 211 powders were calcined at 400 °C-600 °C. No impurity phases were detected as investigated by the X-ray diffraction (XRD). Two kinds of Gd211 powders of 30 mol%, namely i) Ultrasonically pre-treated Gd211 ($Gd211_U$) and ii) normal Gd211 powders were mixed in Gd123 powders. These mixtures were then pressed by a cylindrical mould with an inner diameter of 10 mm and preform pellets were prepared by a uniaxial press. For initiating epitaxially growth, a seed crystal of single crystalline bulk Nd123 with the (00*l*) texture was placed on the upper surface of the precursor pellet, and isothermal experiments were carried out at various growth temperatures from 1035 °C-1015 °C. The heat treatment schedule followed for isothermal under-cooling experiments are shown in Fig. 1 and the sample assembly utilized for the growth experiments is shown as the inset.

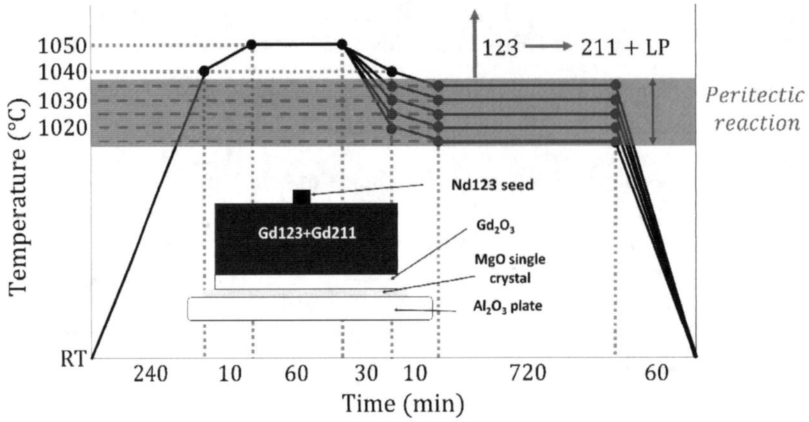

Figure 1. The temperature program of the furnace during isothermal growth of bulk materials. Inset shows the schematic diagram of the precursors of the TSMG method used in this study.

After the growth experiments, sub-specimens were collected from the growth area and oxygen annealing was applied by holding at 450 °C-400 °C for 150 hours with continues flow of 30 m*l*/min. This process changes the crystal structure from the non-superconducting tetragonal to the orthorhombic structure, resulting in the development of the superconductivity. Final samples fabricated by the addition of $Gd211_U$ and isothermal temperature of 1035 °C-1015 °C are referred as Gd-U-1035 °C-Gd-U-1015 °C respectively. The GdBCO sample isothermally prepared at 1025 °C with non-ultrasonicated Gd211 is referred as Gd-1025 °C. Superconducting transition temperature was determined by measuring the dc magnetic susceptibility by applying a magnetic field of 1 mT in the *c*-axis direction employing superconducting quantum interference device (SQUID, Quantum design MPMS-XL). For estimating field dependence of J_c, Magnetic hysteresis loops were measured sweeping the magnetic field up to 7 T in the *c*-axis direction at 77 K. J_c was estimated utilizing the extended Beans critical state model [18-19].

3. Results and discussion

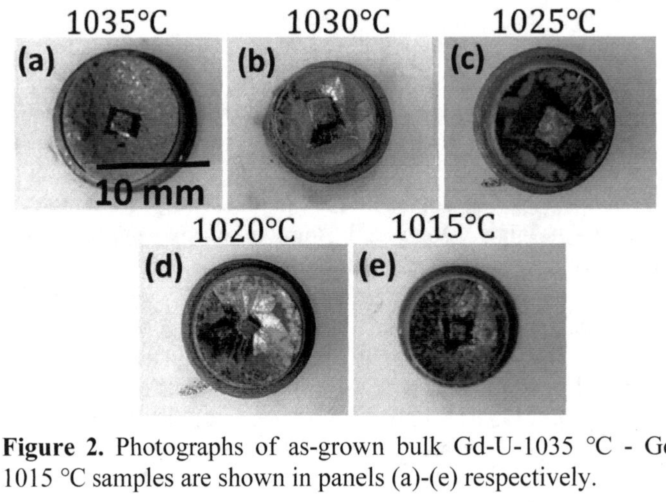

Figure 2. Photographs of as-grown bulk Gd-U-1035 °C - Gd-U-1015 °C samples are shown in panels (a)-(e) respectively.

The photographs of the as-grown bulk GdBCO samples at various isothermal under-cooling temperatures 1035 °C-1015 °Care shown in Fig. 2 (a)-(e), respectively. The epitaxial growth initiated from the Nd123 seed can be observed for Gd-U-1035 °C sample. As the isothermal under-cooling temperature decreased, growth area increased. When the under-cooling temperature was further decreased to 1025 °C, parasitic growth i.e., random nucleation was appeared, especially at the edges of the bulks. In REBCO bulk superconductors, random nucleation could occur due to various reasons such as differences in the temperature especially at edges of the sample, inhomogeneity of the liquid phases' availability for peritectic reaction, etc, [20-21]. When the under-cooling temperature decreased further to 1020 °C, parasitic growth has increased and hence the epitaxial growth area from the seed was inhibited largely. When the isothermal temperature is reduced to 1015 °C, the growth is reduced. From all the present isothermal growth experiments, it can be considered that the effective temperature window for fabricating large GdBCO single grains with the addition of 30 mol% of $Gd211_U$ is 1035 °C to 1015 °C.

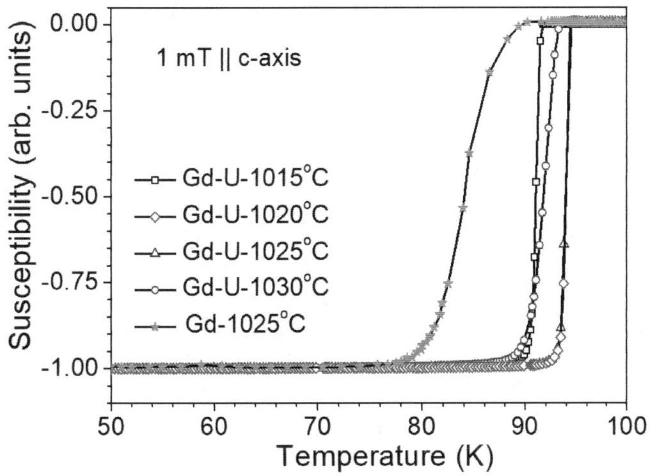

Figure 3. DC magnetic susceptibility curves of isothermal under-cooled GdBCO samples normalized to its maximum superconducting values.

Superconducting transition curves (ZFC curves) of Gd-U-1035 °C-Gd-U-1015 °C samples are shown in Fig. 3 and compared with Gd-1025 °C. All curves were normalized to full dia-magnetic moment with each sample maximum dia-magnetic moment. The determined onset of critical temperature ($T_{c\ on}$) of GdBCO samples are given in Table I and are ranged between 91.5 K – 94.5 K indicating all samples exhibit superconductivity with sharp transition. While the Gd-1025 °C sample superconducting transition curve is broader, as observed in our previous work [21], this is due to the chemical inhomogeneity occurred by the incomplete growth. In the case of Gd-U samples, the low $T_{c\ on}$ values of the Gd-U-1030 °C and Gd-U-1015 °C samples may be due to chemical inhomogeneity and incomplete oxygenation as the transition curves are broader at the tail. The superconducting transition widths (ΔT_c) of these samples are also calculated and the values are given in Table I.

Table 1. Transition temperature and critical current density of bulk Gd-U–*X* °C and Gd–1025 °C superconductors fabricated by isothermal growth at each temperature.

Sample	Transition temperature (K)			J_c (0T, 77 K, A/cm^2)
	$T_{c\ on}$	T_c	ΔT_c	
Gd-U-1030 °C	93.0	90.1	2.9	42700
Gd-U-1025 °C	94.5	93.5	1.0	65100
Gd-U-1020 °C	94.5	93.5	1.0	55000
Gd-U-1015 °C	91.4	90.2	1.2	37200
Gd-1025 °C	87.5	79.2	8.3	29000

The field dependence of J_c curves of Gd-U-1030 °C - Gd-U-1015 °C and Gd-1025 °C superconductors determined at 77 K is compared in Fig. 4. As given in the Table I, the self-field J_c values are of the order of 37200 A/cm^2 for Gd-U-1015 °C sample to 65100 A/cm^2 for Gd-U-1025 °C. As the isothermal temperature reduced, the peritectic reaction and the growth rate were improved. This supported higher J_c's compared to the samples grown at lower and higher isothermal temperatures. If the Gd-1025 °C is compared with the Gd-U-1025 °C superconductor in J_c performance, the latter is superior due to possessing strong pinning centres. Applying thermal gradient method, it may be possible to grow large-sized bulk GdBCO superconductors with superior properties. Addition of dopants may further improve the properties of GdBCO-Ultra superconductors as strong superconducting magnets for high-field applications. These bulk single grain GdBCO superconductors with ultrasonically pre-treated 211 particles are significant in various applications such as generating tesla-scale homogeneous magnetic fields useful for motors and various medical fields, levitation platforms, vibration dampers, etc.

4. Conclusions

Pure phases of $GdBa_2Cu_3O_{7-\delta}$ and Gd_2BaCuO_5 powders were synthesized by conventional solid-state sintering method. High energy ultrasonic radiation of 450 W was employed for 30 mins for producing sharp edged, individual and ~100 nm sized Gd211 particles. By the addition of 30 mol.% of nano Gd211$_U$ in Gd123, isothermal growth experiments were conducted via top-seeded melt growth process. The effective growth temperature window for GdBCO with Gd211$_U$ system is identified to be 1015 °C–1035 °C. Field dependence critical current density of 65100 kA/cm^2 was obtained at the self-field and 77 K in an ultrasonicated GdBCO sample grown at 1025 °C and is more than 2 times larger compared to a conventional GdBCO sample grown at same growth temperature due to improved peritectic reaction.

References

[1] Cardwell D A 1988 *Mater. Sci. Eng. B* **53** 1−10.
[2] Tomita M and Murakami M 2003 *Nature* **421** 517.
[3] Durrell J H, Dennis A R, Jaroszynski J, Ainslie M D, Palmer K G B, et al. 2014 *Supercond. Sci.Technol.* **27** 082001.
[4] Nariki S, Sakai N and Murakami M 2005 *Supercond Sci. Technol.* **18** S126.
[5] Pavan Kumar Naik S, Muralidhar M, Kento T, Jirsa M and Murakami M 2019 *J. App. Phys.* **125** 093907.
[6] Zhang K, Dabrowski B, Segre C U, Hinks D G, Schuller I K, Jorgensen J D and Slaski M 1987 *J. Phys. C: Solid State Phys.* **20** L935-L940.
[7] Kramer M J, Yoo S I, Mc Callum R W, Yelon W B, Xie H and Allenspach P 1994 *Physica C* **219** 145-155.
[8] Cloots R, Koutzarova T, Mathieu J P and Ausloos M 2005 *Supercond. Sci. Technol.* **18** R9-R23.
[9] Pavan Kumar Naik S, Nagaveni K, Missak Swarup Raju P, Muralidhar M, Koblischka M R, Koblischka-Veneva A, Oka T, Iwasa Y, Ogino H, Eisaki H and Murakami M 2019 *Proc. of SPIE* **11054** 110540H-8 pp.
[10] Murakami M, Yamaguchi K, Fujimoto H, Nakamura N, Taguchi T, Koshizuka N and Tanaka S 1992 *Cryogenics* **32** 930–935.
[11] Zmorayova K, Vojtkova L, Hlasek T, Plechacek J and Diko P 2020 *Supercond. Sci. Technol.* **33** 034005.
[12] Devendra Kumar N, Shi Y, Palmer K G, Dennis A R, Durrell J H and Cardwell D.A 2016 *J. Eur. Ceram. Soc.* **36** 615.
[13] Vojtkova L, Diko P, Kovac J and Vojtko M 2018 *Supercond. Sci. Technol.* **31** 065003.
[14] Missak Swarup Raju P, Devendra Kumar N, Pavan Kumar Naik S, Rajasekharan T and Seshubai V 2014 *J. Supercond. Nov. Mag.* **27** 2277-2282.
[15] Pavan Kumar Naik S, Muralidhar M, Koblischka M R, Koblischka-Veneva A, Oka T and Murakami M 2020 *ACS Omega* **5** 6250–6259

[16] Pavan Kumar Naik S, Muralidhar M, Koblischka M R, Koblischka-Veneva A, Oka T and Murakami M 2019 *Appl. Phys. Exp.* **12** 063002.
[17] Arvapalli S S, Miryala M, Sunsanee P, Jirsa M and Murakami M 2021 *Mater. Sci. Eng. B.* **265** 115030
[18] Chen D X and Goldfarb R B 1989 *J. Appl. Phys.* **66** 2489-2500.
[19] Bean C P 1962 *Phys. Rev. Lett.* **8** 250.
[20] Endo A, Chauhan H S, Egi T and Shiohara Y 1996 *J. Mater. Res.* **11** 795-803.
[21] Pavan Kumar Naik S, Muralidhar M, Jirsa M and Murakami M 2017 *J. Appl. Phys.* **122** 193902.

High-energy ultrasonic irradiation effects on the growth and critical current density of top-seeded melt grown single grain $YBa_2Cu_3O_{7-\delta}$ bulk superconductors

Rikako Hagiwara [1], Shinnosuke Ishibashi [1], Sugali Pavan Kumar Naik [1,2,*], Hiraku Ogino [2], Paweł Pęczkowski [3], Shigeyuki Ishida [2], Hiroshi Eisaki [2], Naomichi Sakai [4], Tetsuo Oka [4] and Taichiro Nishio [1,#]

[1] Department of Physics, Tokyo University of Science, 1 Chome-3 Kagurazaka, Shinjuku City, Tokyo, 162-8601, Japan
[2] Research Institute for Advanced Electronics and Photonics, National Institute of Advanced Industrial Science and Technology (AIST), 1-1-1 Central 2 Umezono Tsukuba, Ibaraki, 305-8568, Japan
[3] Institute of Physical Sciences, Faculty of Mathematics and Natural Sciences, School of Exact Sciences, Cardinal Stefan Wyszyński University, K. Wóycickiego 1/3 Street, 01-938, Warsaw, Poland
[4] Shibaura Institute of Technology, 3-7-5 Toyosu, Koto-ku, Tokyo 135-8548, Japan.
*spavankumarnaik@yahoo.in and #nishio@rs.tus.ac.jp

Abstract. The enhancement of a critical current density (J_c) of $REBa_2Cu_3O_{7-\delta}$ (REBCO/YBCO, where RE=rare earth elements and Y) bulk superconductors can be achieved via dispersing normal-conductive RE_2BaCuO_5 (Y211) particles and other defects which are of nano-metric size in the superconducting matrix. Recently, we demonstrated the integration of high-energy ultrasonic irradiation for fabricating high quality YBCO superconductors which effectively improved J_c and trapped fields. In the present work, we investigated the effect of 30 mol.% of Y211 particle produced via high-energy ultrasonication irradiation on the growth and superconducting properties of top-seeded melt grown bulk YBCO single grains. Ultrasonic power of 450 W was employed for 30 minutes to create the sharp-edged, individual and nano-meter sized Y211 phase particles. Systematic isothermal experiments were conducted at various temperatures from 1005 °C – 980 °C. The growth of YBCO grain from a single crystalline NdBCO seed was initiated at 1000 °C and as the isothermal temperature decreased the growth area increased due to the improved peritectic reaction. The field dependence of J_c was improved when the under-cooling isothermal temperature decreased. The present method is reliable, cost-effective, and free from any chemical contamination.

1. Introduction

Copper oxide-based superconductors such as $REBa_2Cu_3O_{7-\delta}$ (REBCO, 123, RE- rare earth elements such as Nd, Sm, Gd, Y, etc), $Bi_2Sr_2Ca_2Cu_3O_{10+x}$, etc., exhibit critical temperature (T_c) exceeding the boiling point of liquid nitrogen, i.e., 77 K [1-3]. In addition, in theses superconductors carriers that are responsible for the superconductivity are not electrons but holes and the coherence length is very short [3]. As a result, T_c is greatly affected by the number of holes and shows the highest value with optimally doping. Among them, the YBCO superconductor which is studied in the present work has a high T_c of 92 K and a high ability in pinning magnetic flux. In this regard, the other families of cuprates such as

Bi-Sr-Ca-Cu-O systems suffer a serious flux creep problem and may not be superior for developing practical applications at higher temperatures [4-5]. Several practical applications out of REBCO are already developed in various fields of science and technology. Namely, energy-storing flywheel, frictionless bearings, superconducting mixer that agitates media without contact, and a purification device that combines magnetic separation using a filter and a superconducting bulk magnet [6-7]. Several methods have already developed for synthesizing single grain REBCO bulk superconductors, namely (i) Melt growth (MG) and (ii) Infiltration growth (IG) [8-10]. Among, the MG method is simpler compared to the IG method.

For developing any applications, improving the critical current density and trapped fields are crucial, and these can be enhanced by improving their microstructures [11-12]. To further improve the microstructure and hence the magnetic field dependence of YBCO superconductors, new methodologies need to be integrated. For improving the J_c of bulk REBCO/YBCO several methods are proposed, mainly to improve the growth, including secondary phases by dopants and fine sized non-superconducting phases such as Y_2BaCuO_5 (Y211) [13]. Among, utilization of smaller sized 211 particles are beneficial for higher growth rates and embedding fine sized non-superconducting particles in Y123 bulk matrix. In this regard, in order to improve the flux pinning of REBCO superconductors, by a team of S. Pavan Kumar Naik et al., for the first time, a novel method of high-energy ultrasonic radiation was employed for pre-treating 211 powders [14-15]. Later this method is applied to other superconductors successfully and improved their performance [16]. In this method, individual 211 particles with sharp edges were produced by high-energy ultrasonic irradiation reducing the micron sized starting powders to ~100 nm size. Employing these pre-treated and new type of 211 phases, single grain YBCO was constructed via the IG process and demonstrated superior J_c and trapped fields improvements, without addition of any dopants or grain-refining agents [12, 17].

However, as discussed above, the IG method is complex compared to the MG method [6]. Thus, it is necessary to employ and study the effect of new type of Y211 particles on the growth and their superconducting properties of YBCO compounds grown via the MG process too, in which larger sized samples can be grown. Therefore, in the present work we synthesize YBCO bulk superconductors with ultrasonically produced Y211 ($Y211_U$) particles using the top-seeded melt growth method. A systematic isothermal experiment at various constant temperatures were employed for studying growth of the single grain. Detailed growth and superconducting properties were investigated.

2. Experimental details

First, the preform powders of Y211 (Y_2O_3:BaO_2:CuO=2:1:1) and Y123 (Y_2O_3:BaO_2:CuO=1:2:3) were produced via a conventional sintering method. Mixed precursor powders were sintered at 860 °C – 880 °C - 900 °C for 12 h with intermediate mixing. Pure phases of respective Y211 and Y123 powders were confirmed by X-ray diffraction (XRD) measurements. After production of Y211, high energy ultra-sonication was carried out employing MITSUI, model UX-600. For this, ethanol was used as a liquid medium and an ultrasonication power of 450 W of energy was employed via pulse mode for 30 min. The mixture of ethanol and Y211 powders were then heated at 110 °C until the evaporation of ethanol. It was then calcined at 400 °C for 12 h and 1 h at 600 °C removing carbon related compounds in the dried powder. The ultrasonically pre-treated Y211 of 30 mol% was then added to Y123 powders. The mixture of the 123 and Y211 powders were compacted into pellets for the MG process at 460 MPa. Bulk preparation was carried out using the MG method. A single crystal of $NdBa_2Cu_3O_{7-\delta}$ (Nd123) was placed as a seed to promote the textured growth of the YBCO bulk. The sample assembly was then transferred to a box furnace. Furnace was then heated to 850 °C in 4 h and the temperature was held for 1 h for preform pellet strengthening. Then the furnace was heated to 1045 °C in 1 h and 1050 °C in 10 minutes and holding for 1 hour and the sample was quenched to 1010°C in 30 minutes and then maintained at several isothermal temperatures from 1005 °C-980 °C with a dwell time of 12 h and then the furnace was switched off for faster cooling. Fig. 1 shows the time- temperature schedule followed for studying the growth of various YBCO samples. The final isothermally top-seeded MG (TSMG) processed YBCO samples are referred as Y-U-X°C, where "X" refers as the isothermal temperature and "U" refers as Ultrasonicated Y211 particles added.

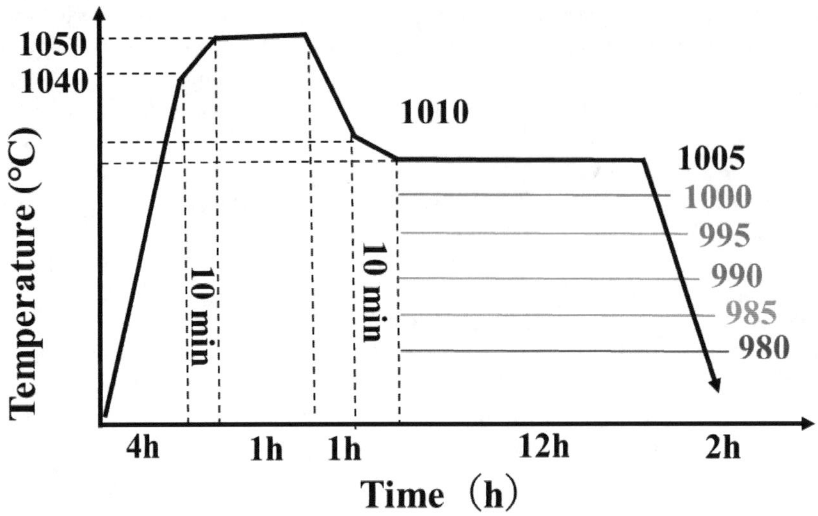

Figure 1. Heating schedule followed for isothermal experiments of ultrasonically pre-treated Y211 added Y123 bulks.

Since the as grown YBCO bulk superconductor will be of tetragonal crystal structure, which is non-superconducting, oxygen annealing was carried out for transitioning the crystal structure to the orthorhombic and superconducting [18]. For this all YBCO samples were oxygenated at 450°C for 150 h with a constant oxygen flow of 0.3 l/min. A small specimen was sliced from each bulk sample ~2 mm below the Nd123 seed for physical properties characterizations. XRD patterns of YBCO samples were investigated by XRD. The superconducting transition and M-H loops were characterized on samples of a typical size ~2 × 1.5 × 0.5 mm^3 by a superconducting quantum interference device (SQUID, MPMS-XL, Quantum design). The critical current density of the YBCO samples was determined utilizing the extended the Beans critical state model [19-20].

3. Results and discussion

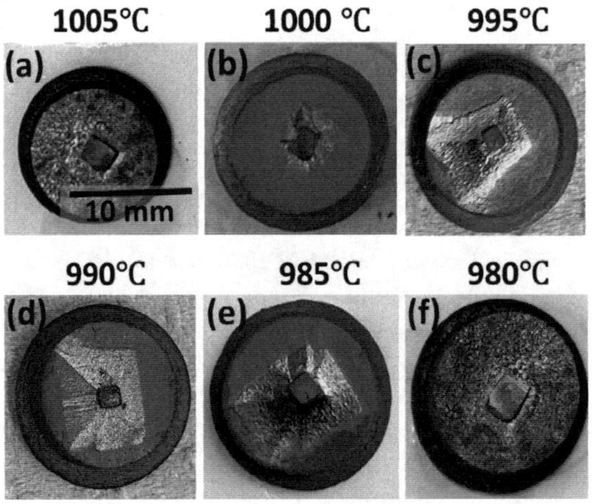

Figure 2. Bulk YBCO material photographs fabricated by the TSMG method and at various isothermal under-cooling temperatures (given on the top of each sample). As the under-cooling temperature decreased, the growth area increased.

Before sonication, Y211 particles were spherical in shape and about 5-10 microns in size and were agglomerated due to the sintering procedure. In contrast, the $Y211_U$ particles had an average particle diameter of ~100 nm along with the sharp edges and surface damages. This increases the surface area, which could result in more opportunity for participating in the growth process while peritectic re-combination reaction. There is also variability in the growth and dispersion of Y211 phase particles in the bulk sample due to the presence of two different types of Y211 particles in the liquid [21-22]. This is beneficial to improve the growth rate as well as flux pinning to increase the J_c of YBCO superconductors.

Fig. 2 (a) to (f) show the photographs of top surfaces of the as synthesized bulk YBCO samples at various isothermal temperatures (1005 °C to 980 °C) respectively. It was confirmed from the appearance that the growth was initiated from the Nd123 seed at 1005 °C, and the area of the growth increased as the isothermal temperature decreased. Furthermore, due to seeded growth, four-fold growth in the form of 'X' can be seen indicating that all samples grew as a single grain. However, when the sample assembly is kept at 980 °C, there was no reaction. As a result of the isothermal experiments, it was found that a temperature range of 1000 °C-980 °C is optimum for growing a high-quality bulk single-grain YBCO superconductors with the addition of $Y211_U$.

Table 1. Superconducting properties of Y-U-X°C and Y-990°C samples.

Sample	Transition temperature (K)			J_c (0T, 77 K, A/cm^{2})
	$T_{c\,on}$	T_c	ΔT_c	
Y-U-985°C	92.7	91.5	1.2	51000
Y-U-990°C	92.4	91.2	1.2	58000
Y-U-995°C	92.4	90.1	2.3	44700
Y-U-1000°C	92.6	89	3.6	31600

The temperature dependence of the superconducting transition curves normalized to each sample maximum dia magnetic moment of Y-U-X°C samples are shown in Fig. 3. The onset of T_c and the width of the superconducting transition (ΔT_c) are determined, as given in Table 1. All samples exhibit diamagnetism below about 92.4-92.7 K, showing that all samples are superconductors. The values of ΔT_c indicate the good quality of the samples.

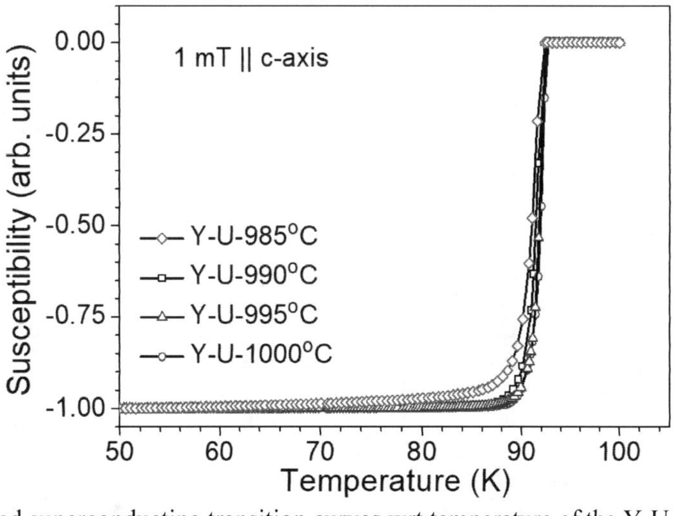

Figure 3. Normalized superconducting transition curves wrt temperature of the Y-U-X°C samples.

Fig. 4 shows the field dependence of J_c (J_c(H)) curves obtained at 77 K for YBCO superconductors fabricated at various isothermal under-cooling temperatures. As given in Table I, the zero field J_c (J_c(0)) of Y-U-985°C, Y-U-990°C, Y-U-995°C and Y-U-1000°C superconductors are determined to be 50100 A/cm^2, 58000 A/cm^2, 45000 A/cm^2, and 31600 A/cm^2 respectively. The present results clearly show that as the isothermal temperature decreased, the J_c(0) value increased. The superior J_c performance of the

Y-U-990°C superconductor could be due to the improved growth due to the presence of tiny size, individual and surface damaged Y211 particles.

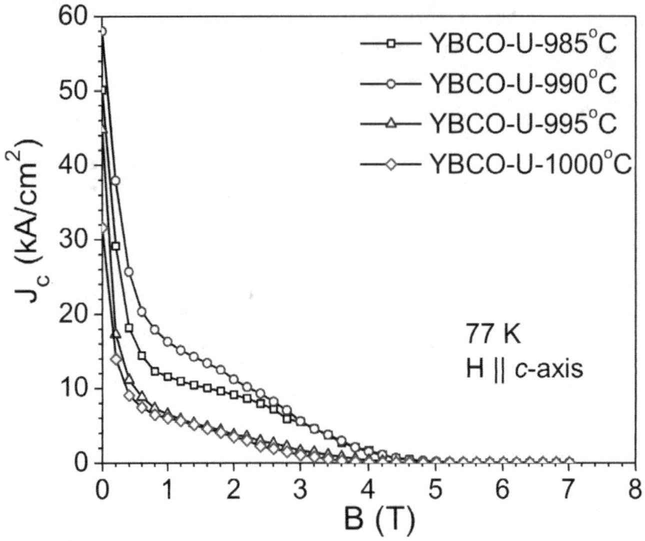

Figure 4. The $J_c(H)$ curves for Y-U-985°C, Y-U-990°C, Y-U-995°C and Y-U-1000°C bulk superconductors determined at 77 K. Magnetic field was applied parallel to the *c*-axis of all the samples.

The present J_c performance of the YBCO superconductors fabricated by the isothermal TSMG process are superior to previous reports [23-24]. We attribute such improvement in J_c performance to the improved growth reaction in the presence of new type of $Y211_U$ particles added for the MG process. Also, it should be noted here that the J_c performance of the TSMG processed YBCO samples can be further enhanced by a fabricating temperature gradient method, which we will be conducting subsequently. In order to determine the irreversibility field (H_{irr}) of the fabricated YBCO superconductor, we defined it, as the magnetic field required reaching the J_c value of 100 A/cm^2. The determined H_{irr} values for Y-U-985°C, Y-U-990°C, Y-U-995°C and Y-U-1000°C superconductors are 4.9 T, 5.2 T, 4.4 T and 4.1 T respectively. This indicates that H_{irr} of the YBCO superconductors is also influenced by the isothermal temperature and improved as the growth was enhanced.

4. Conclusions

Pure phases of $GdBa_2Cu_3O_{7-\delta}$ and Gd_2BaCuO_5 powders were synthesized by conventional solid-state sintering method. High energy ultrasonic radiation of 450 W was employed for 30 mins for producing sharp edged, individual and ~100 nm sized Gd211 particles. By the addition of 30 mol.% of nano $Gd211_U$ in Gd123, isothermal growth experiments were conducted via top-seeded melt growth process. The effective growth temperature window for GdBCO with $Gd211_U$ system is identified to be 1015 °C–1035 °C. Field dependence critical current density of 65100 kA/cm^2 was obtained at the self-field and 77 K in an ultrasonicated GdBCO sample grown at 1025 °C and is more than 2 times larger compared to a conventional GdBCO sample grown at same growth temperature due to improved peritectic reaction.

References

[1] Cardwell D A 1988 *Mater. Sci. Eng. B* **53** 1–10.
[2] Tomita M and Murakami M 2003 *Nature* **421** 517.
[3] Durrell J H, Dennis A R, Jaroszynski J, Ainslie M D, Palmer K G B, et al. 2014 *Supercond. Sci.Technol.* **27** 082001.
[4] Nariki S, Sakai N and Murakami M 2005 *Supercond Sci. Technol.* **18** S126.

[5] Pavan Kumar Naik S, Muralidhar M, Kento T, Jirsa M and Murakami M 2019 *J. App. Phys.* **125** 093907.
[6] Zhang K, Dabrowski B, Segre C U, Hinks D G, Schuller I K, Jorgensen J D and Slaski M 1987 *J. Phys. C: Solid State Phys.* **20** L935-L940.
[7] Kramer M J, Yoo S I, Mc Callum R W, Yelon W B, Xie H and Allenspach P 1994 *Physica C* **219** 145-155.
[8] Cloots R, Koutzarova T, Mathieu J P and Ausloos M 2005 *Supercond. Sci. Technol.* **18** R9-R23.
[9] Pavan Kumar Naik S, Nagaveni K, Missak Swarup Raju P, Muralidhar M, Koblischka M R, Koblischka-Veneva A, Oka T, Iwasa Y, Ogino H, Eisaki H and Murakami M 2019 *Proc. of SPIE* **11054** 110540H-8 pp.
[10] Murakami M, Yamaguchi K, Fujimoto H, Nakamura N, Taguchi T, Koshizuka N and Tanaka S 1992 *Cryogenics* **32** 930–935.
[11] Zmorayova K, Vojtkova L, Hlasek T, Plechacek J and Diko P 2020 *Supercond. Sci. Technol.* **33** 034005.
[12] Devendra Kumar N, Shi Y, Palmer K G, Dennis A R, Durrell J H and Cardwell D.A 2016 *J. Eur. Ceram. Soc.* **36** 615.
[13] Vojtkova L, Diko P, Kovac J and Vojtko M 2018 *Supercond. Sci. Technol.* **31** 065003.
[14] Missak Swarup Raju P, Devendra Kumar N, Pavan Kumar Naik S, Rajasekharan T and Seshubai V 2014 *J. Supercond. Nov. Mag.* **27** 2277-2282.
[15] Pavan Kumar Naik S, Muralidhar M, Koblischka M R, Koblischka-Veneva A, Oka T and Murakami M 2020 *ACS Omega* **5** 6250–6259
[16] Pavan Kumar Naik S, Muralidhar M, Koblischka M R, Koblischka-Veneva A, Oka T and Murakami M 2019 *Appl. Phys. Exp.* **12** 063002.
[17] Arvapalli S S, Miryala M, Sunsanee P, Jirsa M and Murakami M 2021 *Mater. Sci. Eng. B.* **265** 115030
[18] Chen D X and Goldfarb R B 1989 *J. Appl. Phys.* **66** 2489-2500.
[19] Bean C P 1962 *Phys. Rev. Lett.* **8** 250.
[20] Endo A, Chauhan H S, Egi T and Shiohara Y 1996 *J. Mater. Res.* **11** 795-803.
[21] Pavan Kumar Naik S, Muralidhar M, Jirsa M and Murakami M 2017 *J. Appl. Phys.* **122** 193902.

Effect of boron precursor ultra-sonicated in hexane medium on the critical current density of sintered bulk MgB_2

Sai Srikanth Arvapalli[1], Milos Jirsa[2], Masato Murakami[1], Muralidhar Miryala[1]*

[1]Materials for Energy and Environmental Laboratory, Superconducting Materials Group, College of Engineering / Graduate School of Science & Engineering, Shibaura Institute of Technology, 3-7-5 Toyosu, Koto-ku, Tokyo 135-8548, Japan

[2]Institute of Physics ASCR, Na Slovance 2, Praha 8 18221, Czech Republic

*miryala1@shibaura-it.ac.jp

Abstract. The large coherence length of MgB_2 material allows large defects such as grain boundaries to act as strong pinning centres. As an alternative to the high-cost commercial nano precursors, we used a novel ultra-sonication technique to refine a cheap rough boron precursor and avoid the use of the expensive commercial nano amorphous boron powder. Recent novel low-cost ultra-sonication technique proved to improve the critical current density, J_c, of MgB_2. Here, we discuss the consequences of ultra-sonication duration in hexane medium on the boron precursor as well as the superconducting properties of MgB_2 bulks. XRD revealed small peaks corresponding to unreacted Mg. First derivative of susceptibility was plotted as a function of temperature to explain the $T_{c,onset}$ and ΔT_c. J_c was found to grow if the MgB_2 bulk was prepared of boron ultra-sonicated for up to 30 minutes and then reduced with further boron ultra-sonication (60 minutes). SEM micrographs showed that boron powder subjected to longer ultra-sonication resulted in agglomeration of broken boron particles, forming large MgB_2 grains. Based on the comparison of superconducting properties with previous results, we speculate that viscosity of the ultra-sonication medium might have a crucial role in the refinement of boron precursor powder.

1. Introduction

Superconductivity in MgB_2 ($T_c \sim 39$ K) was discovered in 2001 [1], which immediately attracted interest of many researchers. MgB_2 has a hexagonal lattice (P6/mmm) with alternate Mg and B layers and lattice parameters of a = b ~ 3.086 and c ~ 3.524 Å. MgB_2 has some visible advantages compared to other superconductors, especially cuprates, Nb-based ones etc. Irrespective of the moderate critical temperature (it can be cooled using liquid H_2, Ne), the large coherence length allows grain boundaries to be effective flux pinning centers, supporting a high critical current density in polycrystalline bulks. Light weight makes the material attractive for space [2] and portable device applications [3], and the uniform/isotropic trapped field behavior [4] contributes to reliability in application in various fields. One of the issues of MgB_2 is its moderate critical current density, which has been a hot research area since 2 decades. Effective flux pinning centers are needed to enhance J_c in MgB_2. Major flux pinning entities are i) grain boundaries [5,6], ii) lattice defects [7–9] and iii) additives [10–13]. Extensive study of doping has brought amazing results. However, issues such as excessive cost [14] or non-uniform distribution arose. Manipulation with grain boundaries is the easiest and straightforward way to enhance flux pinning. Among other methods, the simplest and cheapest way is high energy ultra-sonication that we have employed recently [15,16]. This technique uses ultrasonic waves generated by metal probe in a solution and causes i) particle-particle collisions, ii) particle-wall collisions and iii) destructive forces generated by collapse of tiny air bubbles, thereby refining the particles. So far we have tried various solvents to optimize the refinement. However, results pointed out that after certain duration of ultra-sonication, the J_c value of corresponding MgB_2 bulk starts to reduce. We explored the boron powder refinement in hexane medium. In this work, based on microstructural observations, we attributed the detrimental effect of long ultra-sonication on the boron powder to a fine particles' agglomeration. We also explored the influence of solvent on grain refinement.

2. Materials and Methods

Cheap commercial boron (300 mesh, Furu-uchi chemicals) is ultra-sonicated in hexane for various durations upto one hour with 15 minute intervals. We applied a power of 150W at 20 kHz frequency. This applied power is 50% of the maximum power of the Ultrasonicator. The ultra-sonicated B powder was then analyzed using X-ray Diffraction and the size reduction was observed using scanning electron microscopy. In-situ MgB_2 sintered bulks were prepared from commercial Mg powders (99.9% purity, 200 meshes, Furu-uchi Chemical Corporation) and boron ultra-sonicated in hexane. Mg and B powders were mixed in the stoichiometric ratio in a glove box containing Ar atmosphere and then pressed into pellets (20 mm diameter, 7 mm thickness). These pellets were wrapped in titanium (Ti) foils and sintered in a tube furnace at optimized conditions such as 775 °C for 3 hours with continuous Ar flow (200-300 ml/min) [6]. More information such as working principle of the ultrasonic treatment can be found in Refs. [15,17].

To characterize the bulk products, we used the high-resolution automated Rigaku smart-lab X-ray powder diffractometer (RINT2200) (step size of 0.01°, Cu-K_α radiation generated at 40 kV and 30 mA), field emission – scanning electron microscope (FE-SEM/JSM-7100F/JEOL) and SQUID magnetometer (Quantum Design, model MPMS5). Critical current densities (J_c) were calculated based on extended Bean critical state model as $J_c = 20\ \Delta m/[a_s^2 c_s\ (b_s\text{-}a_s/3)]$, where a_s and b_s are cross-sectional dimensions, $b_s > a_s$, and c_s is thickness of the SQUID specimen. The typical dimensions of cuboidal specimens a_s, b_s, and c_s for SQUID measurements are approximately 1, 1, and 0.5 mm, respectively. Δm (in emu units, 1 emu = 10^{-3} Am^2) is the difference of magnetic moments during decreasing and increasing magnetic field in the *M-H* loop.

3. Results and discussion

XRD of pristine boron, boron ultra-sonicated for 30 min. (the best sample) and for 60 min. (the longest duration) and MgB_2 prepared from these boron precursors is presented in figure 1. Unlike in ball-milling [18], these boron powders did not contain any boron oxide phase that would encourage formation of non-superconducting MgO phase during Mg and B reaction. In addition, when using ultra-sonication, we avoid the impurities from milling balls, sometimes observed in ball-milled specimens [19]. Very minute MgO and no other impurities were observed in the MgB_2 fabricated from ultra-sonicated powders. Only some unreacted Mg was observed in MgB_2 prepared from boron ultra-sonicated for 60 minutes. In general, unreacted Mg was observed either due to lower reaction temperatures such as below 600°C [20] or when the stoichiometry of Mg and B was offset towards Mg-rich side. Unreacted Mg is also observed when Mg cannot diffuse completely into boron matrix due to large boron particles or not enough reaction time like in case of IG processed MgB_2, where Mg_2B_{25} phase is formed and then progresses towards MgB_2 with time in the initial stages of reaction [21]. On the other hand, unreacted B is often observed in powder-in-tube based MgB_2 bulks or wires, as the Mg vaporizes and offsets the stoichiometry to B-rich [22]. We also focused on the (110) XRD peak of MgB_2 bulks situated at $2\theta \sim 60°$ and found that there was no shift proving that there was no carbon incorporation into the bulk from hexane medium. We chose (110) peak because the boron planes in MgB_2 lie parallel to ab-plane and carbon can only substitute boron, due to similarity in atomic radii. For example, carbon substitution using carbon-coated boron, SiC, toluene etc. can result in shift in the (110) peak, leading to a lattice parameter change [14,23,24]. Usually, the shift is in proportion with the amount of substitution. We calculated the lattice parameter ‘a’ of the MgB_2 specimens using inter-planar distance calculated from Bragg’s law and values were around 3.085, 3.084 and 3.085 Å respectively for 0, 30 and 60 min ultra-sonication. Such values indicate that there is no carbon substitution. In comparison, when we tried to incorporate carbon using carbon-encapsulated boron, we observed a change to 3.078 Å for just 1.5 wt% addition [14]. In case of 5 % nano-C doping, the value was reported to drop to 3.060 Å [25]. Similarly, the lattice parameter ‘c’ also didn’t change. The values of ‘c’ were around 3.521, 3.522 and 3.521 Å respectively for 0, 30 and 60 min ultra-sonication.

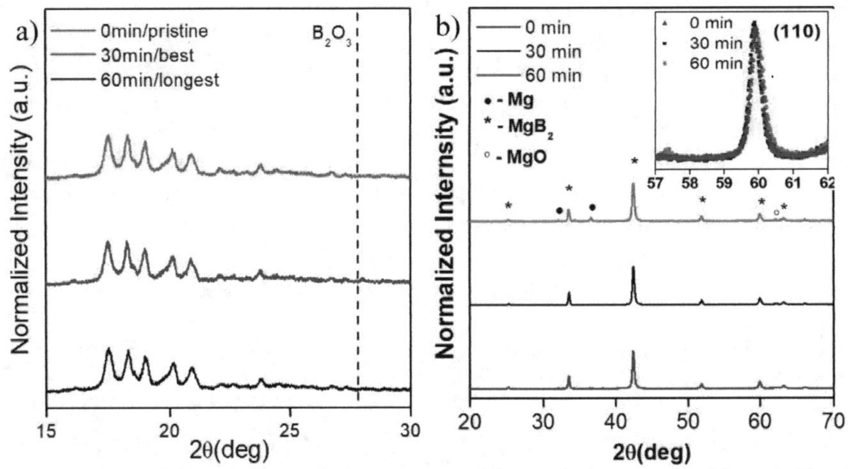

Figure 1. XRD of (a) pristine and ultra-sonicated boron in hexane for 30, and 60 min and (b) MgB_2 sintered bulk prepared from it.

For more compressive understanding, we compared the micrographs obtained using FE-SEM. While refinement of boron can be observed with increase in sonication time from figure 2, we were also able to identify cluster of irregular shaped boron particles interlocked within themselves. Such clusters can be expected as ultrasonic refinement comprises of random collisions of boron particles which generates random shaped boron particles that can entangle with each other as shown in inset of 60 min ultra-sonicated boron. Compared to ball-milling, the ultra-sonication technique much faster and thus more cost efficient. Moreover, it not only refines but also maintains purity of the plain boron. Microscopic observations show a wider size distribution of the refined particles, which is due to the turbulent process. On the other hand, ball-milling can generates a narrower size distribution via a proper choice of milling duration, rotation speed, and milling ball size. The average particle size after subjecting to milling for 4 hours was around 170 nm and the size didn’t change much with further milling, only the size distribution became uniform [26]. Boron particles refined via ultra-sonication for 30 min varied around 30 nm to 620 nm, with an average particle size of around 190 nm. In accordance micrographs of MgB_2 showed refined grains in case of 30 minutes compared to un-ultra-sonicated (0 min) and larger grains in MgB_2 made from 60 min ultra-sonicated boron (see figure 3). We believe that the boron clusters observed in 60 minute ultra-sonicated boron resulted in large MgB_2 grains while also shielding the crust of the cluster from reacting with Mg thereby resulting in unreacted Mg, as depicted in XRD analysis.

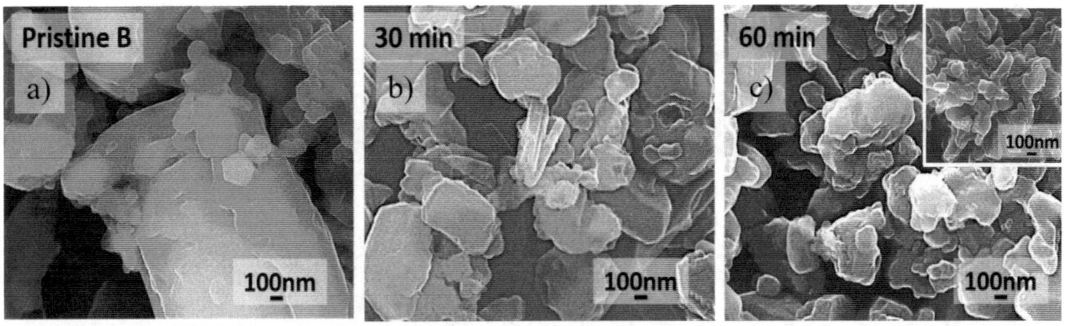

Figure 2. FE-SEM micrographs of (a) pristine boron, boron ultra-sonicated for (b) 30 and (c) 60 minutes in hexane medium. Interlocked clusters can be observed with long ultrasonic treatment.

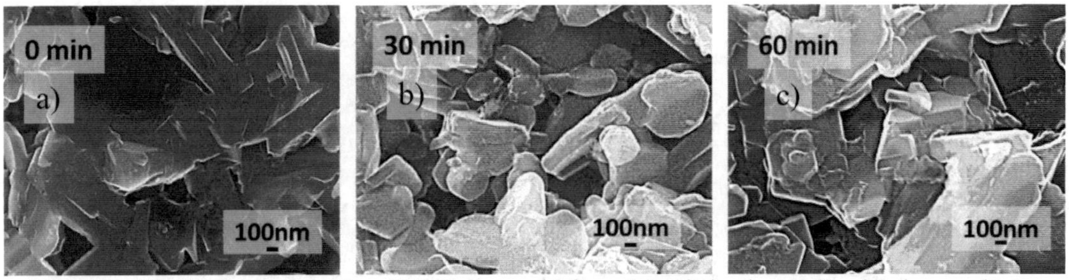

Figure 3. FE-SEM micrographs of MgB_2 prepared from (a) pristine boron (0 min), boron ultra-sonicated for (b) 30 and (c) 60 minutes in hexane medium.

First derivative of susceptibility plotted against temperature in figure 4 (a) shows that the critical onset transition temperature ($T_{c,onset}$) of all specimens are close to 38.5 K with a low transition width (ΔT_c) obtained as the full width of the peak. The ΔT_c value seems to increase with ultra-sonication duration, probably due to increase in unreacted Mg content. The critical current density values at various fields like 0, 0.5 and 1 T at 20 K of MgB_2 specimens prepared from boron ultra-sonicated in hexane for various durations shows (see figure 4 (b)) that 30 minutes was optimum ultra-sonication duration as it results in optimum particle refinement while avoiding clustering of boron particles. Detailed information can be found elsewhere [17]. The dominance of this bulk at extended field range can be attributed to flux pinning from the refined MgB_2 grains. In case of ethanol, the optimum is 15 minutes, however shows better J_c values [15]. To compare the extent of grain refinement and flux pinning from it, we compared the flux pinning force density (F_p) calculated as product of critical current density and field. The maximum pinning force density ($F_{p,max}$) is around 1.2 and 1.4 GN/m^3 in best samples formed from boron ultra-sonicated in hexane and ethanol media as shown in figure 5. This emphasizes that the flux pinning is higher in case of ethanol compared to hexane. It can be due to stronger refinement effect in ethanol medium ($\mu\sim$1.095 cP at 298 K or room temperature) due to its high viscosity compared to hexane ($\mu\sim$0.297 cP at 298 K or room temperature) [27,28]. From stokes law, it is clear that drag force on a particle in a fluid is proportional to the dynamic viscosity of the fluid. Therefore the particles in ethanol during ultra-sonication experience higher drag force compared to that of hexane. This can be explained by the lower average MgB_2 grain sizes in case of ethanol compared hexane. A comprehensive comparison is shown in table 1. Further, one should also take into consideration, the possible pinning from special Mg-B-O phases forming on the pore walls of the bulk MgB_2 matrix, which we have previously observed [29]. To establish a profound relationship we are further working with solvents with different viscosities.

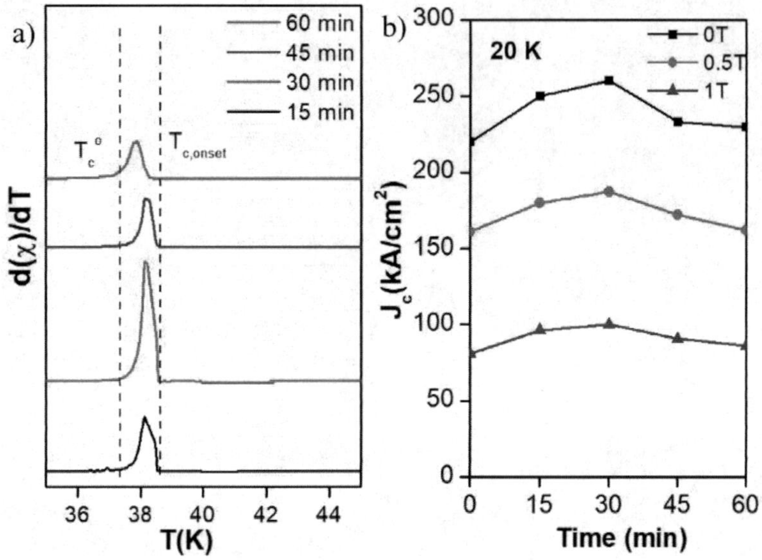

Figure 4. (a) First derivative of susceptibility Vs Temperature and (b) critical current density at various fields of MgB_2 bulks prepared from boron ultra-sonicated in hexane.

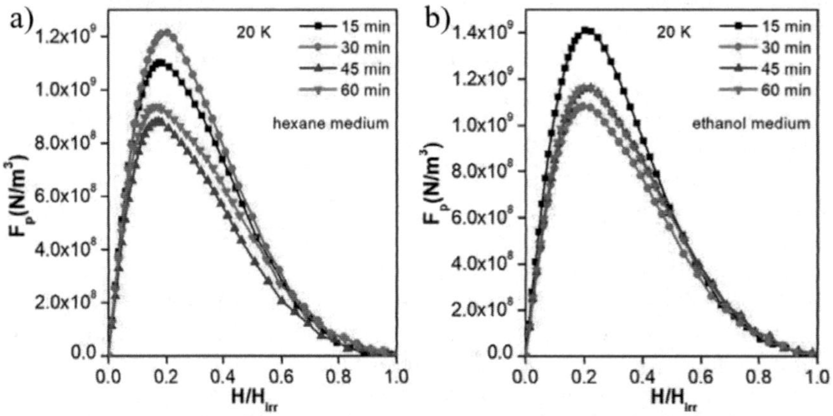

Figure 5. Flux pinning force density (F_p) Vs reduced magnetic field of MgB_2 bulks prepared from boron ultra-sonicated in (a) hexane and (b) ethanol medium.

Table 1: Superconducting properties of MgB_2 prepared from boron ultra-sonicated in different viscous media.

Ultra-sonication Medium	Viscosity(room temp)	J_c at 20K/ 0T (kA/cm^2)	$F_{p,max}$ (GN/m^3)
Ethanol (15 min)	1.095	300 [15]	1.4
Hexane (30 min)	0.297	260 [17]	1.2

4. Conclusion:

Boron was ultra-sonically refined using hexane medium, without any oxide impurity formation. FE-SEM micrographs revealed clustering or entanglement of irregular shaped boron particles in long ultrasonic treated specimens. This clusters led to increased MgB_2 grain size as well as unreacted Mg in the final microstructure, leading to low increase in J_c. 30 minutes of ultra-sonication in hexane can result in optimum J_c by avoiding formation of boron clusters. Flux pinning force density value was highest (1.2 GN/m^3) in the same sample, indicating the strong grain boundary pinning. This value is inferior compared to MgB_2 made from ethanol ultra-sonicated boron which is around 1.4 GN/m^3. High viscosity of ethanol might have led to strong drag force and higher refinement thereby enhancing flux pinning.

Acknowledgements:

This work was partly supported by Shibaura Institute of Technology (SIT) International Research Center for Green Electronics and Grant-in-Aid FD research budget code: 721MA56383. One of the authors (SSA) acknowledges support from SIT for his post-doctoral fellowship. MJ acknowledges support from the program Strategy AV 21-VP3 "Energy storage in flywheels."

References

[1] Nagamatsu J, Nakagawa N, Muranaka T, Zenitani Y and Akimitsu J 2001 Superconductivity at 39K in magnesium diboride *Nature* **410** 63–4

[2] Schlachter S I, Goldacker W, Frank A, Ringsdorf B and Orschulko H 2006 Properties of MgB_2 superconductors with regard to space applications *Cryogenics (Guildf).* **46** 201–7

[3] Takahashi Y, Naito T, Nakamura T and Takahashi M 2021 Detection of 1H NMR signal in a trapped magnetic field of a compact tubular MgB_2 superconductor bulk *Supercond. Sci. Technol.* **34** 06LT02 (6pp)

[4] Yamamoto A, Ishihara A, Tomita M and Kishio K 2014 Permanent magnet with MgB_2 bulk superconductor *Appl. Phys. Lett.* **105** 2–6

[5] Arvapalli S S, Muralidhar M and Murakami M 2019 High-Performance Bulk MgB_2 Superconductor Using Amorphous Nano-boron *J. Supercond. Nov. Magn.* **32** 1891–5

[6] Muralidhar M, Inoue K, Koblischka M R, Tomita M and Murakami M 2014 Optimization of processing conditions towards high trapped fields in MgB_2 bulks *J. Alloys Compd.* **608** 102–9

[7] Serquis A, Zhu Y T, Peterson E J, Coulter J Y, Peterson D E and Mueller F M 2001 Effect of lattice strain and defects on the superconductivity of MgB_2 *Appl. Phys. Lett.* **79** 4399–401

[8] Masui T, Lee S and Tajima S 2004 Carbon-substitution effect on the electronic properties of MgB_2 single crystals *Phys. Rev. B - Condens. Matter Mater. Phys.* **70** 1–7

[9] Zhu Y, Pogrebnyakov A V., Wilke R H, Chen K, Xi X X, Redwing J M, Zhuang C G, Feng Q R, Gan Z Z, Singh R K, Shen Y, Newman N, Rowell J M, Hunte F, Jaroszynski J, Larbalestier D C, Baily S A, Balakirev F F and Voyles P M 2010 Nanoscale disorder in pure and doped MgB_2 thin films *Supercond. Sci. Technol.* **23**

[10] Yang Y, Sumption M D and Collings E W 2016 Influence of Metal Diboride and Dy_2O_3 Additions on Microstructure and Properties of MgB_2 Fabricated at High Temperatures and under Pressure *Sci. Rep.* **6** 1–11

[11] Jirsa M, Rameš M, Miryala M, Svora P, Duchoň J, Molnárová O, Arvapalli S S and Murakami M 2020 Flux pinning and microstructure of a bulk MgB_2 doped with diverse additives *Supercond. Sci. Technol.* **33** 094007 (8pp)

[12] Muralidhar M, Inoue K, Koblischka M R, Murakami A and Murakami M 2015 Effects of Silver Addition on Critical Current Densities and Mechanical Properties in Bulk MgB_2 *Adv. Eng. Mater.* **17** 831–8

[13] Miryala M, Arvapalli S S, Diko P, Jirsa M and Murakami M 2019 Flux Pinning and

Superconducting Properties of Bulk MgB_2 with MgB_4 Addition *Adv. Eng. Mater.* **1900750** 1–7
[14] Muralidhar M, Higuchi M, Jirsa M, Diko P, Kokal I and Murakami M 2017 Improved critical current densities of bulk MgB_2 using carbon coated amorphous boron MgB_2 *IEEE Trans. Appl. Supercond.* **27** 6201104 (4pp)
[15] Arvapalli S S, Miryala M, Jirsa M and Murakami M 2020 Size reduction of boron particles by high-power ultrasound for optimization of bulk MgB_2 *Supercond. Sci. Technol.* **33** 115009 (11pp)
[16] Pinmangkorn S, Miryala M, Arvapalli S S and Murakami M 2020 Enhancing the superconducting performance of melt grown bulk $YBa_2Cu_3O_y$ via ultrasonically refined Y_2BaCuO_5 without PtO_2 and CeO_2 *Mater. Chem. Phys.* **244** 122721 (7pp)
[17] Arvapalli S S, Miryala M, Sunsanee P, Jirsa M and Murakami M 2021 Superconducting properties of sintered bulk MgB_2 prepared from hexane-mediated high-energy-ultra-sonicated boron *Mater. Sci. Eng. B Solid-State Mater. Adv. Technol.* **265** 115030
[18] Kang M O, Joo J, Jun B H, Park S, Kim C S and Kim C 2019 Effect of boron milling on phase formation and critical current density of MgB_2 bulk superconductors *Prog. Supercond. Cryog.* **21** 18–24
[19] Chua C K, Sofer Z, Khezri B, Webster R D and Pumera M 2016 Ball-milled sulfur-doped graphene materials contain metallic impurities originating from ball-milling apparatus: their influence on the catalytic properties *Phys. Chem. Chem. Phys.* **18** 17875–80
[20] Liu Y C, Shi Q Z, Zhao Q and Ma Z Q 2007 Kinetics analysis for the sintering of bulk MgB_2 superconductor *J. Mater. Sci. Mater. Electron.* **18** 855–61
[21] Bhagurkar A G, Yamamoto A, Dennis A R, Durrell J H, Aljohani T A, Nadendla H B and Cardwell D A 2017 Microstructural evolution in infiltration-growth processed MgB_2 bulk superconductors *J. Am. Ceram. Soc.* **100** 2451–60
[22] Ye S, Matsumoto A, Togano K, Zhang Y, Ohmura T and Kumakura H 2014 Fabrication of MgB_2 superconducting wires with a hybrid method combining internal-Mg-diffusion and powder-in-tube *Supercond. Sci. Technol.* **27** 055017 (6pp)
[23] Bohnenstiehl S D, Susner M A, Yang Y, Collings E W, Sumption M D, Rindfleisch M A and Boone R 2011 Carbon doping of MgB_2 by toluene and malic-acid-in-toluene *Phys. C Supercond. its Appl.* **471** 108–11
[24] Li W X, Zeng R, Wang J L, Li Y and Dou S X 2012 Dependence of magnetoelectric properties on sintering temperature for nano-SiC-doped MgB_2/Fe wires made by combined in situ/ex situ process *J. Appl. Phys.* **111** 3–5
[25] Ghorbani S R, Farshidnia G, Wang X L and Dou S X 2014 Flux pinning mechanism in SiC and nano-C doped MgB_2: Evidence for transformation from δtc to δℓ pinning *Supercond. Sci. Technol.* **27** 125003 (7pp)
[26] Xu X, Kim J H, Hossain M S A, Park J S, Zhao Y, Dou S X, Yeoh W K, Rindfleisch M and Tomsic M 2008 Phase transformation and superconducting properties of MgB2 using ball milled low purity boron *J. Appl. Phys.* **103** 023912
[27] Murov S 2020 Properties of Solvents Used in Organic Chemistry *http://murov.info/orgsolvents.htm*
[28] Database of Pure Component Properties Chemical Engineering and Materials Research Information Center(CHERIC) *https://www.cheric.org/research/kdb/hcprop/cmpsrch.php*
[29] Arvapalli S S, Miryala M, Jirsa M, Sakai N and Murakami M 2021 Pinning behavior in bulk MgB_2 prepared using boron powder refined via high-energy ultra-sonication *J. Supercond. Nov. Magn.* **34** 1297–301

Study on performance improvement of superconductive-assisted machining (SUAM) with superconducting tape

S Iwasaki[1], Y Kinoshita[1], H Ishii[1], E S Otabe[1, 3], T Nakasaki[2], K Suzuki[2]

[1] Department of Computer Science and Electronics, Kyushu Institute of Technology, 680 – 4 Kawazu, Iizuka, Fukuoka 820 – 8502, Japan
[2] Department of Mechanical Information Science and Technology, Kyushu Institute of Technology, 680 – 4 Kawazu, Iizuka, Fukuoka 820 – 8502, Japan
[3] Research Center for Neuromorphic AI Hardware, Kyushu Institute of Technology, 2 – 4 Hibikino, Wakamatsu-ku, Kitakyushu, Fukuoka, 808 – 0196, Japan

otabe@phys.kyutech.ac.jp

Abstract. In this study, we calculated the repulsive force of magnetic levitation for superconductive-assisted machining (SUAM) using the finite element method. Conventionally, SUAM utilizes bulk superconductors; here, we propose a magnetic levitation tool using superconducting tape. To obtain a stronger repulsive force, we fabricated various SUAM models and performed calculations. The results showed that we could obtain a strong repulsive force by arranging six superconducting tape pieces radially at equal intervals from the center of the SUAM tool, shortening the distance between the layers, and making the tape trapezoidal in shape. By increasing the number of layers, we can expect to achieve a greater repulsive force than that obtained with bulk superconductors.

1. Introduction

With the progress in industrial product development, the parts used in these products are becoming more complicated. Hollow processing technology is used to process such complex parts. There are two types of hollow processing technologies, one that processes a part before molding and one that processes after molding. When hollow processing is carried out after molding, it can be difficult to internally process an object with a complicated shape. Therefore, superconductive-assisted machining (SUAM) has been proposed in which a permanent magnet is levitated on bulk superconductors to process an object by utilizing the magnetic flux-pinning phenomenon of superconductors [1, 2]. The permanent magnet is a single-sided 4-pole magnet and is levitated by using bulk superconductors. The levitation of the magnet occurs because the magnet is held in the air and cooled in a magnetic field, i.e., field-cooled (FC). Then, the permanent magnet rotates when the superconducting part rotates due to the pinning effect. Therefore, a material can be polished from the inside by the surface of the permanent magnet.

While improving the performance of bulk superconductors [3], research on the fabrication of superconducting tape has progressed, and high-performance tapes are currently being developed [4]. Superconducting tape can work in the same way as bulk superconductors, and by stacking superconducting tape pieces having 130 layers each and then magnetizing them, a magnetic field of 7.9 T could be obtained at 4.2 K [5]. Tape is easy to cut and combine. To design a SUAM tool using superconducting tape, it is necessary to evaluate the parameters affecting the force required to levitate the permanent magnet. In our previous work [6], it was shown that SUAM with superconducting tape shows comparable performance to SUAM with bulk superconductors. For real applications, it is necessary to improve the levitation performance of SUAM with superconducting tape.

In this study, we perform electromagnetic field analysis using a simple model consisting of only permanent magnet and superconducting tape using the finite element method (FEM), and search for a method to obtain stronger repulsive force with various shapes of superconducting tape pieces and configurations of SUAM. Finally, a comparison of the results for bulk superconductors and superconducting tape is given.

2. Calculation Methods

In this study, the finite element method (FEM) was used to analyze the electromagnetic field of the superconductive-assisted machining (SUAM) tool. For this purpose, JMAG manufactured by JSOL was utilized for the finite element method calculation. For the magnetic field dependence of the critical current density and *E-J* characteristics used in the calculation, the experimental results of $GdBa_2Cu_3O_{7-\delta}$ superconducting tape [7] were used. The critical current of the superconducting tape is 1.7×10^{10} A/m2 at 77.3 K and 0 T. The critical temperature of the tape is 93 K. The size of the $GdBa_2Cu_3O_{7-\delta}$ superconducting tape pieces was 20 mm long, 12 mm wide, and 1 μm thick for the superconducting layer. Six pieces were employed in each model, and models with two different configurations were fabricated. The arrangement of the superconducting tape pieces and permanent magnet is shown in figure 1. The model with six tape pieces placed azimuthally at equal intervals from the center of the SUAM is designated Type A, and the model with six tape pieces placed radially is designated Type R. In addition, we created up to four layers similar to the upper layer at intervals of 50 μm, and analyzed repulsive force for Type A and Type R. In order to investigate the change in repulsive force as a function of the distance between the upper and lower layers, four layers similar to the upper layer were created using the Type A model at intervals of 50 μm, 100 μm, 300 μm, and 500 μm. In addition, the size of the superconducting tape in Type A was changed to a trapezoidal shape with an upper base of 20 mm and a lower base of 34 mm to try to reduce the gap between the tape pieces. The arrangement using the trapezoid-shaped superconducting tape pieces is shown in figure 2. The 4-pole permanent magnet was placed so that the center of the permanent magnet coincided with the center of the tape pieces. We used a permanent magnet with an inner diameter of 20 mm, outer diameter of 59 mm, thickness of 10 mm, and a magnetic field of 450 mT at the magnet surface. These values are the same as the 4-pole permanent magnet used in reference [1].

The superconducting tape pieces were magnetized with field-cooled magnetization (FC) at the magnetizing position of 10 mm between the upper part of the tape and the 4-pole permanent magnet. The cooling temperature is 77.3 K of liquid nitrogen. Then, the repulsive force on the permanent magnet

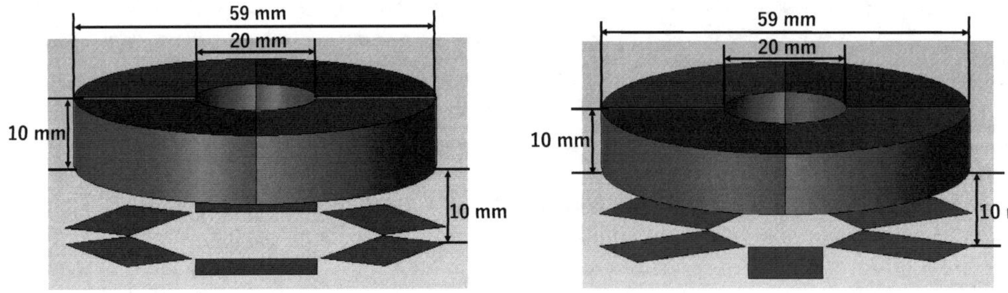

(a) Type Azimuthal (Type A) (b) Type Radial (Type R)

Figure 1. Superconductive-assisted machining (SUAM) with different superconducting tape configurations. The model with six pieces placed azimuthally at equal intervals from the center of the SUAM is designated Type A, and the model with six sheets placed radially is designated Type R.

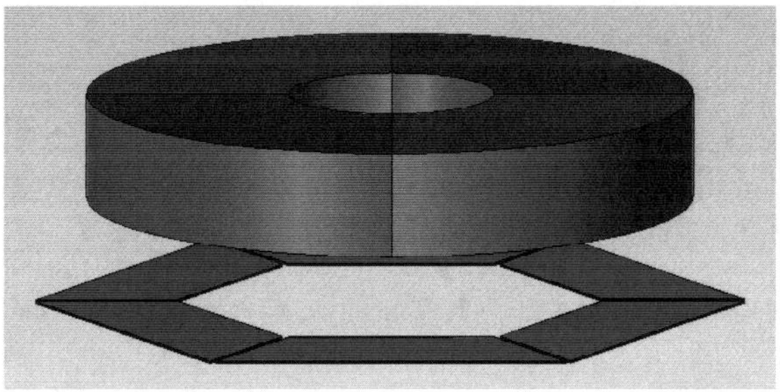

Figure 2. Superconductive-assisted machining (SUAM) with trapezoid-shaped superconducting tape. The superconducting tape pieces in Type A were changed to a trapezoidal shape with an upper base of 20 mm and a lower base of 34 mm.

generated by the superconducting tape was calculated when the permanent magnet was moved perpendicularly to the superconducting tape pieces at a speed of 0.1 mm/s. The distance between the permanent magnet and the superconducting tape pieces was decreased from 10 mm to 1 mm. In this paper, we only focus on the repulsive force, since the attractive force and torque are almost proportional to the results for the repulsive force, and thus it is clear that if the repulsive force is improved, the attractive force and torque are also improved.

3. Results and discussion

Figure 3 shows the FEM calculation results for the two models when the number of layers of superconducting tape was increased from 1 to 4. The difference between Type A and Type R was also calculated and is shown in figure 3. As can be seen in figure 3, both repulsive forces are inversely proportional to the distance and increase with the number of layers. It was confirmed that the force is zero at a distance of 10 mm, which is the magnetizing position, and that the repulsive force emerges when the permanent magnet is brought closer to the tape from the magnetizing position, and grows larger with shorter distance. Furthermore, it can be seen that Type A has a larger repulsive force than Type R when the distance is closer than 5 mm with two or more layers. This is because Type A can obtain a larger repulsive force in the tape pieces at the boundary between the north and south poles of the permanent magnet. In addition, the area of shielding current inside the tape pieces for Type A is larger than that for Type R, since the tape pieces are located at the edge of the permanent magnet. Therefore, if the superconductive-assisted machining (SUAM) tool is configured according to Type A and the number of layers is increased, the repulsive force will increase. Thus, if we increase the number of layers, we can expect a larger repulsive force than that obtained with bulk superconductors.

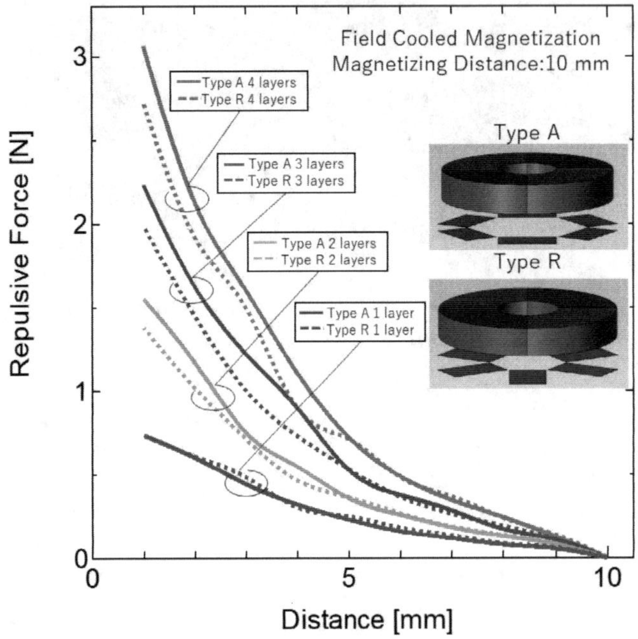

Figure 3. Repulsive force when magnetizing distance is 1 – 10 mm for Type A and Type R. Magnetizing position is 10 mm with field-cooled magnetization.

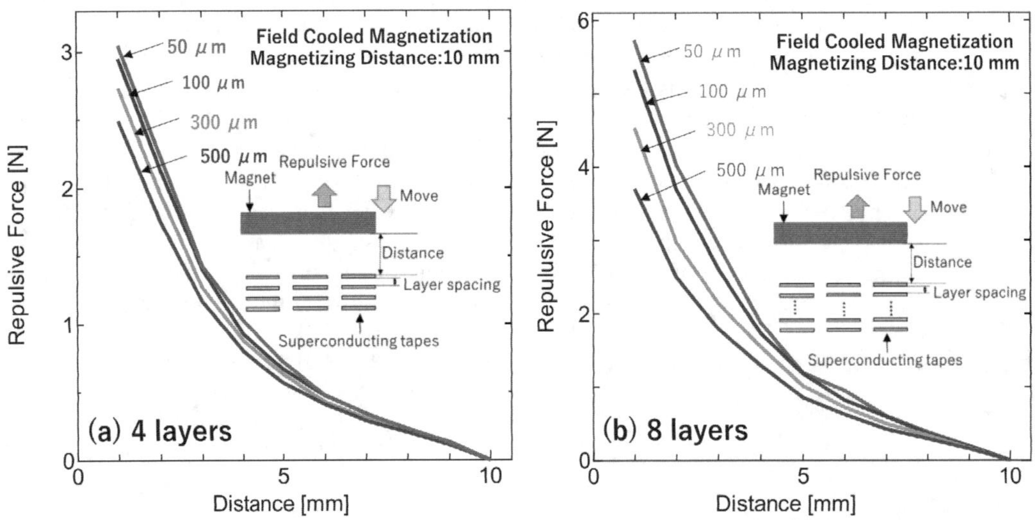

(a) (b)

Figure 4. Repulsive force when magnetizing distance is 1 – 10 mm for various distances between layers, (a) 4 layers and (b) 8 layers of superconducting tape.

Figure 4(a) shows the results of FEM calculations for four layers of superconducting tape with different layer spacings. Figure 4(b) shows the results of FEM calculations for eight layers of superconducting tape with different layer spacings. From figure 4(a), the repulsive force increases by 10% when the distance between layers is 300 μm compared to 500 μm, and by 23% when the distance between layers is 50 μm. Furthermore, figure 4(b) shows that the repulsive force increases by 40% at

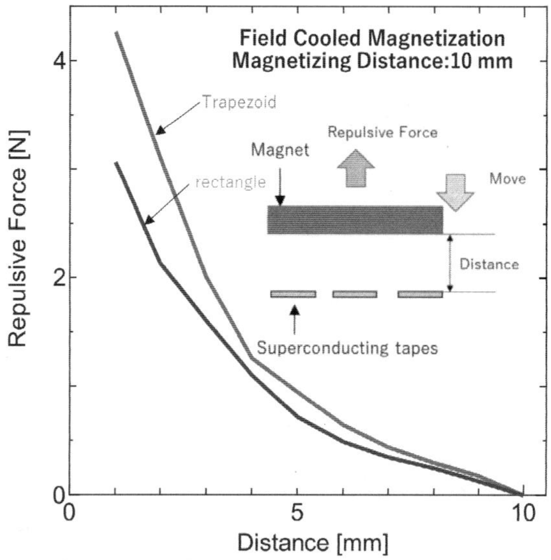

Figure 5. Repulsive force when magnetizing distance is 1 – 10 mm for Type A and model of Type A superconducting tape with the size changed to a trapezoidal shape with 20 mm upper side and 34 mm lower side.

50 μm compared to 500 μm. Therefore, it can be seen that the repulsive force is improved more when the distance between the layers is shortened and the number of layers is 8 than when the number of layers is 4. This is because as the number of layers increases, the total pinning force becomes greater. From these results, it can be concluded that the shorter the distance between layers, the greater the repulsive force.

The results of FEM calculations are shown in figure 5 for Type A and Type A with the tape shape changed to a trapezoid with an upper base of 20 mm and a lower base of 34 mm. From figure 5, the repulsive force increases by 40% when the shape of the superconducting tape is changed to a trapezoid. This large increase in repulsive force is due to the fact that the number of tape pieces between the N and S poles of the 4-pole magnet is greater than before. Then, a strong repulsive force can be obtained with the tape between the N and S poles of the 4-pole magnet. Therefore, when six superconducting tape pieces are arranged radially, a strong repulsive force can be obtained by changing the tape to have a trapezoidal shape.

In our previous work [6], it was shown that the performance of SUAM in terms of the repulsive force can be improved by increasing the number of superconducting tape pieces in the layer. Therefore, combining our previous work with the present results, we found that if the shape is tailored and the layers are properly arranged, repulsive force greater than that generated by bulk superconductors can be obtained easily and cheaply. The reason why stacked superconducting tapes perform better than bulk superconductors is that superconducting tape has a high critical current density and is thin compared with bulk superconductors.

4. Conclusion

In this study, in order to efficiently obtain a large repulsive force, various models for superconductive-assisted machining (SUAM) were fabricated and analyzed. As a result, it was found that it is possible to obtain a larger repulsive force by fabricating a SUAM tool using the Type A configuration shown in figure 1, shortening the distance between superconducting layers, and making the tape trapezoidal in shape. In addition, by increasing the number of layers, we can expect to achieve a larger repulsive force than that obtained in the bulk. In future work, it is necessary to develop a technology to reduce the thickness of the support layer in the superconducting tape, such as Hastelloy, in order to shorten the distance between layers.

Acknowledgements

The present study was supported in part by JSPS KAKENHI Grant Number 19H00771.

References

[1] Nakashima H, Nakasaki T, Tanaka T, Kinoshita Y, Khajornrungruang P, Otabe E S and Suzuki K 2021 *Int. J. of Automation Technology* **15** 234 – 242
[2] Hiramatsu Y, Takahashi Y, Otabe E S, Suzuki K, Tanaka Y and Kiuchi M 2017 *J. Phys.: Conf. Ser.* **871** 012048
[3] Murakami M 2007 *Int. J. Appl. Ceram. Technol.* **4** 225 – 241
[4] Lao M, Willa R, Meledin A, Rijckaert H, Chepikov V, Lee S, Petrykin V, Van Driessche I, Molodyk A, Holzapfel B 2019 *Supercond. Sci. Technol.* **32** 094003
[5] Tamegai T, Hirai T, Sun Y and Pyon S 2016 *Physica C* **530** 20 – 23
[6] Kinoshita Y, Zhang R, Otabe E S, Suzuki K, Tanaka Y, Nakashima H, Nakasaki T 2020 *J. Phys.: Conf. Ser.* **1590** 012023
[7] Inoue M, Tanaka K, Imamura K, Higashikawa K, Awaji S, Watanabe K, Taneda T, Yoshizumi M, Izumi T and Kisu T 2015 *Physics Procedia* **67** 903 – 907

PySQIF, a Statistical Analysis Tool for Bi-SQUID Magnetometers

A Akgun[1] and S Razmkhah[1]

[1] Electrical and Electronics Engineering Department, TOBB University of Economics and Technology, Ankara, Turkey.

Email: ali.akgunphys@gmail.com

Abstract. Superconductor loops with Josephson junctions such as SQUIDs and Bi-SQUID have unique properties that can be used as quantum accurate magnetometers, amplifiers, and many more applications. However, due to quantum interference between junctions and their nonlinear nature, their interaction with magnetic fields is hard to analyze. In this work, we introduce an open-source library titled PySQIF that determines the voltage response of the SQUIDs and Bi-SQUIDs in the presence of the external magnetic flux and can determine the design parameter margins by a statistical approach. This library was used to design and fabricate a Bi-SQUID magnetometer.

1. Introduction

Nowadays, Superconductor Quantum Interference Devices (SQUIDs) are extensively used in many applications such as sensitive magnetometers, magnetic microscopy, amplifiers for cryogenic readout electronics [1], nondestructive tests [2], and biomagnetism [3]. DC SQUID consists of two parallel Josephson Junctions in the superconductive loop. Figure 1 (a), shows the structure of a simple DC SQUID. DC SQUID's voltage response versus the external applied magnetic field is non-linear, which makes the readout circuit that much more complicated and expensive. Bulky and expensive readout electronics limit the application capabilities, and will affect the intrinsically high dynamic range and speed of the sensor. Therefore, researchers investigate SQUID-like circuits, which are more linear than conventional SQUIDs such as Bi-SQUID, arrays of SQUIDs, Superconducting Quantum Interference Filters (SQIFs), and so on [4–6]. Bi-SQUIDs are one of the alternative solutions. Bi-SQUID is designed by adding a parallel Josephson junction to a typical DC SQUID as demonstrated in Figure 1 (b). The addition of the new junction in the loop for Bi-SQUID sensors makes the voltage versus flux response of the circuit much linear compared to DC SQUID [7].

Bi-SQUID response to applied external magnetic flux is characterized by a set of differential equations. Unfortunately, there is no analytic solution to this set of equations [8]. Therefore, numerical analysis plays a critical role in these types of systems. Modeling and simulation tools can support design studies by using numerical methods. Today, several superconductor circuit analysis tools exist for SQUIDS/Bi-SQUIDs, such as PSCAN2[9] and WRSpice [10,11]. These are great tools, and they proved their capability in the field. However, their software architecture complicates the development and integration stages. Thus, we developed a flexible statistical analysis tool for Josephson-based circuits, such as SQUIDs and Bi-SQUIDs named "PySQIF". PySQIF library architecture simplifies development stages compared to other superconductivity circuit analysis tools.

PySQIF is an open-source and user-friendly statistical analysis tool [12]. Our Python library calculates the average voltage response of SQUID/ Bi-SQUID for each corresponding normalized applied external magnetic flux as an output. The normalized applied external magnetic flux range can be determined by the user. Moreover, our simulation tool provides multiple runs for the statistical analysis of Bi-SQUID. Users can determine the margin for the critical current of the shunt junction parameter, and the tool generates Gaussian distributed random numbers in a specific margin for the parameter. After that, the voltage response in an external applied magnetic flux is achieved in the defined margin. The output set provides a wide range of design options to the user who can easily observe the reliable working condition of Bi-SQUID circuits. The user can optimize Bi-SQUID design problems and determine the margin and voltage response of the sensor based on the output dataset received from the simulation.

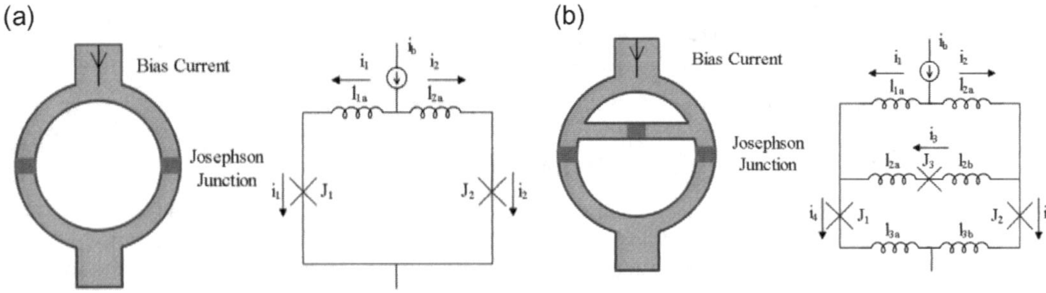

Figure 1 The conventional representation of DC SQUID, with its schematic is in part (a) and the Bi-SQUID representation and schematic are seen in part (b)

2. Theoretical Model and Implementation Methodology

2.1. Theoretical Model

Figure 1 (b) shows Bi-SQUID circuit representation with applied external magnetic field and normalized bias current ($i_b = I_b / I_c$). We assumed that the circuit is symmetric, which means the two Josephson junctions have the same critical current ($I_c = I_{c1} = I_{c2}$). The normalized currents that flow through Josephson junctions are determined by using the RCSJ model and Kirchhoff laws[8,13].

$$
\begin{aligned}
&i_b = i_1 + i_2\,, && i_2 = i_3 + i_5\,, && i_4 = i_1 + i_3 \\
&i_4 = \dot{\varphi}_1 + \sin(\varphi_1(\tau))\,, && i_5 = \dot{\varphi}_2 + \sin(\varphi_2(\tau))\,, && i_3 = \dot{\varphi}_3 + i_{c3}\sin(\varphi_3(\tau))
\end{aligned}
\tag{1}
$$

Here, i_{c3} is the normalized critical current equals to I_{c3} / I_c, and dot states the differentiation with normalized time step. φ_i represents phase difference across i. The other parameters that will be used are represented in Table 1.

Table 1 Definition of the parameters and their equations.

Magnetic flux quantum	$\Phi_0 = \hbar(2e)^{-1}$	**Characteristic frequency**	$\omega_c = 2\pi V_c(\Phi_0)^{-1}$
Normalized time step	$\tau = \omega_c t$	**Characteristic voltage**	$V_c = I_c R_n$

Applying magnetic flux quantization to the Bi-SQUID circuit results in a phase conservation equation [8]:

$$\varphi_1 + i_1 l_{1a} + i_4 l_{2a} = 2\pi\varphi_{ext}\beta + \varphi_2 + i_2 l_{1b} + i_5 l_{2b} \tag{2}$$

Here, $l_i = 2\pi(\Phi_0)^{-1} L_i I_c$ represents normalized inductances, and β controls the size of the upper loop. Re-writing equation (2) gives us to Bi-SQUID systems of differential equations [8]:

$$(l_1 + l_{2a})\dot{\varphi}_1 - l_{2b}\dot{\varphi}_2 - l_1\dot{\varphi}_3 = l_{1b}i_b + \varphi_2 - \varphi_1 + 2\pi\varphi_{ext}\beta + l_1 i_{c3}\sin\varphi_3 + l_{2b}\sin\varphi_2 - (l_1 + l_{2a})\sin\varphi_1 \quad (3a)$$

$$l_{2a}\dot{\varphi}_1 - (l_1 + l_{2b})\dot{\varphi}_2 - l_1\dot{\varphi}_3 = -l_{1a}i_b + \varphi_2 - \varphi_1 + 2\pi\varphi_{ext}\beta + l_1 i_{c3}\sin\varphi_3 - l_{2a}\sin\varphi_1 + (l_1 + l_{2b})\sin\varphi_2 \quad (3b)$$

$$l_{2a}\dot{\varphi}_1 - l_{2b}\dot{\varphi}_2 - (l_{3a} + l_{3b})\dot{\varphi}_3 = \varphi_2 - \varphi_3 + \varphi_1 - (l_{3a} + l_{3b})i_{c3}\sin\varphi_3 - l_{2a}\sin\varphi_1 + l_{2b}\sin\varphi_2 \quad (3c)$$

Where, $l_1 = (l_{1a} + l_{1b})$, equations (3a) to (3c) represent a set of differential equations for Bi-SQUID. These equations can be represented with matrix representation. Equations (4a) and (4b) show a numerically solvable row-echelon form of the first two rows of the matrix [14]:

$$\dot{\varphi}_1 = \left(1 + \frac{l_{12d}}{dl_{12s}}\right)\left(\frac{i_b}{2}\right) - \frac{1}{dl_{12s}\beta}(\varphi_1 - \varphi_2 - 2\pi\varphi_{ext}\beta) + \frac{1}{d}\left(1 + \frac{l_{23s}}{dl_{12s}}\right)i_{c3}\sin(\varphi_2 - \varphi_1) - \frac{1}{2}\left(1 + \frac{1}{d}\right)\sin(\varphi_1) - \frac{1}{2}\left(1 - \frac{1}{d}\right)\sin(\varphi_2) \quad (4a)$$

$$\dot{\varphi}_2 = \left(1 + \frac{l_{12d}}{dl_{12s}}\right)\left(\frac{i_b}{2}\right) + \frac{1}{dl_{12s}\beta}(\varphi_1 - \varphi_2 - 2\pi\varphi_{ext}\beta) - \frac{1}{d}\left(1 + \frac{l_{23s}}{dl_{12s}}\right)i_{c3}\sin(\varphi_2 - \varphi_1) - \frac{1}{2}\left(1 - \frac{1}{d}\right)\sin(\varphi_1) - \frac{1}{2}\left(1 + \frac{1}{d}\right)\sin(\varphi_2) \quad (4b)$$

$l_{12s} = l_{1a} + l_{1b} + l_{2a} + l_{2b}$, $l_{12d} = l_{1b} - l_{1a} + l_{2b} - l_{2a}$, $l_{23s} = l_{2a} + l_{2b} + l_{3a} + l_{3b}$, $d = 3 + 2l_{23s}(l_{12s})^{-1}$.

Normalized average voltage across the Bi-SQUID, can be represented as $\frac{(\dot{\varphi}_1 + \dot{\varphi}_2)}{2}$ [8,13,14]. The numerical solutions of equation (4a) and (4b) gives us the voltage response of Bi-SQUID for the applied external magnetic flux.

2.2. PySQIF Implementation Methodology

We chose Python as a programming language for this project. PySQIF is an object-oriented Python library[12]. Python has rich scientific libraries (NumPy, SciPy, Pandas…, etc.) and a large community. These advantages made the development stage of PySQIF easier.

PySQIF consists of several classes, and each of them has separate tasks. This object-oriented library architecture simplifies the lifetime development stages of the project and provides a user-friendly structure to the users. Figure 2, shows an association diagram of the PySQIF library.

Definition of the PySQIF classes:

- InputData class forms input-protocol with a user, namely InputData class reads simulation parameters from input CSV (Comma Separated Values) file.
- InductanceConstants takes input and then calculates normalized inductance constants of a set of differential equations (4a).
- TimeStep class generates normalized time steps for solving a set of differential equations (4).
- PhiExtArray generates normalized applied external magnetic flux in the defined range.
- Solver class solves Bi-SQUID systems of differential equations (4).
- Voltage class calculates normalized voltage response of Bi-SQUID for each defined normalized external magnetic flux.
- YieldIc3 class calculates voltage response for defined ic3 margin.
- Lastly, PySQIF class provides an input-output communication protocol between other library classes.

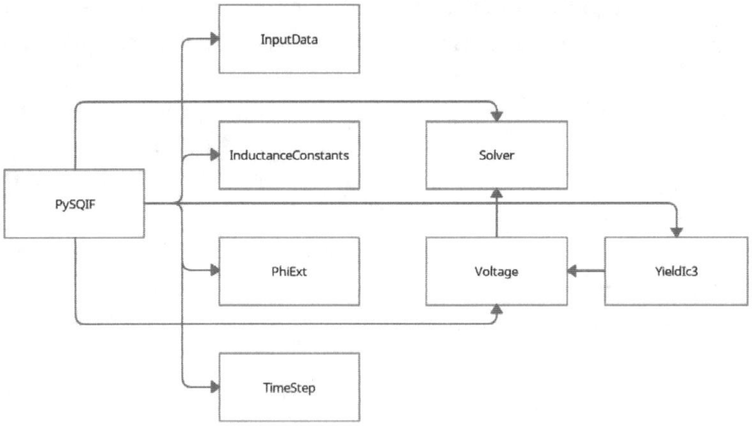

Figure 2 PySQIF Object-Oriented Organization.

2.3. *PySQIF Algorithm*

The algorithm of PySQIF consists of three steps. First, PySQIF generates a Bi-SQUID circuit by using a user-defined input-text file. Second, PySQIF generates proper time steps and normalized external magnetic flux arrays. And finally, PySQIF calculates the normalized average voltage response of Bi-SQUID for each defined applied external magnetic flux. Flow-chart of PySQIF algorithm shown in Figure 3.

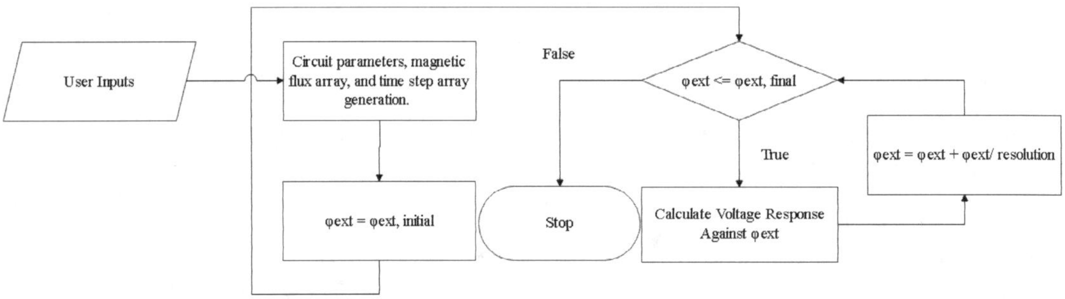

Figure 3 PySQIF flowchart.

3. PySQIF Features

PySQIF is an open-source project with a well-documented user manual available on its GitHub page [12]. We optimized the program structure for providing a good user experience. The users don't have to be good at any programming language to be able to use this library. Moreover, our object-oriented library structure makes integration and lifetime development phases easier. The object-oriented structure of PySQIF, makes integration with other simulation tools easier and developers can use its classes in other projects. However, the working environment should have all the dependencies for working with our library; these dependencies are SciPy, NumPy, Matplotlib, and Pandas [15–18].

PySQIF has a CSV input text file format for simulation inputs which consists of the circuit and simulation-based parameters. (Figure 4 (a) and Figure 4 (b)) Users can freely edit these parameters according to their objectives and it makes it more convenient as a plugin library rather than standalone code. Users can define both Bi-SQUID and SQUID circuits by changing normalized inductance parameters, i_{c3} (normalized critical current of the shunted junction), and i_b (normalized bias current) from an input CSV file. As a library, PySQIF can work for the general cases viz, users can define their circuits at both symmetric and asymmetric variations. Furthermore, our simulation tool provides multiple runs for the statistical analysis of Bi-SQUID. Output set of multiple runs, provide a wide range of design options to the user who can easily observe a reliable working range of Bi-SQUID circuits.

4. Results

We designed a Bi-SQUID circuit by using our tool. The input values for normalized inductance parameters are, $l_{1a} = l_{2a} = 0.05$ and $l_{2a} = l_{2b} = l_{3a} = l_{3b} = 0$. Other circuits parameters are $i_b = I_b / I_c = 2$, and $\beta = 1$. We used our library in multiple run mode to reach the voltage response of our circuit against the defined i_{c3} margin. We determined the mean of the $i_{c3} = I_{c3} / I_c$ distribution as 3 and the standard deviation as 0.5. When we run this simulation, our tool generates Gaussian distribution for i_{c3} (Figure 4 (c)). Subsequently, our tool calculates the voltage response of the defined circuit for each i_{c3} parameter in the defined normalized external magnetic flux range. Our circuit model, input file, critical current of shunted junction distribution (i_{c3}), and voltage response are shown in Figure 4 (d). Figure 5 demonstrates the voltage versus flux response for two extremum values, as the i_{c3} value of the circuit increases, the linearity of voltage response decreases. When i_{c3} approaches zero, the circuit will behave like a SQUID, so this result seems meaningful.

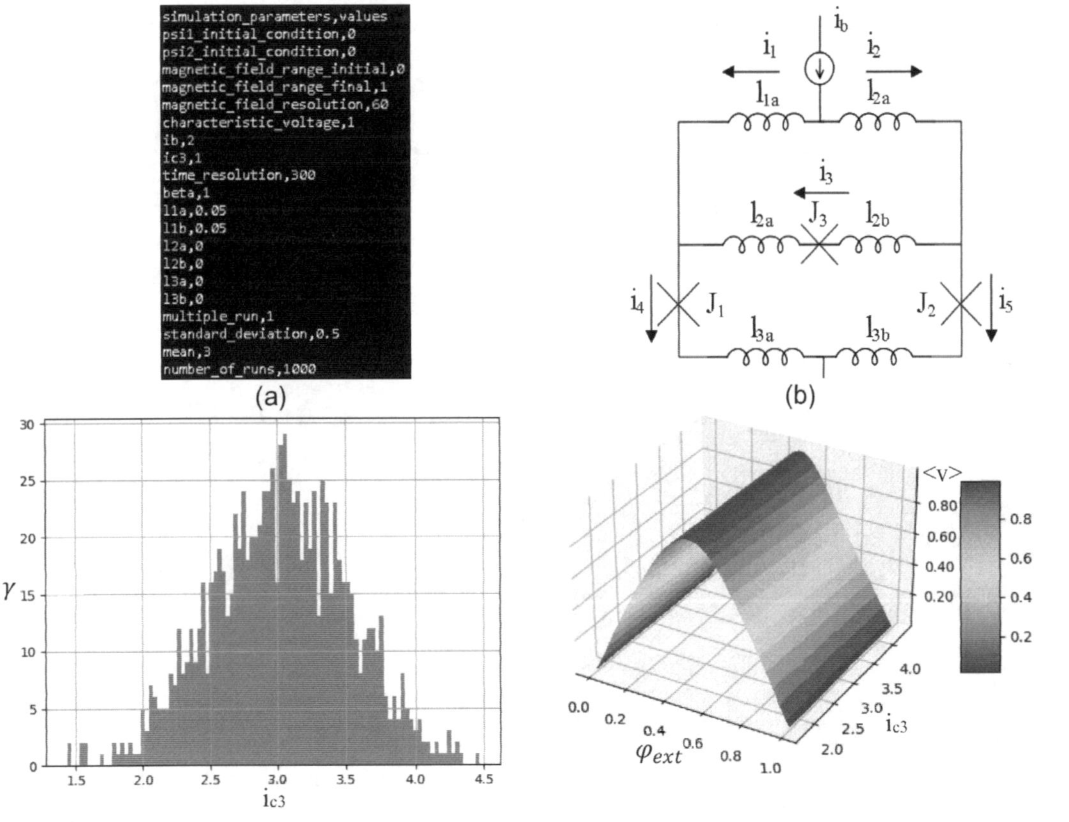

Figure 4 (a) İnput file for multiple runs, (b) Circuit model, (c) ic3 distribution, (d) Voltage response surface plot.

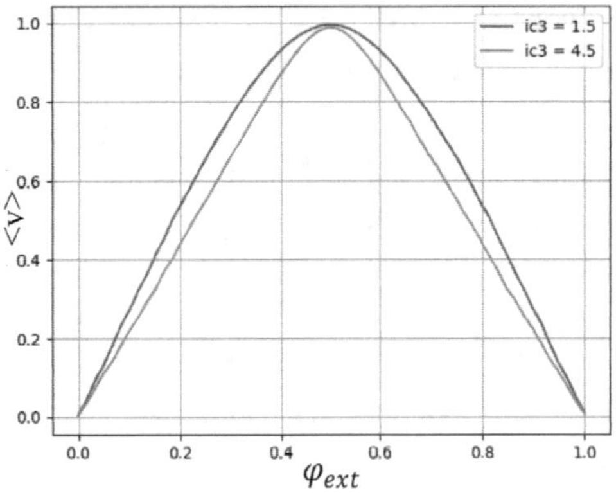

Figure 5 Voltage Response for minimum (1.5) and maximum (4.5) i_{c3} value.

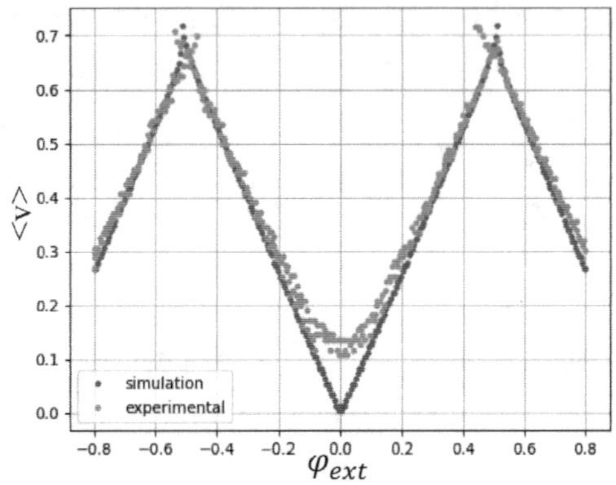

Figure 6 Comparing simulation result from PySQIF with experimental data measured by Kornev et al [7].

To confirm the simulation results, we compare them with the measurement values in Figure 6. The measurements for the Bi-SQUID were published by Kornev et al [7]. Using the same parameters for our simulator, we see a great agreement between both data. However, when the external field is small, the results start to drift apart. This is due to the thermal fluctuation noise seen in experimental results. In future works, we will update the junctions to include thermal noise.

Using the values we get from the simulations and margin analysis of the PySQIF, we have designed a Bi-SQUID magnetometer based on the four-layer HSTP Niobium fabrication [19]. There is a ±5% minimum parameter margin calculated with a statistical analysis tool. The circuit is designed in Klayout [20] opensource layout editor and then the inductances and couplings were extracted with Inductex [21] tool. The layout is demonstrated in Figure 7.

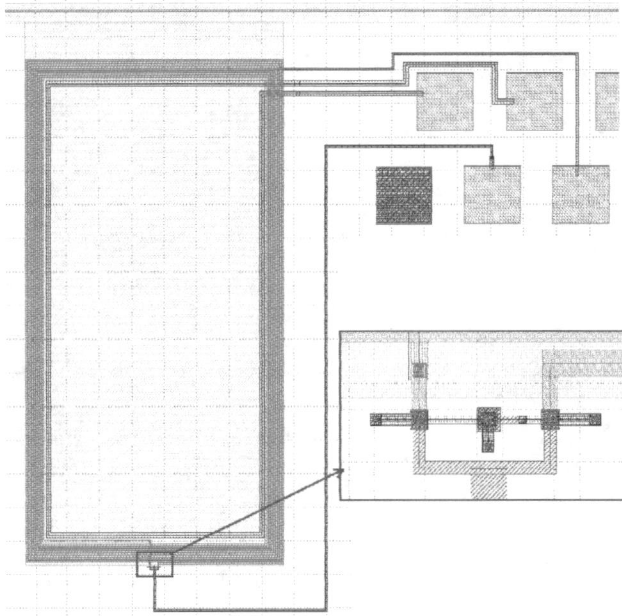

Figure 7 Bi-SQUID layout designed with PySQIF open source Python library.

The layout is fabricated but hasn't been tested yet. However, we have confirmed the results of the PySQIF simulations with experimental results of the literature [6,22] as seen in Figure 6.

5. Conclusion and future works

PySQIF is an open-source python library capable of design and analysis of the SQUIDs and Bi-Squid's netlists. PySQIF outputs support design studies for generalized situations, hence researchers are not concerned with minor details. Later, we will improve our library for Bi-SQUID arrays modeling and simulation so PySQIF will be expanded to be able to solve systems of parallel or series arrays of SQUID and Bi-SQUID. PySQIF results are in good agreement with the literature. Since experimental results include a noisy environment Josephson junction noise will be added to the PySQIF library in the future. Today, users can use PySQIF by using programming editors or in the terminal across multiple platforms. In the future, we will develop desktop, mobile, and web-based GUI for PySQIF.

6. Acknowledgement

The authors want to thank Dr. Pascal Febvre from Université Savoie Mont Blanc, IMEP-LAHC, CNRS France for his valuable guidance toward this work.

References

[1] Razmkhah S, Bozbey A and Febvre P 2021 A compact high frequency voltage amplifier for superconductor–semiconductor logic interface *Supercond. Sci. Technol.* **34** 045013

[2] Sarreshtedari F, Razmkhah S, Hosseini N, Schubert J, Banzet M and Fardmanesh M 2011 An efficient SQUID NDE defect detection approach by using an adaptive finite-element modeling *J Supercond Nov Magn* **24** 1077–81

[3] Fagaly R L 2006 Superconducting quantum interference device instruments and applications *Review of Scientific Instruments* **77** 101101

[4] Kornev V K, Soloviev I I, Klenov N V and Mukhanov O A 2007 Development of SQIF-based output broad band amplifier *IEEE transactions on Applied Superconductivity* **17** 569–72

[5] Mitchell E E, Hannam K E, Lazar J, Leslie K E, Lewis C J, Grancea A, Keenan S T, Lam S K H and Foley C P 2016 2D SQIF arrays using 20 000 YBCO high R n Josephson junctions *Supercond. Sci. Technol.* **29** 06LT01

[6] Kornev V K, Kolotinskiy N V, Bazulin D E and Mukhanov O A 2018 High-linearity bi-SQUID: Design Map *IEEE Transactions on Applied Superconductivity* **28** 1–5

[7] Kornev V K, Soloviev I I, Klenov N V and Mukhanov O A 2009 Bi-SQUID: a novel linearization method for dc SQUID voltage response *Supercond. Sci. Technol.* **22** 114011

[8] Longhini P, Berggren S, de Escobar A L, Palacios A, Rice S, Taylor B, In V, Mukhanov O A, Prokopenko G, Nisenoff M, Wong E and De Andrade M C 2014 Voltage response of non-uniform arrays of bi-SQUIDs *Int. Conf. on Theory and Application in Nonlinear Dynamics (ICAND 2012)* (*Understanding Complex Systems*) ed V In, A Palacios and P Longhini (Cham: Springer International Publishing) pp 77–90

[9] Polonsky S, Shevchenko P, Kirichenko A, Zinoviev D and Rylyakov A 1997 PSCAN'96: new software for simulation and optimization of complex RSFQ circuits *IEEE Transactions on Applied Superconductivity* **7** 2685–9

[10] Whiteley S R 1991 Josephson junctions in SPICE3 *IEEE Transactions on Magnetics* **27** 2902–5

[11] Anon Whiteley Research Incorporated. Available: http://wrcad.com/

[12] aakgn 2021 *PySQIF.* Available: https://github.com/aakgn/PySQIF

[13] Barone A and Paternò G 1982 *Physics and applications of the Josephson effect* (New York: Wiley)

[14] Longhini P, Berggren S, Palacios A, In V and Leese de Escobar A 2012 Coupled non-uniform bi-squid: A numerical investigation *Advances In Cryogenic Engineering: Transactions of the Cryogenic Engineering Conf. - CEC*, Volume 57 (Spokane, Washington, USA) pp 1167–74

[15] Virtanen P *et al.* 2020 SciPy 1.0: fundamental algorithms for scientific computing in Python *Nat Methods* **17** pp 261–72

[16] Harris C R *et al.* 2020 Array programming with NumPy *Nature* **585** 7825 pp 357–62

[17] Hunter J D 2007 Matplotlib: A 2D graphics environment *Comput. Sci. Eng.* **9** pp. 90–5

[18] McKinney W *et al.* 2010 Data structures for statistical computing in python *9th Python in Science Conference* **445** pp 51–6

[19] Nagasawa S and Hidaka M 2013 Design instruction for AIST standard process(AIST-STP2) National Institute of Advanced Industrial Science and Technology (AIST)

[20] *klayout*. KLayout EDA Tool, 2021. Accessed: Sep. 09, 2021. [Online]. Available: https://github.com/KLayout/klayout

[21] Fourie C J 2015 Full-gate verification of superconducting integrated circuit layouts with InductEx *IEEE Transactions on Applied Superconductivity* **25** pp 1–9

[22] Kornev V K, Kolotinskiy N V, Bazulin D E and Mukhanov O A 2017 High-inductance Bi-SQUID *IEEE Transactions on Applied Superconductivity* **27** pp 1–4

Guided Wave Transceive Technique Using Magnetostrictive Effect of STPG370 Pipe itself and High-Temperature Superconductor SQUID

K Watanabe[1], K Okada[1], S Masumitsu[1], T Munkhnyam[1], W Sun[1], Y Hatsukade[1], Y Shoyama[2] and A Tomotoshi[3]

[1]Kindai University, 1 Umenobe, Takaya, Higashi-Hiroshima, Hiroshima 739-2116, Japan

[2]Chugai Technos, 9-12 Yokogawashimmachi, Nishi-ku, Hiroshima, Hiroshima 733-0013, Japan

[3]NIHON SEALAKE, 1-6-11 Tomonishi, Asaminami-ku, Hiroshima, Hiroshima 731-3169, Japan

Abstract. We developed a non-contacting ultrasonic guided wave testing technique using magnetostrictive effects and high-temperature superconductor (HTS) SQUID for STPG370 steel pipes. A pair of C-shaped electromagnets was used to magnetize the STPG370 pipe while rotating the pipe, in order to give uniform remnant fields as bias fields. Conditions to give larger remnant field in the pipe were investigated using an electromagnetic field simulation. The simulation results indicated that C-shaped pb permalloy iron cores with an aperture angle of 20 degree and a diameter of 20 mm should be utilized for the magnetization. An excitation coil was wound around one of the magnetized parts, which is used as a transmitter. One cycle sine current of 9 A_{pp} at 50 kHz was repeatedly supplied to the coil to generate T(0,1) mode guided waves in the pipe. The other magnetized part, which was used as a receiver, was combined with the HTS-SQUID gradiometer, which was located above the pipe with a lift-off of 4 mm. In measurements, the pipe was rotated to perform all-round measurements. A contour map of the time-series signal distribution of T(0, 1) mode guided wave was obtained around the pipe.

1. Introduction

For many years, accidents caused by corrosion under insulation (CUI) of steel pipes in social infrastructures have been problems [1]. Non-destructive inspection (NDI) of piping is conducted to prevent the accidents by detecting the CUI. There are various techniques for the NDI of piping. Ultrasonic guided wave testing (UGWT) is one of the techniques for the NDI, which can inspect a large area (several tens meters) at a time. T(0,1) mode guided waves are mainly used in the UGWT because the mode has a constant group velocity and no velocity dispersion [2-4].

One of the methods of transmitting and receiving the guided waves is the magnetostrictive sensor (MsS) technology [5] that uses magnetostrictive effects, which refer to Joule effect and its inverse effect, the Villari effect. This technique is conventionally done by placing a ferromagnetic material with high magnetostriction, such as nickel, in contact with the transmitting and receiving parts of the pipe. However, most of the pipes are wrapped with heat insulating materials. Therefore, it is necessary to remove the heat insulator from the measurement area once for inspection and to reinstall it. This procedure is very costly and time consuming. In order to solve this problem, a technology that can transmit and receive the guided waves from outside the heat insulator is needed. However, it has not been developed yet.

We have noticed that the magnetism and magnetostrictive effects of the steel pipes, which are used in many facilities, can be utilized to transceive the guided waves without removing the heat insulators in place of the nickels. If the magnetostrictive effects of the steel pipes itself can transceive the guided waves, there is a possibility that guided waves can be transceive from outside the heat insulators. Principle of this method is firstly to magnetize the steel pipe from outside the heat insulator to provide remnant fields that serves as bias fields, secondly to generate the guided waves by providing excitation magnetic fields to the magnetized transmitting part, and finally to receive the guided waves using the Villari effect at the magnetized receiving part. At the receiving part, some magnetic devices such as induction coils or magnetic sensors can be used.

In our previous work, two rectangular air-core electromagnets were used sandwiching a Ni pipe with the large magnetostrictive effects, and two parts of the pipe were magnetized by the electromagnets while rotating the pipe or the electromagnets. It was demonstrated that the circumferential remnant magnetizations were given in the Ni pipe and these parts of the pipe could be used to transceiver the T(0,1) mode guided waves without contacting the pipe [6]. In that study, we used an HTS-SQUID gradiometer [7-8] as the receiving device to measure magnetic signals derived from the guided waves. By rotating the pipe under the SQUID to measure the signal distributions around the pipe, it was demonstrated that we could transceive the approximately uniform guided waves in the Ni pipe using this method. It is assumed that the above method can be applied to the steel pipes also. For this purpose, it is necessary to clarify the magnetic property (B-H curve) of the steel pipes, and parameters to magnetize the pipes that give suitable remnant magnetization to occur the magnetostrictive effects on the pipes. It is also presumed that the magnetostrictive effects of the commonly-used steel pipes should be smaller than that of materials such as Ni. Therefore, it is considered that the use of a highly sensitive receiving device, such as the HTS-SQUID gradiometer, is necessary also.

The purpose of this study is to develop a novel non-contacting UWGT method for the steel pipes in the heat insulators using the T(0,1) mode guided waves, which has excellent performance in localization of defects of the pipes under test by using its own magnetostrictive effects. First, we measure the magnetic properties of a STPG370 pipe, which is most commonly used in this field. Next, we clarify how much remnant magnetization can be given by the above-mentioned rotational magnetization method by an electromagnetic field analysis using the measured magnetic property. In this analysis, we introduce the hysteretic B-H curve of the STPG370 pipe in order to optimize geometric parameters such as size and aperture of the C-shaped Permalloy iron cores used for the electromagnets to magnetize the STPG370 pipe. We demonstrate the feasibility of the method to transceive the T(0,1) mode guided waves on the STPG370 pipe using the magnetization parameters derived from the simulation results. The uniformity of the transceived guided waves was investigated by measuring the entire circumference magnetic signal distribution around the STPG370 pipe using the HTS-SQUID gradiometer.

2. Electromagnetic field analysis for magnetization of STPG370 pipe

A sheet with a thickness of 0.5 mm made from the STPG370 material was cut by a laser into small pieces of 30 mm × 60 mm, and the B-H curve measurements were carried out on one of the pieces. The measured B-H curves are shown in figure 1. According to the specifications of the equipment used in the measurements, the maximum magnetic field H that can be applied to the small piece of the STPG370 was about 7.5 kA/m. The saturation field was about 1.7 T and the remnant field was about 1.0 T.

The measured B-H curves of the magnetization characteristics were introduced into the electromagnetic field analysis using JMAG [9]. By changing the parameters such as the shape of the C-shaped electromagnets and the magnetizing current, we investigated to optimize the parameters to obtain sufficient remnant field that causes the magnetostrictive effects. The magnetization characteristics of the STPG370 suggest that it is necessary to provide a stronger magnetic field than 1.0 T. For this reason, the electromagnets with the C-shaped iron cores of pb permalloy with a saturation flux density of about 1.4 T were used to magnetize the STPG370 pipes. The C-shaped cores were used with consideration to reduce the magnetoresistance of the magnetization circuit. The parameters of electromagnets and magnetization method to provide a strong and uniform remnant field in the circumference of the STPG370 pipe were investigated.

In the simulation, a model of the STPG370 pipes have an outer diameter of 34 mm and a thickness of 1 mm. Two iron cores of the C-shaped permalloy electromagnets were modelled. One has an outer diameter of 20 mm and a thickness of 3 mm, and another has an outer diameter of 36 mm and a thickness of 3 mm. The electromagnets were coiled with 100 turns around the iron cores. The respective pair of the C-shaped pb permalloy electromagnets were assumed to be placed at 1 mm from each side of the magnetized part of the STPG370 pipe. We assumed the position above the top of the circumference of the STPG370 pipe as an angle of 0 degree, and the central positions of the apertures of the C-shaped electromagnets were set at 90 degree and -90 degree, as shown in figure 2. The analysis method is a two-dimensional electromagnetic field transient analysis, which allows the introduction of the hysteresis magnetic characteristics of the STPG370 material. For reasons of analysis cost and ease of convergence, the finite elements of the models were divided into triangle meshes with one side of 0.5 mm. The time division was set at 0.05 s. We set a total analysis time of 7.0 s with 141 steps with the time division. In the analysis, a magnetic field was generated by applying a DC current to the electromagnet, which magnetized the STPG370 pipe for 6 seconds. The STPG370 pipe was rotated three times during the magnetization at a constant speed of 0.5 rps. The current for the magnetic field was rapidly reduced to 0 A at 6.0 s ~ 6.05 s, corresponding to that the magnetic field was reduced to 0 T in the duration. The rotation of the pipe was stopped at 6 s and the simulation is continued for 1 s after stopping the rotation. Firstly, to determine the aperture of the C-shaped magnets, the remnant field in the pipe was calculated while varying the aperture from 20 degrees to 180 degrees with a step of 20 degrees. In this simulation, a constant magnetizing current of 10 A was applied to the C-shaped electromagnet with the outer diameter of 20 mm to give a remnant field in the pipe. A contour map of the magnetic field of the pipe model at the last step 141 is shown in Figure 2 in the case of the aperture angle of 20 degrees. The maximum, minimum and average values of the magnetic field distribution are 0.38 T, 0.23 T and 0.28 T, respectively. The relationship between change in the aperture angle and averaged remnant field in the pipe is shown in figure 3. The remnant field has a peak value at aperture angles of 20 and 40 degrees, and tends to decrease with increasing aperture from 40 to 80 degrees, reaching the minimum value of 0.04 T. We also performed the analysis with the distance of 1 mm between the C-shaped electromagnets and the pipe, as shown in figure 4, while changing the positions of the coils on the electromagnets, compared with those in figure 2. The obtained average remnant field in the pipe was 0.24 T in this case. This result showed that it is possible to magnetize the STPG370 pipe using this method without contacting the pipe.

Secondly, the remnant fields with the different two sizes of the C-shaped electromagnet was investigated. The pair of the C-shaped electromagnets with the outer diameter of 20 mm and the aperture of 20 degrees, and the pair of the C-shaped electromagnets with the outer diameter of 36 mm and the aperture of 20 degrees were used. The analysis was carried out in the same way as before, with the magnetization current to the C-shaped magnets varying from 1 to 9 A to obtain the optimum value of the remnant field. The relationships between magnetizing fields generated by the pair of the electromagnets with the different diameters and averaged remnant field in the pipe is shown in figure 5. The horizontal axis of the graph was converted into the magnetizing field generated from the current applied to the electromagnet. For the electromagnets with the outer diameter of 20 mm, maximum averaged remnant field of 0.57 T was obtained at the magnetization field of about 180 mT, which corresponded to the feed current of 4 A. The remnant magnetic field in the pipe had a minimum value of 0.53 T and a maximum value of 0.60 T. This distribution of the remnant field was the best uniform and strong distribution obtained in the simulation. On the other hand, the maximum averaged remnant field of about 0.48 T was obtained with the electromagnets with the outer diameter of 36 mm.

3. Magnetization of STPG370 pipe and UGWT system using HTS-SQUID

3.1. Magnetization of STPG370 pipe

As a sample, STPG370 pipe of 34 mm in outer diameter, 1.0 mm in thickness, and 1000 mm in length were prepared. According to the simulation results, a pair of the C-shaped Permalloy cores with the outer diameter of 20 mm, the thickness of 3 mm and the aperture of 20 angle was prepared. The electromagnets were made by winding enamelled wire with 0.8 mm in a diameter around the cores. The electromagnets were coiled with 74 turns. The transmitting and receiving parts of the guided waves were located at 200 mm from the both ends of the sample pipe, and the Litz wire was wound around the magnetized transmitting part with 21 turns. As shown in figure 6, the transmitter and the receiver were magnetized with the pair of electromagnets, which were located at the distance of 1 mm between the pipe and the cores, as in the analysis, while feeding a DC current to the electromagnets 7 A, which corresponded to the magnetizing field of 121 mT, where the almost largest magnetization was obtained as shown in figure 5. During the magnetization, the pipe was rotated three times manually at roughly 0.5 rps, also as in the analysis.

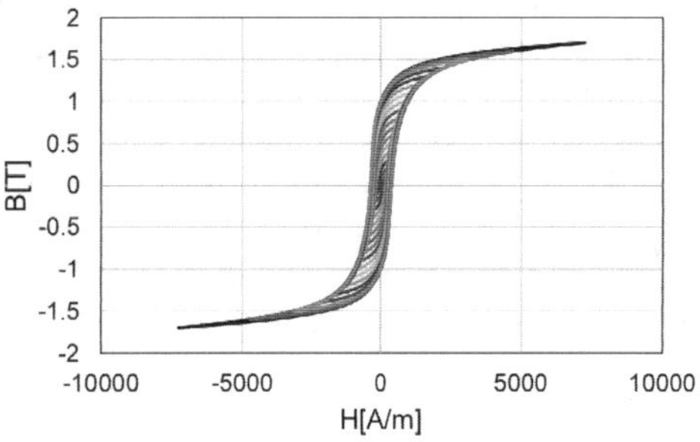

Figure 1. B-H curves of STPG370 material

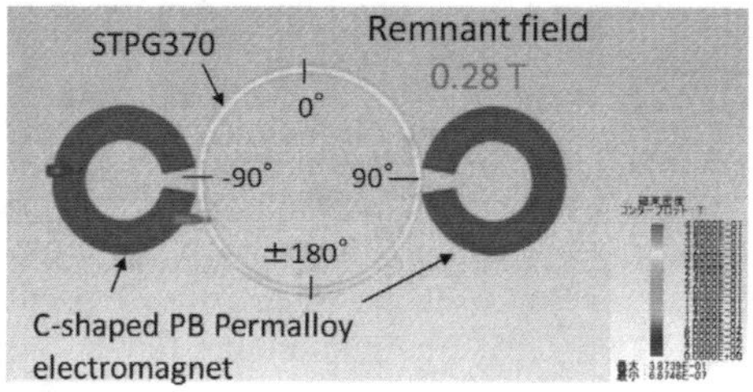

Figure 2. Remnant field inside the pipe after magnetization by electromagnets with an outer diameter of 20 mm and magnetizing current of 10 A.

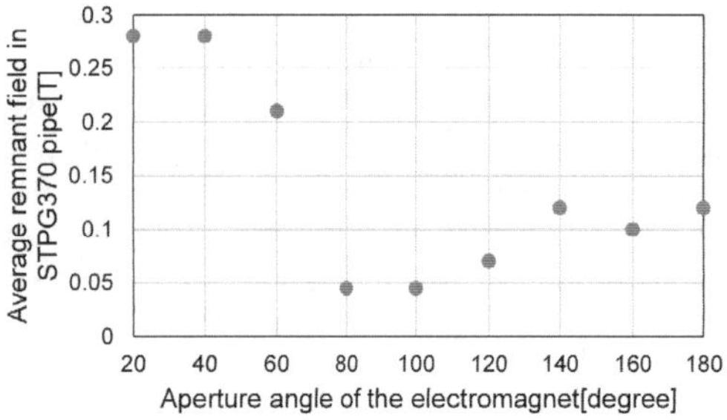

Figure 3. Relationship between aperture angle of a 20-mm-diameter C-shaped electromagnets and average remnant field inside the pipe.

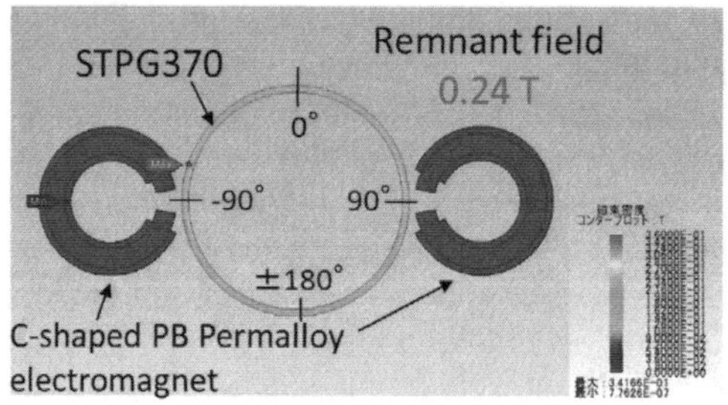

Figure 4. Remnant field inside the pipe with the distance of 1 mm between the STPG370 pipe and the electromagnets.

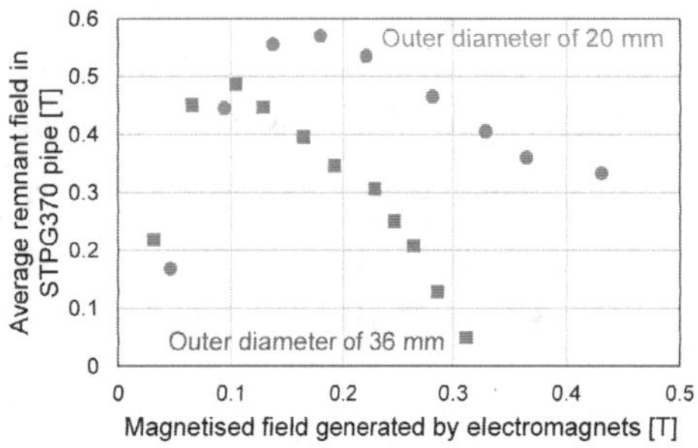

Figure 5. Relationship between magnetization field generated by the electromagnets and average remnant field in STPG370 pipe.

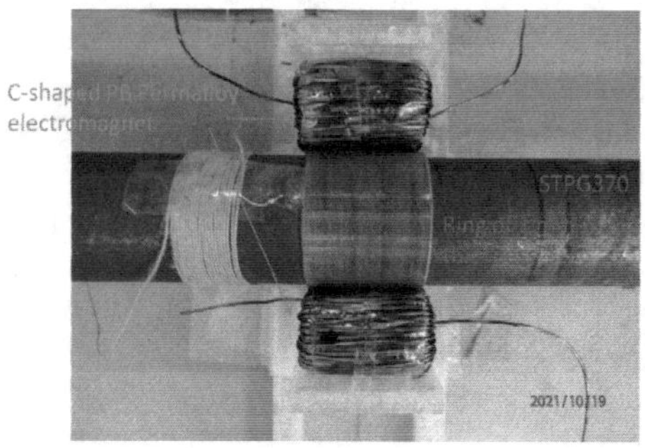

Figure 6. Non-contacting magnetization of the STPG370 pipe using C-shaped electromagnets.

3.2. UGWT system using HTS-SQUID

We constructed the measurement system for the STPG370 pipes based on the magnetostrictive effects using the HTS-SQUID gradiometer. A schematic diagram of the system is depicted in figure 7 (a) with predicted times of T(0,1) mode and L(0,2) mode guided waves to be measured in the figure 7 (b), which was calculated with the group velocity of about 3180 mm/ms and 5000 mm/ms in the STPG370. We notice that non-uniform magnetization would cause the L(0, 2) mode guided waves. The system consisted of the HTS-SQUID gradiometer, a pulse tube refrigerator with a cryostat, a SQUID electronics, function generators, a power amplifier, an oscilloscope, a DAQ module and a PC. The transmitter of the guided wave consisted of the magnetized part on the pipe, an excitation coil, and the current source. One cycle of sine wave voltage at 50 kHz from the function generator was amplified into current by the power amplifier, and the current was repeatedly applied to the coil of the transmitter to generate the pulse excitation magnetic field. In principle, with the uniform circumferential magnetization in the part of the pipe, the T(0, 1) mode guided waves would be generated by the excitation fields. The guided waves would propagate to the both sides of the pipe. The receiver consisted of the other magnetized part with the HTS-SQUID gradiometer, which was cooled at about 73.5 K in the coaxial pulse tube refrigerator with the cryostat. The HTS-SQUID gradiometer was placed above the pipe at 0 degree to detect the magnetic signals, which would be derived from the guided waves at the receiving part. The HTS-SQUID gradiometer with ramp-edge Josephson junctions and a planar first-order differential pickup coil made by former SUSTERA [10] was used. The flux noise level of the device when operated in a normal environment using the SQUID electronics was about 20 $\mu\Phi_0/\mathrm{Hz}^{1/2}$ in a frequency range from 10 to 10 kHz. The HTS-SQUID gradiometer was located at a lift-off of about 4 mm between the surface of the pipe and the SQUID so that the spatial gradient of the vertical x-direction magnetic field (B_x) in the longitudinal z-direction of the pipe, dB_x/dz, would be measured. The guided waves generated by the transmitter were measured by the SQUID gradiometer and the output was acquired using the DAQ module with 510 samples, a sampling rate of 500 kHz and an additive averaging of 256 times, and then, the SQUID outputs were recorded in the PC.

4. All-round inspection of T(0, 1) mode guided waves using HTS-SQUID

4.1. Measurements on defect-free STPG370 pipe

In order to evaluate feasibility of the method and the uniformity of the T(0,1) mode guided waves, time-series distribution of magnetic signal due to the guided waves around the entire circumference of the

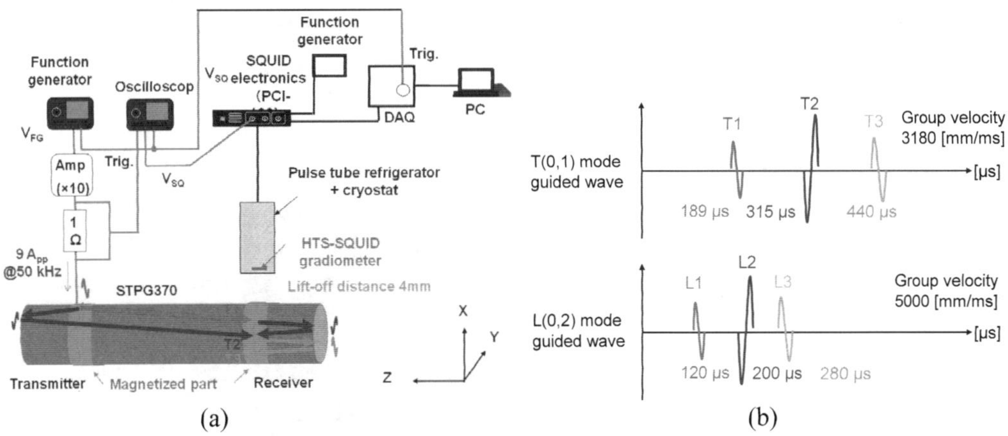

Figure 7. HTS-SQUID-based UGWT system.
(a) UGWT system using HTS-SQUID gradiometer. (b) Predicted times of T(0, 1) mode and L(0, 2) mode guided waves to be measured. It is known that non-uniform magnetization would cause the L(0, 2) mode guided waves.

magnetized pipe was measured using the HTS-SQUID gradiometer. One cycle sine wave current of 9 A_{pp} at 50 kHz was repeatedly applied to the excitation coil of the transmitter to generate the guided waves. Magnetic signals from the guided waves were detected by the gradiometer while rotating the pipe for 360° degrees with steps of 15 degrees.

One of the measured waveforms at the rotation angle of -90 degree is shown in figure 8 (a). There is a large drift of the signal between 0 to 200 μs. So far, we have noticed that if the uniformity of the remnant bias field in the target pipe is not enough, such kind of large drift in the signal just after the generation of the guided wave have been seen. In the drift signal, a signal was detected from about 110 μs. This smaller signal than the later T(0,1) mode guided waves should be the L(0,2) mode guided wave as shown in figure 7 (b), which is probably caused by the non-uniformity of the remnant bias field in the pipe. In spite of these unexpected noise, magnet signals due to the T(0, 1) mode guided waves were clearly detected from about 200 μs and 320 μs respectively, which are approximately agreed with those in figure 7 (b). Time-series distribution of the magnetic signals around the pipe is shown as a colour contour map in figure 8 (b). The horizontal axis shows the measurement time and the vertical axis shows the angle of the pipe. The contour map shows that the T(0,1) mode guide waves such as T1 and T2, due to the incident wave and the reflected waves from the both ends of the pipe, were well detected all around the pipe. The T2 distribution is roughly uniform.

4.2. Measurement on the STPG370 pipe with artificial defect

Next, a semicircle flange was attached on the STPG370 pipe at 800 mm from the left end of the pipe as an artificial defect. The transmitter and the receiver parts were magnetized with the same magnetization method at 200 mm and 500 mm from the left end of the pipe. One cycle sine wave current of 5 A_{pp} at 50 kHz was applied and the lift-off was 5 mm in this experiment. The measurement of the T(0,1) mode guided wave on the pipe was done with the same method as 4.1.

Figures 9 (a) and (b) show the measured waveform and scalogram of the waveform at 60 degree. The latter was obtained by applying a wavelet transformation to the waveform. A signal was detected from about 100 μs in the both graphs. The gravity point of the signal may be around 40 kHz as shown in the scalogram. From the timing and the distance between the transmitter and the receiver, this signal is the T(0, 1) mode incident wave. A signal from about 220 μs is the T(0, 1) mode wave reflected at the left end of the pipe. Signals around 75 μs and 175 μs, which are clearer in figure 9 (b), should be the L(0, 2) mode guided waves, because the gravity points of the signals are higher than those of the T(0, 1) mode waves. A small signal at about 300 μs is the T(0, 1) mode wave reflected at the artificial defect. These results indicate the possibility of this non-contacting method to detect defects on the STPG370 pipes.

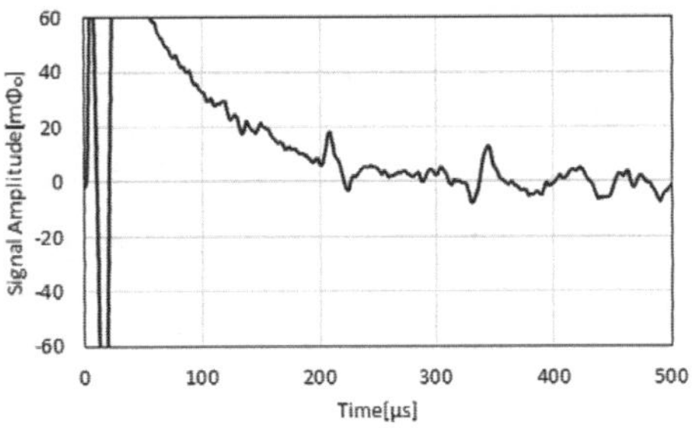

(a)

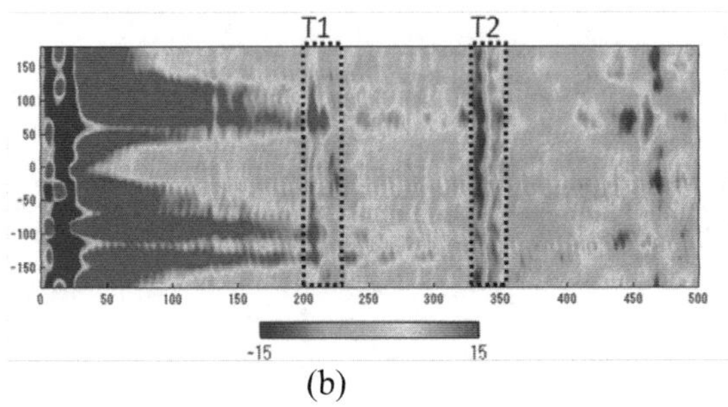

(b)

Figure 8. Measurement results of all-round inspection of T(0, 1) mode guided waves using HTS-SQUID gradiometer. (a) Waveform at -90 degree of the pipe. (b) Contour map of time-series distribution of the magnetic signals around the pipe.

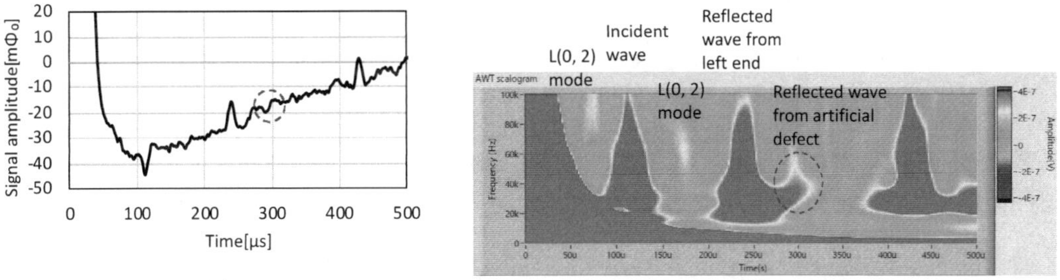

Figure 9. Measurement results of T(0, 1) mode guided waves using HTS-SQUID gradiometer at 60 degree. (a) Time waveform. (b) Scalogram of the waveform obtained by applying the wavelet transformation to the waveform.

5. Conclusion

In this study, we developed the completely non-contacting ultrasonic guided wave testing technique using the magnetostrictive effects and the HTS-SQUID gradiometer for the STPG370 pipes. By the simulation which varied the magnetization parameters of the core geometry and the magnetization current, we have clarified the parameters that give a maximum remnant field of 0.57 T inside the STPG370 pipe with the electromagnets with C-shaped permalloy cores of the outer diameter of 20 mm and the aperture angle of 20 degrees. In the experiments, the pipe was magnetized by the electromagnets, which yielded the magnetizing field of 121 mT. One cycle sine current of 9 A_{pp} at 50 kHz was applied to the excitation coil. While rotating the pipe to perform the all-round measurement, the time-series signal distribution with the T(0, 1) mode guided wave signals were measured around the entire circumference of the pipe at the approximate times. Furthermore, the possibility of defect detection was shown by measuring the reflected wave from the artificial defect on the pipe.

References

[1] Winnik S 2016 *Corrosion Under Insulation (CUI) Guidelines* (Cambridge: Woodhead Publishing)
[2] Nagai T et al. 2002 *Journal of JSNDI* **51** 622 (in Japanese)
[3] Rose J L 1999 *Ultrasonic Waves in Solid Media* (Cambridge: Cambridge University Press)
[4] Rose J L 2014 *Ultrasonic Guided Waves in Solid Media* (New York: Cambridge University Press)
[5] Kwun H and Bartels K A 1998 *Ultrasonic* **36** 171
[6] Hatsukade Y, Watanabe K, Masumitsu S and Munkhnyam T 2021 *IEEE Trans. Appl. Supercond.* **31** 1600805.
[7] Hatsukade Y, Kobayashi T, Nakaie S, Masutani N and Tanaka Y 2017 *IEEE Trans. Appl. Supercond.* **27** 1600104.
[8] Hatsukade Y, Mastani N, Teranishi S, Masamoto K, Kanenaga S, Adachi S and Tanabe K 2017 *J. Phys.: Conf. Ser.* **871** 012072.
[9] https://www.jmag-international.com/
[10] Adachi S, Tsukamoto A, Oshikubo Y, Hato T, Ishimaru Y and Tanabe K 2011 *IEEE Trans. Appl. Supercond.* **21** 12022963.

Narrow-area Bragg-edge transmission of iron samples using superconducting neutron sensor

The Dang Vu[1,2], Hiroaki Shishido[3,4], Kazuya Aizawa[2], Takayuki Oku[2], Kenichi Oikawa[2], Masahide Harada[2], Kenji M. Kojima[4,5], Shigeyuki Miyajima[6], Kazuhiko Soyama[2], Tomio Koyama[4], Mutsuo Hidaka[7], Soh Y. Suzuki[8], Manobu M. Tanaka[9], Alex Malins[10], Masahiko Machida[10] and Takekazu Ishida[1,4]

[1]Division of Quantum and Radiation Engineering, Osaka Prefecture University, Sakai, Osaka 599-8570, Japan
[2]Materials and Life Science Division, J-PARC Center, Japan Atomic Energy Agency, Tokai, Ibaraki 319-1195, Japan
[3]Department of Physics and Electronics, Graduate School of Engineering, Osaka Prefecture University, Sakai, Osaka 599-8531, Japan
[4]NanoSquare Research Institute, Osaka Prefecture University, Sakai, Osaka 599-8570, Japan
[5]Centre for Molecular and Materials Science, TRIUMF, 4004 Wesbrook Mall, Vancouver, BC V6T 2A3, Canada
[6]Advanced ICT Research Institute, National Institute of Information and Communications Technology, 588-2 Iwaoka, Nishi-ku, Kobe, Hyogo 651-2492, Japan
[7]Advanced Industrial Science and Technology, Tsukuba, Ibaraki 305-8568, Japan
[8]Computing Research Center, Applied Research Laboratory, High Energy Accelerator Research Organization (KEK), Tsukuba, Ibaraki 305-0801, Japan
[9]Institute of Particle and Nuclear Studies, High Energy Accelerator Research Organization (KEK), Tsukuba, Ibaraki 305-0801, Japan
[10]Center for Computational Science & e-Systems, Japan Atomic Energy Agency, 178-4-4 Wakashiba, Kashiwa, Chiba 277-0871, Japan.

E-mail: vuthedang@omu.ac.jp

Abstract. This study investigates a current-biased kinetic inductance detector (CB-KID) performance by investigating Bragg-edge spectra from the restricted-area neuron transmission of materials. Iron samples with a size of $5 \times 5 \times 2$ mm^3 were used as typical test materials. The ergodic theorem was used to obtain a visible transmission spectrum so that a long-time averaging of a transmission spectrum can alternatively be evaluated using a space average of independently selected area spectra with the same ensemble size. The most visible edges were observed with a limited area sample of 0.43 mm^2 using a minimum time bin of 25 μs in a time-of-flight (ToF) spectrum or a wavelength resolution of 0.0007 nm of each neutron pulse at beamline BL10 of the Japan Proton Accelerator Research Complex (J-PARC) center. The main Bragg edge of iron as a sum of random 100 ensembles (with an ensemble size of 3.1×2.3 μm^2) thus obtained has a distinctive signal-to-noise ratio and can be fitted well with the Rietveld Imaging of Transmission Spectra (RITS) program with Miller indices. We consider that our CB-KID system is, in principle, able to analyze the Bragg edge of samples as small as 3.1×2.3 μm^2.

1. Introduction

Since neutrons directly interact with nuclei instead of the electron cloud of constituent elements, they have a different characteristic compared with X-ray and other probes [1]. As a matter of fact, substances become visible with neutron beams even though it is almost impossible to see them using X-ray and gamma-ray imaging techniques. A neutron imaging technique provides a useful means of investigating test samples. Moreover, in an energy-dispersive analysis, various material properties are investigated with the aid of the wave nature of neutron beams. Intense pulsed neutrons are powerful in efficiently obtaining wavelength-sensitive properties of substances together with the time-of-flight (ToF) method. Therefore, neutron beams have been recognized as one of the most powerful tools for conducting the nondestructive inspection of a wide range of materials [2]. A neutron detector has three important parameters in the application of neutrons in material science, i.e., spatial resolution, neutron detection efficiency, and energy resolution. We first proposed the idea of a superconducting neutron sensor called a current-biased kinetic inductance detector (CB-KID) [3] and recently reported systematic investigations on its characteristics to optimize the operating conditions [4,5]. We can improve the detection efficiency of the detector by increasing the operating temperature of the superconducting detector to a critical temperature [6]. A delay-line technique of the four-terminal CB-KID successfully achieved neutron imaging with a spatial resolution of 16 μm [5]. We observed neutron transmission images of practical Gd islands with various sizes and thicknesses with CB-KID and demonstrated that the neutron transmission images of various Gd islands were in good agreement with the scanning electron microscope (SEM) images [7]. Since thermal and cold neutrons behave like waves, they inform the wavelength-sensitive interference properties of materials through the Bragg diffraction, i.e., the Bragg edges in the neutron transmission spectra can be observed by coherent scattering in crystalline materials. This enables us to conduct nondestructive studies of phase, texture, and residual stress in the test materials [8,9]. Table 1 presents a comparison of the specifications of three different neutron detectors, namely, CB-KID, microchannel plate (MCP) detector [10], and micropixel chamber-based neutron imaging detector (μNID) [11]. Scintillation-type neutron detectors are widely used, and good review articles are available [12,13].

Table 1. Comparison of three different neutron detectors, i.e., current-biased kinetic inductance detector (CB-KID), microchannel plates (MCPs), and micropixel chamber-based neutron imaging detector (μNID) in several specifications, i.e., operating principle, temporal resolution, sensing area, efficiency for thermal neutrons and spatial resolution. Cited values are not the best but representative (see references).

	CB-KID [5,6,7]	**MCPs** [10]	**μNID** [11]
Operating principle	Superconducting meanderlines	Microchannel plates	Micropatterned electrode
Temporal resolution	1 ns	1 μs	0.25 μs
Sensing area	15x15 mm^2	14x14 mm^2	100x100 mm^2
Efficiency (at 25.3 meV)	1%	78%	26%
Spatial resolution	16 μm	55 μm	0.1mm

This work investigates the Bragg edges in the neutron transmission spectra from restricted areas of the test samples to show the energy-selective wave nature of the materials with our superconducting sensor. For this study, we used an iron test sample because it provides visible Bragg edges with assigned Miller indices as a wavelength function. We analyzed neutron images of the test sample using the ToF method with a time bin of 25 μs (or a wavelength resolution of 0.0007 nm) of each neutron pulse. This study aimed to answer how narrow we can select a testing area for a sample to recover the Bragg edges using our high-resolution superconducting neutron sensor. We systematically investigated the size of the testing area from a few millimeters square to a restricted area of 3.1×2.3 μm^2. Lack of statistics in neutron transmission intensity is crucial in the investigation of a restricted area as it is a trade-off relation between the arriving neutron intensity and the analyzing area size under the condition of fixed neutron flux intensity. Thus, we attempted to realize transmission spectra with good statistics to replace a long-time average spectrum with an ensemble-averaged spectrum based on the ergodic theorem [14]. We performed an ensemble averaging of randomly selected test areas with identical sizes (average pixel size of 3.1×2.3 μm^2) to obtain a visible transmission spectrum. We discuss the potential capabilities of the CB-KID sensor in treating fine structural features of various materials when the neutron intensity increases drastically to be available for researchers in the future.

2. Experimental details

2.1. Superconducting Neutron Sensor

One of the best superconducting neutron sensors was used for this experiment. The characteristics of the sensor were systematically investigated in a previous study [4,6]. The structure of a superconducting neutron sensor consists of a pair of meanderlines (X and Y sensors) [15–17] with an orthogonal stacking and a ^{10}B layer deposited on top of the X and Y meanderlines to convert neutrons into charged particles to make it easier to sense an electroneutral particle. The sensitive area of the sensor used was 15 × 15 mm· The nanowire width of the X (or Y) Nb meanderline was 900 nm, and the space between two neighboring nanowire segments was 600 nm. These values gave a 1.5-μm repetition period with 10,000 repetitions to reach an effective width of 15 mm and a total length ℓ of 151 m of the meanderline. These components were stacked in a thermally oxidized Si substrate with eight layers from the bottom: (1) 300-nm-thick SiO_2, (2) 300-nm-thick Nb ground plane, (3) 350-nm-thick SiO_2 layer, (4) 50-nm-thick Nb Y meanderline, (5) 150-nm-thick SiO_2 layer, (6) 50-nm-thick Nb X meanderline, (7) 50-nm-thick SiO_2 layer, and (8) 70-nm ^{10}B neutron conversion layer on top of the CB-KID. The superconducting neutron sensor used in this study was designed using a computer-aided design software (LayoutEditor) to fabricate the device at the clean room for analog-digital superconductivity (CRAVITY) at the National Institute of Advanced Industrial Science and Technology (AIST).

2.2. Measurement System

Figure 1 presents the schematic of the measurement system used with CB-KID. To refrigerate the superconducting neutron sensor down to an operating temperature of 7.9 K, the neutron sensor and test samples were installed on an antivibration cold stage of a Gifford–McMahon (GM) refrigerator. Long-time measurements of CB-KID were performed at a stable cryogenic temperature without feeding liquid helium. The temperature of the superconducting sensor was controlled and monitored using two temperature controllers (Cryocon Model 44) with electric resistance heaters and Cernox thermometers, installed around the neutron sensor and the cold stage of the GM refrigerator with a LabVIEW program. The X and Y meanderlines were fed with direct current (DC) bias currents in series with two DC adjustable voltage sources and bias resistors. Four voltage signals from a single neutron event on the meanderlines were amplified using low-noise amplifiers (NF SA-430F5), and they were divided by four splitters to input both a 32-channel time-to-digital converter (TDC) module for the Kalliope-DC readout circuit [18] and a four-channel 2.5-GHz sampling digital oscilloscope (Teledyne LeCroy HDO4104-MS). The former has a time resolution of 1 ns, whereas the latter has 0.4 ns. The alternating current (AC) power supplied to the above instruments was provided by two stabilized AC power sources (NF EC1000SA) to avoid notable environmental noises provided through the AC power line of the large-scale facility. The amplitudes of neutron signals were tuned by adjusting the amplitudes of the bias currents to the X and Y meanderlines while monitoring them by the high-speed digital oscilloscope under irradiation of neutron beams to the superconducting neutron sensors at a cryogenic operating temperature. Subsequently, the signal amplitudes of the four channels were adjusted by setting the 16-bit variable attenuators (ATT, Hoshin Electronics N032) placed in front of the TDC inputs of the Kalliope-DC readout circuit. Semirigid cables (50 Ω), microminiature coaxial (MMCX) connectors, and subminiature version A (SMA) connectors were used to transmit a high-frequency signal pulse from the cryogenic temperature CB-KID to the room temperature readout instruments inside the cryostat to reduce noises. Neutron irradiation was performed for 141 h under a proton power of 502 kW at the J-PARC to obtain the transmission images and transmission spectra of the test samples.

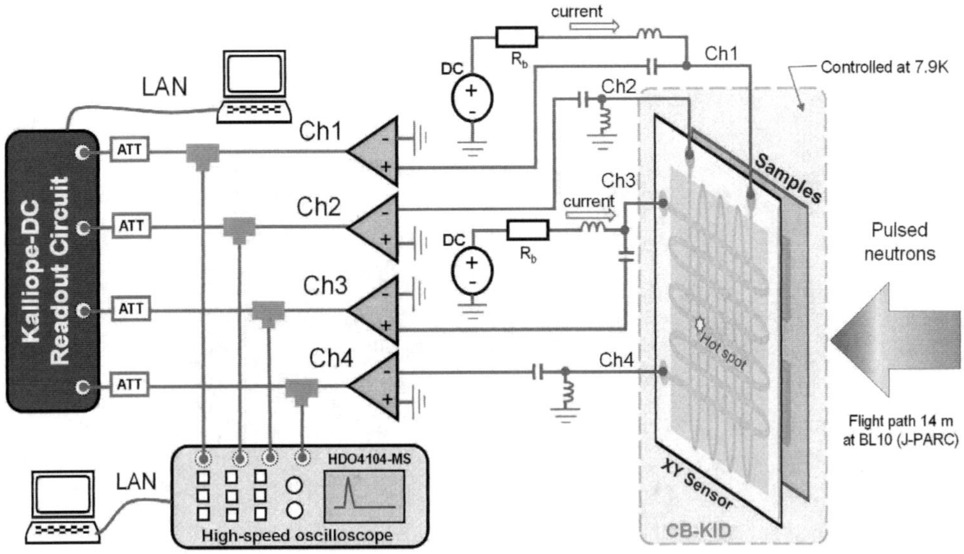

Figure 1. Schematic diagram of the superconducting neutron imaging system. Pulsed neutrons were incident on the substrate side of CB-KID after passing through the test samples through the 14-m long beam-line (BL10) of Material and Life Science Experimental Facility (MLF) at J-PARC. The detector and samples were cooled down to a cryogenic temperature at 7.9 K using a Gifford–McMahon cryocooler. The neutron detector consists of the X and Y-meanderlines and a ^{10}B neutron conversion layer. Neutron signals arising from both ends of the two meanderlines were amplified by ultra-low noise amplifiers to feed into a Kalliope-DC readout circuit via adjustable attenuators (ATT) and an oscilloscope. The system was controlled by two computers for the Kalliope readout circuit and the oscilloscope connected with local area network (LAN) cables.

2.3. Test Samples

Four test samples (5 × 5 mm, 2 mm in thickness) of iron-containing 0.13 wt.% of C were prepared, containing 0.21 wt.% of Si, 1.44 wt.% of Mn, 0.56 wt.% of Mo, and 0.02 wt.% of Nb as impurities. Although the impurities contained were not for this study but for other purposes, such as probing nuclear resonance absorption dips in neutron transmission spectra at higher neutron energies, which have little effect in observing the Bragg edges from iron samples. These samples were fixed with a tiny drop of epoxy adhesive on a 0.1-mm-thick Al plate. Figure 2(a) presents an optical photo of these samples (B1, B2, B3, and B4) on the Al plate, where black lines and a red dot were drawn using a marker ink to discriminate the sample position. The samples on the Al plate were then placed on an Al alloy mount with a milling hole of 13 × 13 mm with a 3.7-mm depth. Finally, the Al plate was located at a 0.7-mm distance from the CB-KID meanderlines. The iron samples were cooled down to the CB-KID temperature to place them at a 0.8-mm distance from the CB-KID sensor. A neutron transmission experiment was conducted at beamline BL10 in the J-PARC, where the pulsed neutrons were used at a 25-Hz repetition employing the ToF method at the beamline and length of 14 m. Neutron transmission imaging with high spatial resolution and wavelength (λ)-sensitive transmission spectroscopy, such as a Bragg-edge transmission with a wavelength resolution of $\Delta\lambda/\lambda = 0.33\%$, were conducted [19].

3. Results and discussion

This work intends to answer how narrow we can select a testing area with a visible Bragg edge using a high-spatial-resolution neutron sensor. The characteristics of iron in a neutron transmission have been thoroughly established in both experiments and simulations [2,20], where multiple Bragg edges appear in a wide range of wavelengths rather visibly shorter than 0.4 nm. Therefore, an iron sample was considered to be a good candidate as a test material for examining the CB-KID performance for revealing the Bragg-edge spectra manifested from a transmission spectrum by selecting a restricted area of materials.

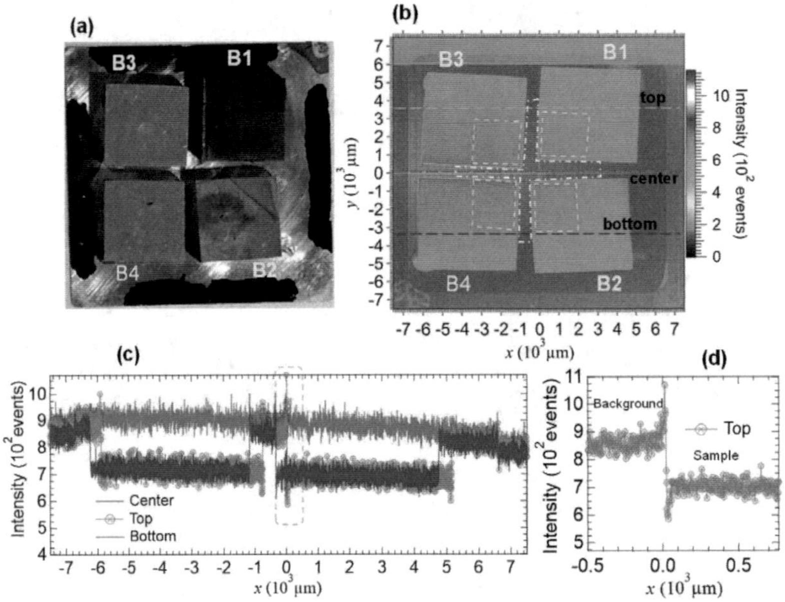

Figure 2. **(a)** Optical photo of four iron test samples (B1, B2, B3, B4) used for the present study. Samples were fixed on the 0.1-mm thick Al plate mounted on an Al-alloy sample mount. **(b)** Neutron transmission image of iron test samples (B1, B2, B3, B4). **(c)** Horizontal line profiles correspond to the top (in green), center (in red) and bottom (in violet) positions in neutron image; **(d)** A zoomed line profile of neutron transmission near the left-hand-side edge of test sample B2. This explains an enhanced contrast of the images at the sample edge.

Figure 2(b) presents the neutron transmission images, where four test samples (B1, B2, B3, and B4) clearly appeared at the corresponding positions of samples in figure 2(a). Furthermore, it is evident that the appearance of epoxy resin was used to fix the iron sample at a corner under the sample (at a sample corner close to the corner of the Al plate). It can be noted that resin adhesives were not so visible in the photo (see figure 2(a)) because they were located under the 2-mm-thick iron samples. On the contrary, the black lines at the four sides and red dots at the corner of figure 2(a) did not appear in figure 2(b). However, the 13 × 13-mm milling hole of the Al alloy mount can be seen in figure 2(b). This is due to a difference in the Al alloy thickness present and the diffraction/reflection edge effect (see below). The appearance of an image contrast above $y > 6$ mm indicates the existence of defect(s) in the superconducting sensor. This indication is a characteristic feature of neutron beams in probing samples compared with optical observations. After careful inspection of the neutron image in figure 2(b), it can be seen that the contrast in neutron intensity at the borders of the samples is somewhat exaggerated in neutron transmission images compared with other parts of the iron samples. To highlight the intensity contrast, we took the horizontal line profiles (see figure 2(c)) at the three different positions: the upper side (along a yellow dashed line in figure 2(b)), the central area (along a red dashed line in figure 2(b)), and the bottom side (along a blue dashed line in figure 2(b)). Negative peaking was present in the neutron intensity at the sample side and positive peaking at the empty side (background-area side). These phenomena are explained by the refraction/reflection and scattering of neutron beams at the sample edges when neutron beams arrive at the test samples in a way that is highly parallel to their surfaces. However, when neutron beams propagate from vacuum to iron along the flat surface area of the sample, refraction/reflection may force back neutron beams from the inside of the iron sample to the vacuum environment. These phenomena should occur at the four horizontal surfaces of each iron sample. Therefore, positive and negative peakings occur in the line profiles in figure 2(c). The dashed-line area in figure 2(c) is zoomed in figure 2(d) to explain the phenomena more clearly. Before this study, Tremsin *et al.* [21] reported similar phenomena in neutron imaging. The negative and positive peakings were disturbing with regard to the purpose of this study, but we only focused on the neutron transmission spectrum inside the samples in our analyses.

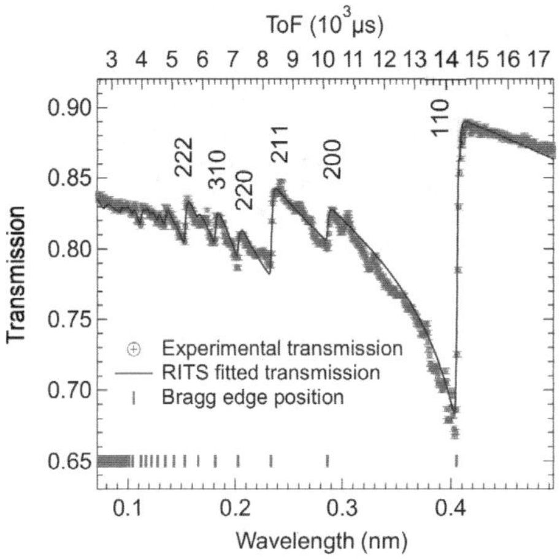

Figure 3. A typical neutron transmission spectrum of the iron sample obtained by the present experiment. The transmission was calculated by the ratio of sum spectrum of four sample areas (indicated by yellow dashed-line areas of 23.64 mm^2 in figure 2(b)) and the spectrum of the background area (indicated by a white dot-and-dash-line area of 10.68 mm^2 in figure 2(b)) for the purpose of reducing a difference in sampling areas. Most of typical Bragg edges of the iron samples can clearly be identified by eyes in the wavelength range from 0.1 nm to 0.45 nm. The transmission spectrum fitted with RITS code was overlaid on red data points as a blue solid line.

We discuss the Bragg-edge profile in the neutron transmission spectra of the iron samples. Figure 3 presents a typical neutron transmission of the sample in a wide area of 23.64 mm^2, which is wide enough and almost the same as that of the single sample. The transmission was calculated by taking the ratio of the neutron intensity in the sample area, which is indicated by the dashed yellow lines in figure 2(b) and the intensity in the area without the iron samples (denoted as the background area) indicated by the white dot and dashed line in figure 2(b). The procedure for taking the ratio between the sample and background areas was helpful in compensating for the differences in the beam intensities as wavelength functions. The neutron transmission data in this study were analyzed using the ToF method from 25 μs to 20 ms or a wavelength of 0.0007 to 0.57 nm in a time bin of 25 μs, giving the best wavelength resolution of 0.0007 nm [19]. A neutron image was constructed with a resolution of an area of elementary pixel size (x and y direction size of 3.1×2.3 μm^2). We found that the experimental transmission had good statistics to be able to see most of the typical Bragg edges in the wavelength of 0.15 to 0.4 nm. Figure 3 presents the transmission of the iron samples evaluated as a function of wavelength or ToF. Many steps can be seen in the experimental spectrum in this figure. We performed fitting of the experimental spectrum using a RITS program [20], where the major Miller indices of 110, 200, 211, 220, 310, and 222 were labeled in the transmission spectrum of the iron samples (see figure 3). We confirmed that the experimental transmission spectrum was in good agreement with the simulation spectrum obtained from the RITS code.

The superconducting neutron sensor in this study exhibited a high spatial resolution [5,7]. This study aimed to determine how narrow we can select a testing area to obtain a visible Bragg edge using an iron test sample with CB-KID. There was a trade-off relation between the choice of the measurement area and time. Although the capability of the facility limits the neutron flux density, it was unrealistic to be assigned a very long beam time in any available facilities. The ensemble average equals the time average from the ergodic theorem in statistical mechanics [14]. In other words, a sufficiently large collection of random ensembles from a sample system can represent the average statistical properties of the entire process of an isolated small ensemble. The long-time averaged behavior of the narrow area of the test sample was examined by considering an ensemble average over many identical ensembles randomly selected from the inside of the test sample.

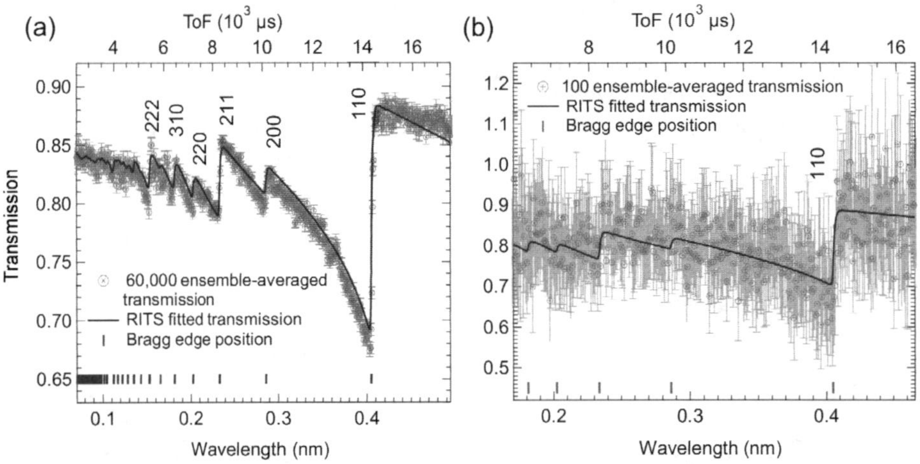

Figure 4. (a) Neutron transmission spectrum of iron test samples obtained by using a limited area of 0.43 mm^2 in total with summing randomly-chosen 60,000-pixel ensembles from the sample area. The transmission spectrum shows a good signal-to-noise ratio while error bars and a blue fitting curve are shown in the figure. This is almost identical with features of the visible Bragg edges obtained by using a wider-sample area 23.64 mm^2 as showed in figure 2(b). (b) The experimental transmission spectrum composed from randomly-chosen 100 pixels (713 μm^2 in total). Although a signal-to-noise ratio in red data points is moderate, a blue RITS fitting curve is shown. The major Bragg edge at around 0.4 nm can be discernably recognized.

To reveal the minimum size of the sample area enough for visualizing Bragg edges in the transmission spectra, the main 110 Bragg edge was particularly important in the discussion of this issue. For this purpose, based on the ergodic theorem [14], we took an ensemble average for several ensembles, of which the area was randomly taken from the inside of the test samples to evaluate the transmission spectrum of each area. This yielded a virtual transmission spectrum that could be obtained by conducting a very long-time average of the transmission spectrum through a single narrow ensemble area of the test samples. The space average of independently selected area spectra with the same ensemble size can be evaluated. Figure 4(a) presents a neutron transmission spectrum of the iron test sample in a restricted area, which was obtained by summing up the spectra from randomly chosen 60,000 pixels of the 3.1 × 2.3 μm^2 area from the test sample area. We observed that the total area consisting of the 60,000 ensembles was considerably smaller than the sample size. Furthermore, we carefully removed the ensemble case whenever the selected area overlapped with the areas of other ensembles in choosing randomly chosen ensembles in analyses. This ensures that any area can be independent with each other so that the ergodic theorem can be applied appropriately. The same number of constituent pixels from the background area (outside the test iron samples) from close locations to the test samples were used. We found that the signal-to-noise ratio was not so good as in the transmission spectrum obtained from the wider area in figure 3, but it was good enough to discriminate several different Bragg edges in the spectrum (see labels in figure 4(a)). We concluded that all Bragg edges are still clearly visible, even employing the restricted area of 0.43 mm^2. This conclusion would become important as it is, in principle, possible to obtain the Bragg-edge profile for a small restricted area of a test material when we use a high-spatial-resolution neutron sensor with plenty of long beam time.

We further attempted to reduce the number of ensemble pixels for studying the minimum area to obtain a neutron transmission spectrum with the discernible main Bragg edge exhibiting a reasonable signal-to-noise ratio. This attempt helped estimate the data acquisition time to obtain a discernible Bragg edge within the limitation of beam-time allocation in actual planning. Figure 4(b) presents the neutron transmission spectrum obtained with 100 ensembles (713 μm^2 in total area) with error bars, in which we noticed that the signal-to-noise ratio was not so good compared with that in figure 4(a). However, the eye still clearly identified the main Bragg edge (110) in figure 4(b), where it was helpful to overlay the transmission spectrum obtained by fitting with the RITS program on the experimental data. We ensured that all analyses in this study were conducted using raw-spectrum data; however, the signal-to-noise ratio can be improved by averaging over serval pixels image or averaging over a time bin in the ToF spectrum. We interpret that this is the minimum area size (3.1 × 2.3 μm^2) at which a visible Bragg edge can be obtained if a beam time is available hundred times longer than the present measurements. Our sensor is suitable to study the properties of the material with a size larger than 31 × 23 μm^2 (10 × 10 pixels) with the beam time of the present study. Regarding the wavelength resolution, Tremsin *et al.* [10] observed the neutron transmission of an Al cylinder with a diameter of 50 mm; they conducted this observation by using an MCP neutron detector with a 14 × 14 mm^2 sensitive area with the Timepix readout. They succeeded in observing a neutron transmission spectrum with main Bragg-edge in an area of 165 × 165 μm^2 (3 × 3 pixels) with a minimum of wavelength bin of 0.0004 nm. The spectra were built with 55-μm shifts, which is one-third of the averaging width.

4. Summary

It is a reasonable hypothesis that a high-spatial-resolution Bragg-edge analysis is possible if a high-spatial-resolution neutron detector is used. To confirm this hypothesis, we analyzed neutron transmission spectra with a restricted area down to spatial resolution (3.1 × 2.3 μm^2) and a minimal wavelength window (0.0007 nm). This is to determine the minimum area size required for recovering clear Bragg edges of the 2-mm-thick iron test samples using our superconducting neutron sensor. Instead of very long-time measurements required to achieve transmission spectra with good statistics, we performed ensemble averaging of the transmission spectra of randomly chosen restricted areas based on the ergodic theorem. The multiple Bragg edges of the iron test samples were nicely reproduced *via* ensemble averaging of 60,000 identical ensembles with an area size of 3.1 × 2.3 μm^2, where the analyzed area reached 0.43 mm^2 in total within the test samples. As a reference, we demonstrated a clear neutron transmission spectrum with a single wider area of 23.64 mm^2 of the test samples. We found that our sensor can still detect the main 110 Bragg edge of iron beyond the noise floor even when the number of ensembles was limited to 100 with the smallest ensemble size (3.1 × 2.3 μm^2). Several visible edges clearly appeared, with a total sample size of 0.43 mm^2 as a sum of all ensembles. These Bragg edges were visible with a good signal-to-noise ratio. The Miller indices were assigned well with the RITS program [20]. This was useful to estimate the required beam time to obtain a visible transmission spectrum with a long data acquisition time. We simulated a transmission spectrum using the RITS program to compare it with the experimental neutron transmission spectrum. It can be concluded that a high-spatial-resolution Bragg edge analysis can be conducted using our superconducting neutron sensor. Our finding would be more important when a neutron source becomes much more intense by three orders of magnitude in the field of inertial fusion energy facilities in the future [22].

Acknowledgment

The authors are grateful to Y. Su for her kind suggestions for using the RITS program to fit our experimental neutron transmission spectra. This work is partially supported by Grant-in-Aid for

Scientific Research (Nos. JP16H02450, JP21H04666, JP21K14566) from Japan Society for the Promotion of Science (JSPS) and the project program of MLF (Nos. 2018P0201, 2021P0501).

References

[1] Tengattini A, Lenoir N, Andò E and Viggiani G 2021 Neutron imaging for geomechanics: A review *Geomech. Energy Environ.* **27** 100206

[2] Kardjilov N, Manke I, Woracek R, Hilger A and Banhart J 2018 Advances in neutron imaging *Mater. Today* **21** 652–72

[3] Ishida T, Yoshioka N, Narukami Y, Shishido H, Miyajima S, Fujimaki A, Miki S, Wang Z and Hidaka M 2014 Toward mega-pixel neutron imager using current-biased kinetic inductance detectors of Nb nanowires with ^{10}B converter *Journal of Low Temperature Physics* **176** 216–221

[4] Vu T D *et al.* 2019 Temperature dependent characteristics of neutron signals from a current-biased Nb nanowire detector with ^{10}B converter *Journal of Physics: Conference Series* **1293** 012051

[5] Iizawa Y, Shishido H, Nishimura K, Vu T D, Kojima K M, Koyama T, Oikawa K, Harada M, Miyajima S, Hidaka M, Oku T, Soyama K, Aizawa K, Suzuki S Y and Ishida T 2019 Energy-resolved neutron imaging with high-spatial resolution using a superconducting delay-line kinetic inductance detector *Supercond. Sci. Technol.* **32** 125009

[6] Vu T D *et al.* 2020 Kinetic inductance neutron detector operated at near critical temperature *Journal of Physics: Conference Series* **1590** 012036

[7] Vu T D *et al.* 2021 Practical tests of neutron transmission imaging with a superconducting kinetic-inductance sensor *Nucl. Instrum. Meth. Phys. Res. A* **1006** 165411

[8] Santisteban J R, Edwards L, Steuwer A and Withers P J 2001 Time-of-flight neutron transmission diffraction *J. Appl. Cryst* **34** 289–97

[9] Steuwer A, Withers P J, Santisteban J R and Edwards L 2005 Using pulsed neutron transmission for crystalline phase imaging and analysis *J. Appl. Phys.* **97** 074903

[10] Tremsin A S, McPhate J B, Kockelmann W, Vallerga J V., Siegmund O H W and Feller W B 2011 High resolution Bragg edge transmission spectroscopy at pulsed neutron sources: Proof of principle experiments with a neutron counting MCP detector *Nucl. Instrum. Meth. Phys. Res. A* **633** S235–S238

[11] Shinohara T *et al.* 2020 Development of Energy-Resolved Neutron Imaging Detectors at RADEN *Rev. Sci. Instrum.* **91** 043302

[12] Rhodes N J 2006 Scientific Reviews: Status and Future Development of Neutron Scintillation Detectors *Neutron News* **17**:1 16-18

[13] Rhodes N J 2012 Scintillation detectors *Neutron News* **23**:4 26-30

[14] Moore C C 2015 Ergodic theorem, ergodic theory, and statistical mechanics *Proc. Natl. Acad. Sci. U. S. A.* **112** 1907–11

[15] Ishida T *et al.* 2008 Superconducting MgB_2 thin film detector for neutrons *J. Low Temp. Phys.* **151** 1074–1079

[16] Shishido H, Miyajima S, Narukami Y, Oikawa K, Harada M, Oku T, Arai M, Hidaka M, Fujimaki A and Ishida T 2015 Neutron detection using a current biased kinetic inductance detector *Appl. Phys. Lett.* **107** 232601

[17] Yoshioka N, Yagi I, Shishido H, Yotsuya T, Miyajima S, Fujimaki A, Miki S, Wang Z and Ishida T 2013 Current-biased kinetic inductance detector using MgB_2 nanowires for detecting neutrons *IEEE Trans. Appl. Supercond.* **23** 2400604

[18] Kojima K M *et al.* 2014 New μSR spectrometer at J-PARC MUSE based on Kalliope detectors *Journal of Physics: Conference Series* **551** 012063

[19] Oikawa K *et al.* 2008 Design and application of NOBORU-NeutrOn Beam line for Observation and Research Use at J-PARC *Nucl. Instrum. Meth. Phys. Res. A* **589** 310–317

[20] Sato H 2018 Deriving quantitative crystallographic information from the wavelength-resolved neutron transmission analysis performed in imaging mode *J. Imaging* **4** 7

[21] Tremsin A S, Lehmann E H, Kardjilov N, Strobl M, Manke I, McPhate J B, Vallerga J V, Siegmund O H W and Feller W B 2012 Refraction contrast imaging and edge effects in neutron radiography *J. Instrum.* **7** C02047

[22] Taylor A, Dunne M, Bennington S, Ansell S, Gardner I, Norreys P, Broome T, Findlay D and Nelmes R 2007 A Route to the Brightest Possible Neutron Source? *Science* **315** 1092–1095

Neutron Imaging toward Epithermal Regime using a Delay Line Current-Biased Kinetic-Inductance Detector

Hiroaki Shishido[1,2,3], The Dang Vu[4,5], Kazuya Aizawa[5], Kenji M. Kojima[4,6], Tomio Koyama[4], Kenichi Oikawa[5], Masahide Harada[5], Takayuki Oku[5], Kazuhiko Soyama[5], Shigeyuki Miyajima[7], Mutsuo Hidaka[8], Soh Y. Suzuki[9], Manobu M. Tanaka[10], Shuichi Kawamata[2,4] and Takekazu Ishida[2,4]

[1]Department of Physics and Electronics, Graduate School of Engineering,Osaka Prefecture University, Sakai, Osaka 599-8531, Japan
[2]NanoSquare Research Institute, Osaka Prefecture University, Sakai, Osaka 599-8570, Japan
[3]The Center for Research and Innovation in Electronic Functional Materials, Osaka Prefecture University, Sakai, Osaka 599-8570, Japan
[4]Division of Quantum and Radiation Engineering, Osaka Prefecture University, Sakai, Osaka 599-8570, Japan
[5]Materials and Life Science Division, J-PARC Center, Japan Atomic Energy Agency, Tokai, Ibaraki 319-1195, Japan
[6]Center for Molecular and Materials Science, TRIUMF and Stewart Blusson Quantum Matter Institute, University of British Columbia, Vancouver, BC, V6T 2A3 and V6T 1Z4, Canada
[7]Advanced ICT Research Institute, National Institute of Information and Communications Technology, 588-2 Iwaoka, Nishi-ku, Kobe, Hyogo 651-2492, Japan
[8]Advanced Industrial Science and Technology, Tsukuba, Ibaraki 305-8568, Japan
[9]Computing Research Center, Applied Research Laboratory, High Energy Accelerator Research Organization (KEK), Tsukuba, Ibaraki 305-0801, Japan
[10]Institute of Particle and Nuclear Studies, High Energy Accelerator Research Organization (KEK), Tsukuba, Ibaraki 305-0801, Japan

E-mail: shishido@pe.osakafu-u.ac.jp

Abstract. Wide energy range neutron imaging and spectroscopy in a single experiment without changing the setup is available by a combination of a high-power pulsed neutron source and fast operating neutron detector. The delay-line current-biased kinetic-inductance detector (CB-KID) is a two-dimensional superconducting detector with high spatial and temporal resolutions, and multi-hit tolerance. We demonstrate that the delay-line CB-KID with ^{10}B neutron conversion layer is applicable for neutron imaging and spectroscopy up to the neutron energy of 100 eV. High spatial neutron transmission image for a Sm-Sn mixed metal is successfully constructed below 0.3 eV. Concomitantly, we examine the neutron transmission up to 100 eV and identify observed resonance dips by Sm nuclei.

1. Introduction

Superconducting detectors are one of the most successful applications among in the superconducting devices [1, 2, 3, 4]. A discontinuous transition in the resistivity at the superconducting transition temperature T_c is applied to sensitive detectors as superconducting nanowire single photon detectors (SNSPD) [4] and transition edge sensors (TES) [1]. Microwave kinetic inductance detectors (MKID) [2, 3] work in the superconducting state, and are based on the detection principle of changes in the kinetic inductance induced by the Cooper pair breaking.

While main applications of these superconducting detectors are detection of electromagnetic waves, superconducting detectors can detect other beams such as neutron beams. There are several reports of applying superconducting detectors to neutron detection by combining a neutron absorption layer. The combination of the TES and a B layer [5], and the combination of superconducting tunnel junctions and $Li_2B_4O_7$ single crystal [6, 7] were reported as a superconducting neutron detector.

We proposed a superconducting neutron detector with the combination of the current-biased kinetic-inductance-detector (CB-KID) and a ^{10}B neutron capture layer [8, 9]. The CB-KID keeps in the superconducting state, and detects a neutron incident event via a transient change of the kinetic inductance induced by the Cooper pair breaking. Therefore, the CB-KID has the advantage of high speed operation over a wide temperature range below T_c. We reported that a high spatial resolution of 16 μm in neutron transmission imaging by the combination of delay-line method (delay-line CB-KID) [10], the energy resolved neutron imaging and transmission measurements up to 10 eV [11, 12] via the the time-of-flight (ToF) method in the pulsed neutron source. The delay-line method was also applied to a single photon imager based on the SNSPD [13]. A dip structure induced by resonance absorption appears in the neutron transmissions as a fingerprint of an isotope. Neutron resonance transmission imaging (NRTI), which is an elemental imaging technique based on resonance dip, was proposed [14]. We also succeeded in achieving Gd [11] and Sm [12] selective NRTI and transmission measurements below 10 eV.

In the present work, we demonstrate that the delay-line CB-KID is applicable to a neutron transmission and imaging in the regime of epithermal neutrons. Dips induced by nuclear resonance absorption of natural Sm nuclei were observed. We also succeeded in acquiring neutron transmission images even with high neutron energy in the range of 10 to 100 eV.

2. Experimental methods

2.1. Delay-line current-biased kinetic inductance detector

As schematically illustrated in Fig. 1, the main part of the delay-line CB-KID consists of two mutually orthogonal superconducting Nb meanderlines. Those meanderlines of 0.9 μm width, 0.6 μm spacing, and segment length h = 15.1 mm act not only as a detecting area but also as a delay line. Thanks to the combination of delay-line method and high-speed time-to-digital converter, high spatial resolved neutron transmission image can be obtained with only four signal detecting channels. The meanderlines with the total length l=151 m were folded 10,000 times, and the electromagnetic pulse wave can propagate from end to end thanks to a stripline structure with a Nb superconducting ground plane [15]. This device was fabricated in the Clean Room for Analog-Digital Superconductivity (CRAVITY) at the National Institute of Advanced Industrial Science and Technology (AIST). A ^{10}B layer with 70 nm thick was deposited by an electron beam deposition technique under ultra high vacuum.

Charged particles (Li-ion and α-particle) are emitted by nuclear reaction between a ^{10}B nucleus at the neutron capture layer and an incident neutron. If one of charged particles hits both X and Y meanderlines, a transient reduction in the Cooper pair density n_s occurs locally at the hot spot, and induces a transient change in local kinetic inductance. A voltage V across the hot spot is generated by the transient change of local kinetic inductance with a DC bias current I_b as

$$V = I_b \frac{dL_0}{dt} \simeq I_b \frac{dL_k}{dt} = -\frac{m_s \Delta l I_b}{n_s^2 q_s^2 S} \frac{dn_s}{dt}, \quad (1)$$

where, L_0 is the total inductance; L_k is the kinetic inductance; Δl and S are the hot spot length and cross-sectional area of the superconducting wire, respectively; and m_s and q_s are the effective mass and electric charge of the Cooper pair, respectively. Voltage signals propagate toward both ends through meanderlines as electromagnetic waves.

The hot spot positions x and y are determined as follows:

$$x = \text{ceil}\left[\frac{(t_{\text{Ch4}} - t_{\text{Ch3}})v_x}{2h}\right] p, \qquad y = \text{ceil}\left[\frac{(t_{\text{Ch2}} - t_{\text{Ch1}})v_y}{2h}\right] p, \tag{2}$$

where t_{Ch1}, t_{Ch2}, t_{Ch3}, and t_{Ch4} are the corresponding time stamps of the signals received at Ch1, Ch2, Ch3, and Ch4; and $v_x = 5.981 \times 10^7$ m/s and $v_y = 4.544 \times 10^7$ m/s are signal propagation velocities for X and Y meanderlines, respectively at T=7.9 K in the present CB-KID.

A hot spot emerging time t_0 can be estimated from the combination of pair channels (t_{Ch1} and t_{Ch2} for Y meanderline, and t_{Ch3} and t_{Ch4} for X meanderline) independently, as follows:

$$t_0 = \frac{t_{\text{Ch4}} + t_{\text{Ch3}}}{2} - \frac{l}{2v_x}, \qquad t_0 = \frac{t_{\text{Ch2}} + t_{\text{Ch1}}}{2} - \frac{l}{2v_y}. \tag{3}$$

These t_0 estimated for the X and Y meanderlines should coincide with each other if the selected signal quartet originates from the same event. The delay-line CB-KID acts as a multi-hit tolerant detector because the coincidence check for t_0 is a useful criterion to identify a

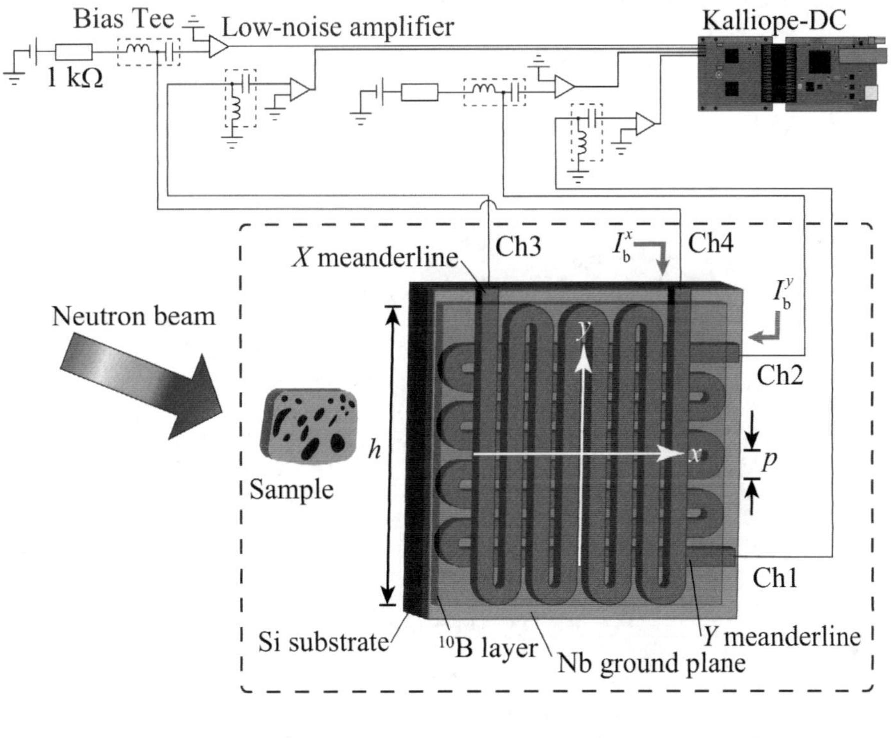

Figure 1. Stacked layers of a superconducting Nb ground plane, X- and Y-superconducting Nb meanderlines, and a ^{10}B neutron capture layer are schematically shown. Neutron beams were irradiated from Si substrate side through test sample. Objects surrounded by the dotted line were cooled down to 7.9 K. DC bias currents I_b^x and I_b^y were fed to meanderlines, and signals from both ends of meanderlines were detected by a Kalliope-DC board.

signal quartet discriminatively from several different signals simultaneously produced on the meanderlines.

2.2. Preparation for a Sm-Sn mixed metal

We grew $SmSn_3$ single crystals by using Sn-self flux method. Starting materials of Sm and Sn with the atomic ratio of 1:9.2 were sealed in a quartz tube under Ar atmosphere. This tube was heated up to 950 °C and cooled down to 200 °C with the cooling rate of 10 °C/h. The obtained ingot, in which $SmSn_3$ single crystals distributed in Sn metal, was cut into a platelet with a thickness of ~0.5 mm for the neutron imaging experiments. $SmSn_3$ is an antiferromagnet with the Néel temperature of 12 K [16, 17].

2.3. Cryogenic and signal measurement systems

The delay-line CB-KID system is schematically shown in Fig. 1. The delay-line CB-KID was cooled down to 7.9 K by using a Gifford–McMahon (GM) refrigerator to made it in a superconducting state, and was held in vacuum surrounded by aluminum-alloy vacuum can. Test sample was also cooled down to 7.9 K because it was placed at a distance of 0.8 mm from the detector. The detector temperature was controlled by a temperature controller (Model 44 by Cryogenic Control Systems Inc.). Neutron beams were irradiated to the substrate side of the detector through the sample.

The DC bias currents for X and Y meanderlines were fed at $I_b^x = I_b^y = 30\,\mu$A. Positive (or negative) signals from Ch2 and Ch4 (or Ch1 and Ch3) were directly (or inversely) amplified by differential ultralow-noise amplifiers (SA-430 F5 by NF Corporation), as shown in Fig. 1. Amplified signals were received by the Kalliope-DC readout circuit, which is a 1 ns-sampling multichannel (16 Ch × 2) time-to-digital converter (TDC) [18, 19], and a 2.5 GHz sampling digital oscilloscope (Teledyne LeCroy HDO4104-MS) simultaneously.

2.4. Neutron imaging

Neutron imaging experiments were performed with pulsed neutrons at BL10 of the Material and Life science experimental Facility (MLF), J-PARC [20]. Neutrons of various energies are created by the nuclear spallation reaction, and travel from a moderator to the detector through the beam line with the length L_{path}=14 m at BL10. Nuclear spallation reactions are initiated every 40 ms and synchronized trigger signals is generated, while the trigger signal contain finite but a non-negligible delay time t_{delay}. ToF through the beam line is proportional to the neutron wavelength λ as

$$\lambda = \frac{h}{m_{\text{n}} L_{\text{path}}} (t_0 - t_{\text{delay}}), \tag{4}$$

where m_{n} is a neutron mass. The neutron energy E is also represented as

$$E = \frac{m_{\text{n}}}{2} \left(\frac{L_{\text{path}}}{t_0 - t_{\text{delay}}} \right)^2. \tag{5}$$

Imaging experiments performed with beam power of 920 kW using a rotary-collimator device located at 8.04 m position. Design of the rotary-collimator is described in the literature [21]. Currently, the"Large" aperture D having $31.6 \times 31.6\,\text{mm}^2$ has been replaced by the "Medium" aperture D having $17.8 \times 17.8\,\text{mm}^2$. Therefore, the collimation ratio L/D was 335 because the distance between the detector and the rotary-collimator is L=5960 mm.

3. Results and discussion

3.1. Detected voltage signals

Figure 2 represents a typical detected signal quartet induced by a intermediate neutron with $E = 63\,\mathrm{keV}$. Signal amplitude and width are not affected by the incident neutron energy. A sharp positive peak, which induced by transient reductions of the Cooper pair density caused by neutron incident events, followed by a wider negative peak appears in each channels. Negative peaks may correspond to recovery of Cooper pairs by recombination, and are well fitted by an exponential relaxation function $A\exp\{-(t-t_\mathrm{p})/\tau\}$ with τ of 0.12-0.21 μs.

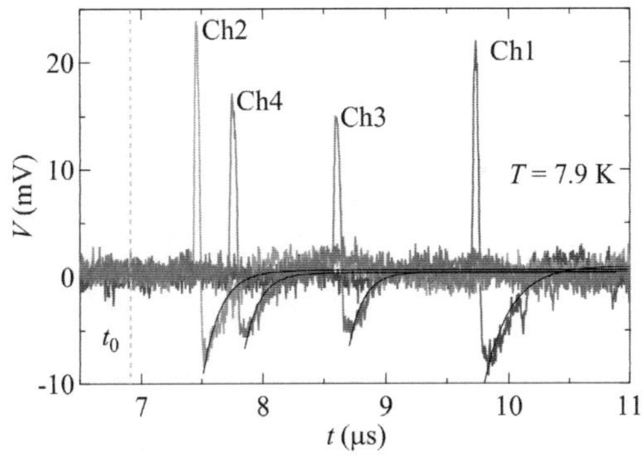

Figure 2. A typical signal quartet detected at 7.9 K. The hot spot emerging time t_0 was calculated by Eq. (3) (see a light-blue dashed line).

The neutron incident event occurred at t_0 and voltage pulses traveled to both ends through meanderlines. An incident neutron energy corresponding to a signal quartet was determined not by a signal detection time for each channel but by t_0 to compensate signal propagation times in CB-KID.

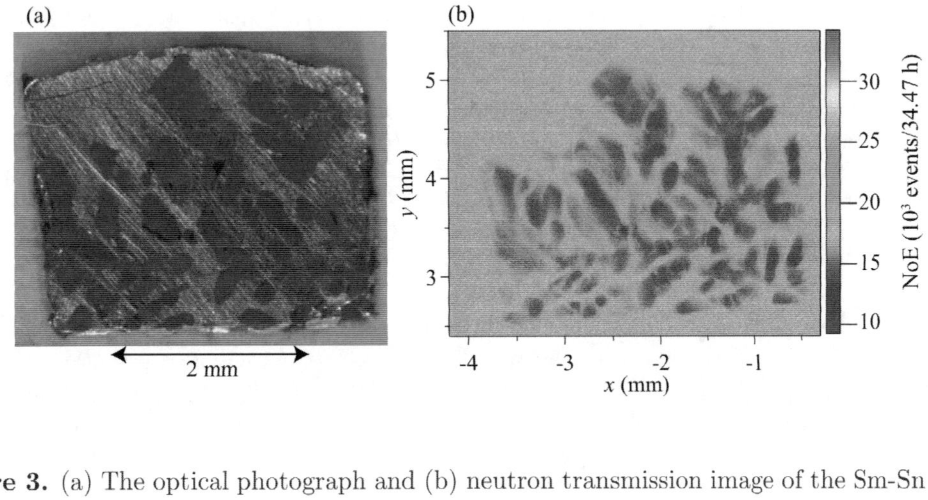

Figure 3. (a) The optical photograph and (b) neutron transmission image of the Sm-Sn mixed metal sample.

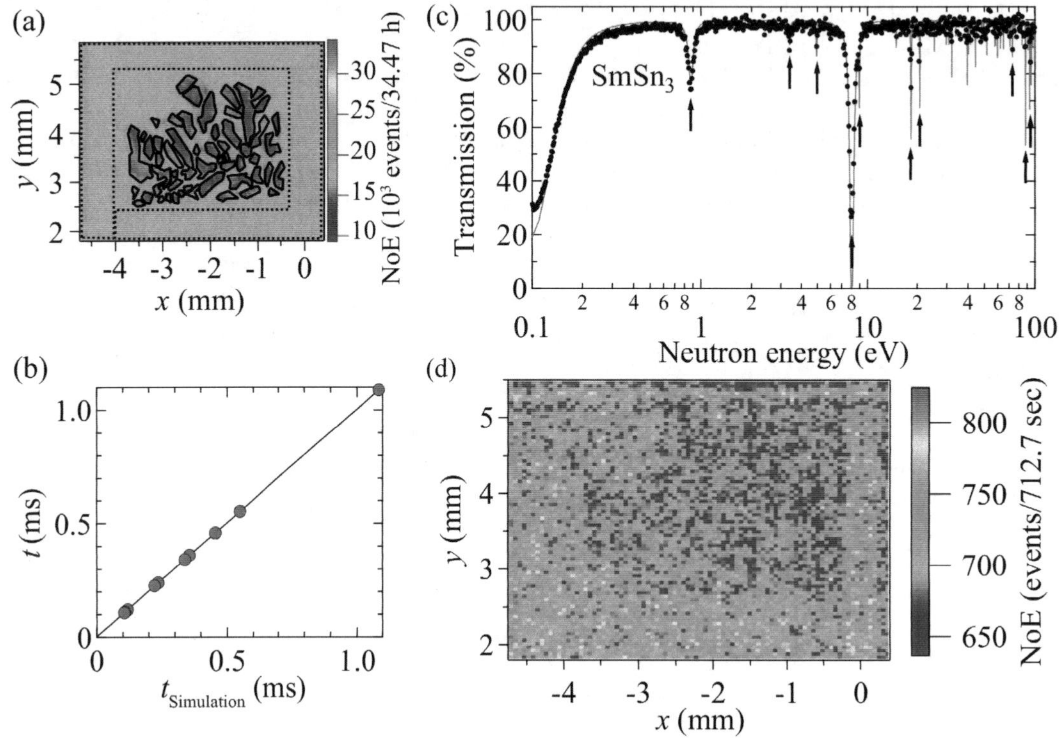

Figure 4. (a) Zoomed out image of Fig. 3(b). (b) Comparison between detected dip energies and simulated ones. (c) Transmission of a $SmSn_3$ single crystals. (d) Neutron transmission image with the neutron energy ranging from 10 eV to 100 eV.

3.2. Neutron transmission image of $SmSn_3$

Figure 3(a) shows an optical photograph of the Sm-Sn mixed metal sample. $SmSn_3$ single crystals are seen as dark gray regions on the surface in a matrix of metallic colored Sn ingot. The neutron transmission image of the sample taken for a range of $0.6 \leq E \leq 300$ meV is shown in Fig. 3(b) with the color scale of number of events (NoEs) collected for 34.47 hours. The $SmSn_3$ single crystals are clearly identified from Sn ingot as the blue colored areas, in which NoEs are much less than green, yellow, and red regions thanks to large neutron absorption cross section of Sm nuclei. Even though Sn is a heavy metal, the neutron beam penetrates well because of low neutron total cross section of Sn. While most of pixels have a size of $3 \times 3\,\mu m^2$, a few pixels have a size of 1.5×3, 3×1.5, and $1.5\times1.5\,\mu m^2$. Resulting from ununiformity in pixel sizes, a thin grid pattern is visible in the Sn area.

Figure 4(a) shows a zoomed out image of Fig. 3(b) in a reduced size, where $SmSn_3$ single crystals are surrounded by solid lines. The peripheral area surrounded by the dotted line contains neither $SmSn_3$ single crystals nor other sample. To estimate the neutron transmission through the $SmSn_3$ single crystals and Sn, NoEs for each neutron energy range were summed for each of the areas surrounded by solid lines and that surrounded by the dotted line, and then we divided the former by the later. The neutron energy dependence of neutron flux, detection efficiency of CB-KID, and neutron absorption by vacuum can and sample holder made by the aluminum alloy are normalized by division, and as a result the neutron transmission is obtained.

The neutron transmission exhibits several dips induced by resonance absorption due to the nuclear reaction between Sm nuclei and neutrons. One can correct a trigger delay time by comparing the experimental results of resonance absorption with those available in the database. Figure 4(b) represents the comparison between experimental ToF dips for resonance absorption and those in the database [22]. A linear relationship holds well between them as shown by a good fitting to a linear line, and the trigger delay is estimated to be 2.88 μs from the intercept of a linear fit. We corrected incident neutron energy by using equation (5) with $t_{\text{delay}} = 2.88\,\mu$s.

The corrected neutron energy dependence of the transmission for $SmSn_3$ single crystals is shown in Fig. 4(c) with arrows indicating the positions of dips induced by resonance absorption. Solid line represents a simulated transmission for natural Sm nuclei. Dip energies and width of resonance absorption are well reproduced by simulations.

We demonstrate epithermal-region neutron imaging of the sample at $10 \leq E \leq 100\,\text{eV}$, as shown in Fig. 4(d), with the color scale of NoE collected for 712.7 seconds. Pixel sizes are 60 $\times$ 60 μm^2 or 60 $\times$ 58.5 μm^2. The outer shape of the crystal is barely discriminated as a blue region thanks to resonance absorption despite the basically high transmittance in the epithermal region. These results are indicative of the capability for neutron transmission and imaging with epithermal region by the delay-line CB-KID.

4. Summary

We demonstrated that the delay-line CB-KID can be used for neutron transmission measurements and imaging with epithermal neutrons. Observed dips in the neutron transmission were well identified by peaks induced by resonance absorption, in the total neutron cross-section of natural Sm. Combination of high temporal resolution and wide time window in the delay-line CB-KID with Kalliope-DC readout circuit ensures that the combination of the delay-line CB-KID and pulsed neutron source can simultaneously measure neutron transmission and imaging over a wide energy range of neutrons up to epithermal region.

Acknowledgments

This work is partially supported by Grant-in-Aid for Scientific Research (Nos. JP16H02450, JP21H04666, JP21K14566) from JSPS. The neutron irradiation experiments at the Materials and Life Science Experimental Facility (MLF) of the J-PARC were conducted under the support of MLF project program (No. 2020P0201).

References

[1] Irwin K D 1995 *Appl. Phys. Lett.* **66** 1998
[2] Day P K, LeDuc H G, Mazin B A, Vayonakis A and Zmuidzinas J 2003 *Nature* **425** 817
[3] Zmuidzinas J 2012 *Annu. Rev. Condens. Matter Phys.* **3** 169
[4] Natarajan C M, Tanner M G and Hadfield R H 2012 *Supercond. Sci. Technol.* **25** 063001
[5] Merlo V, Salvato M, Cirillo M, Lucci M, Ottaviani I, Scherillo A, Celentano G and Pietropaolo A 2015 *Appl. Phys. Lett.* **106** 113502
[6] Nakamura T, Katagiri M, Ukibe M, Ikeuchi T and Ohkubo M 2004 *Nucl. Instrum. Methods Phys. Res. A* **520** 67
[7] Nakamura T, Katagiri M, Chen Y E, Ukibe M and Ohkubo M 2006 *Nucl. Instrum. Methods Phys. Res. A* **559** 766
[8] Ishida T, Yoshioka N, Narukami Y, Shishido H, Miyajima S, Fujimaki A, Miki S, Wang Z and Hidaka M 2014 *J Low Temp Phys* **176** 216
[9] Shishido H, Miyajima S, Narukami Y, Oikawa K, Harada M, Oku T, Arai M, Hidaka M, Fujimaki A and Ishida T 2015 *Appl. Phys. Lett.* **107** 232601
[10] Iizawa Y, Shishido H, Nishimura K, Vu T D, Kojima K M, Koyama T, Oikawa K, Harada M, Miyajima S, Hidaka M, Oku T, Soyama K, Aizawa K, Suzuki S Y and Ishida T 2019 *Supercond. Sci. Technol.* **32** 125009
[11] Shishido H, Nishimura K, Vu T D, Aizawa K, Kojima K M, Koyama T, Oikawa K, Harada M, Oku T, Soyama K, Miyajima S, Hidaka M, Suzuki S Y, Tanaka M M, Kawamata S and Ishida T 2021 *IEEE Trans. Appl. Supercond.* **31** 9
[12] Shishido H, Vu T D, Aizawa K, Kojima K M, Koyama T, Oikawa K, Harada M, Oku T, Soyama K, Miyajima S, Hidaka M, Suzuki S Y, Tanaka M M, Kawamata S and Ishida T 2021 *J. Phys.: Conf. Ser.* **1975** 012023

[13] Zhao Q-Y, Zhu D, Calandri N, Dane A E, McCaughan A N, Bellei F, Wang H-Z, Santavicca D F and Berggren K K 2018 *Nat. Photonics* **11** 247
[14] Festa G, Cippo E P, Martino D D, Cattaneo R, Senesi R, Andreani C, Schooneveld E, Kockelmann W, Rhodes N, Scherillo A, Kudejova P, Biro K, Duzs K, Hajnalf Z and Gorini G 2015 *J. Anat. At. Spectrom.* **30** 745
[15] Swihart J C 1961 *J. Appl. Phys.* **32** 461
[16] Tsuchida T and Wallace W E 1965 *J. Chem. Phys.* **43** 3811
[17] Ito T U, Higemoto W, Ninomiya K and Suzuki H S 2011 *Phys. Rev. B* **84** 064411
[18] Kojima K M, Murakami T, Takahashi Y, Lee H, Suzuki S Y, Koda A, Yamauchi I, Miyazaki M, Hiraishi M, Okabe H, Takeshita S, Kadono R, Ito T, Higemoto W, Kanda S, Fukao Y, Saito N, Saito M, Ikeno M, Uchida T and Tanaka M M 2014 *J. Phys.:Conf. Ser.* **551** 012063
[19] Shishido H, Miki Y, Yamaguchi H, Iizawa Y, Vu T D, Kojima K M, T. Koyama, Oikawa K, Harada M, Miyajima S, Hidaka M, Oku T, Soyama K, Aizawa K, Suzuki S Y and Ishida T 2018 *Phys. Rev. Appl.* **10** 044044
[20] Oikawa K, Maekawa F, Harada M, Kai T, Meigo S, Kasugai Y, Ooi M, Sakai K, Teshigawara M, Hasegawa S, Futakawa M, Ikeda Y and Watanabe N 2008 *Nucl. Instrum. Methods Phys. Res. A* **589** 310
[21] Harada M, Oikawa K, Ooi M, Kai T, Shinohara T, Sakai K and Maekawa F 2014 *JPS Conf. Proc.* **1** 014015
[22] Shibata K, Iwamoto O, Nakagawa T, Iwamoto N, Ichihara A, Kunieda S, Chiba S, Furutaka K, Otuka N, Ohsawa T, Murata T, Matsunobu H, Zukeran A, Kamada S and Katakura J 2011 *J. Nucl. Sci. Technol.* **48** 1

Implementing High Q-Factor HTS Resonators to Enhance Probe Sensitivity in ^{13}C NMR Spectroscopy

J N Thomas[1], T L Johnston[1], I M Litvak[1], V Ramaswamy[2], M E Merritt[3], J R Rocca[3], A S Edison[4] and W W Brey[1]

[1]National High Magnetic Field Laboratory, Tallahassee FL, USA
[2]Bruker Biospin, Faellanden, Switzerland
[3]University of Florida, Gainesville FL, USA
[4]University of Georgia, Athens GA, USA

jnt15@my.fsu.edu (J N Thomas), tjohnston@fsu.edu (T L Johnston), litvak@magnet.fsu.edu (I M Litvak), vijaykumar.ramaswamy@bruker.com (V Ramaswamy), matthewmerritt@ufl.edu (M E Merritt), jrrocca@ufl.edu (J R Rocca), aedison@uga.edu (A S Edison), wbrey@magnet.fsu.edu (W W Brey)

Abstract. Nuclear magnetic resonance spectroscopy (NMR) probes using thin-film high temperature superconducting (HTS) resonators provide exceptional mass sensitivity in small-sample NMR experiments for natural products chemistry and metabolomics. We report improvements in sensitivity to our 1.5 mm ^{13}C-optimized NMR probe based on HTS resonators. The probe has a sample volume of 35 microliters and operates in a 14.1 T magnet. The probe also features HTS resonators for ^{1}H transmission and detection and the ^{2}H lock. The probe utilizes a ^{13}C resonator design that provides greater efficiency than our previous design. The quality factor of the new resonator in the 14.1 T background field was measured to be 4,300, which is over 3x the value of the previous design. To effectively implement the improved quality factor, we demonstrate the effect of adding a shorted transmission line stub to increase the bandwidth and reduce the rise/fall time of ^{13}C irradiation pulses. Initial NMR measurements verify ^{13}C NMR sensitivity is significantly improved while preserving detection bandwidth. The probe will be used for applications in metabolomics.

1. Background

Nuclear magnetic resonance (NMR) is a powerful molecular characterization technique useful for the study of solution samples in biology, biochemistry, and organic chemistry. Yet, the detection sensitivity of the NMR at temperatures suitable for biological samples is still too low to solve a variety of important biological problems [1]. High Q-factor high temperature superconducting (HTS) resonators have demonstrated remarkable detection sensitivity in comparison to normal-metal resonators [2–6]. We report here the in-field resonator performance and initial validation of a higher sensitivity 1.5 mm all-HTS NMR probe optimized for ^{13}C detection. A 1.5 mm ^{13}C-optimized all-HTS probe [2] of similar design (but also including a ^{15}N channel) previously developed in our lab has shown to be a powerful tool for metabolomics applications [7–11]. In development of the new probe, advances in coil design and probe construction technique led to significantly higher resonator Q-factors [12,13]. However, the leap in resonator Q-factor has led to the encounter of new obstacles to overcome in their implementation. The purpose of this work is to first outline these obstacles and discuss techniques used to overcome them. The combined results of innovative HTS probe technology and the new implementation techniques discussed here have produced a ^{13}C NMR probe with sensitivity improvements over the previous probe while preserving signal bandwidth.

While high-Q resonators are crucial to increasing probe sensitivity, they also have inherent drawbacks to NMR. The first drawback is a long ring-down and ring-up time. The decay of the oscillation amplitude of an RF pulse generated in a coil follows an exponential law $A(t) = A_0 \exp(-t/\tau)$ where A is the amplitude of the signal and τ is the ring-down time constant. Because an HTS coil presents little loss, the RF current tends to keep circulating in the resonator after the irradiation pulse. Specifically, the ring-down time constant is inversely related to the Q-factor by $\tau = 2Q/\omega_0$ where ω_0 is the NMR frequency [14]. A long ring-down time results in distorted pulse shapes and can significantly increase the NMR dead time following a pulse before reception can begin [15].

Similarly, the Q-factor is inversely related to the resonator bandwidth [16]. A high Q-factor reduces noise by rejecting unwanted signals outside the small bandwidth of the receiver. Yet, there may be an opposing demand for larger band coverage depending on the spectral range of the NMR nucleus being detected. For high-field NMR, too small of a bandwidth can produce poor quantization of peak heights and reduce S/N (signal to noise) in parts of the NMR spectrum. For nuclides such as ^1H with a narrow (~12 ppm) spectral range, the reduction in bandwidth characteristic to HTS resonators has not proven problematic [17]. For the ~200 ppm spectral range of ^{13}C, the application of HTS resonators is more challenging. Since biological NMR samples almost always contain carbon, ^{13}C represents a relevant example where a wide band is important [18]. However, ^{13}C has a low gyromagnetic ratio and is only present at a 1.1% natural abundance, which makes high sensitivity essential. The previous ^{13}C optimized 1.5 HTS probe developed by our lab [2] featured a ^{13}C in-field Q-factor of 1300 (measured at the 3 dB points) and produced an unparalleled ^{13}C NMR sensitivity without adverse bandwidth related effects. However, at the ^{13}C Larmor frequency a loaded Q-factor of 5000 corresponds to a bandwidth of 28 kHz or 186 ppm [15]. The newly designed HTS resonators utilized in our new ^{13}C optimized probe exhibit a zero-field Q-factor of 6000 [13]. Therefore, for ^{13}C NMR, RF coil loss has been sufficiently decreased to put further sensitivity improvements at odds with the spectral range. Utilizing higher Q-factors to gain sensitivity requires the development of new techniques to balance the contradictory incentives for high detection sensitivity and wide bandwidth.

2. Description of NMR Signal and RF Pulse Bandwidth Broadening Techniques for HTS NMR Probes

An NMR transceiver system utilizes a transmit-receive switch (TR switch) to allow the same sample coil to be used to both irradiate the sample and detect the precessing magnetization. A block diagram of the scheme is shown in Figure 1. In HTS NMR probes, the two planar HTS resonators are tuned to approximately the same frequency and function as a Helmholtz-like coil pair to generate a uniform RF magnetic field over the sample volume. The TR switch is coupled to the HTS coils through a matching loop inductor which is adjusted to provide impedance matching [2,3,17,19–22]. The use of inductive coupling is an established approach in NMR spectrometers and has been shown to reduce both electric losses and frequency shift when compared to wired connections [23–28]. During the probe's "load phase", the RF power amplifier (RFPA) and matching loop induce a resonant co-rotating current in the HTS coil pair. An RF pulse produced by the HTS then excites the spin system within the NMR sample. In the probe's "source phase", the precessing magnetization of the sample induces a current in the HTS pair which causes a voltage to appear in the same matching loop. The TR switch is flipped to direct the incoming signal to the low noise amplifier (LNA). In both cases, the loaded Q-factor depends on the coupling coefficient (k) between the HTS coil and the inductive matching loop, as well as the Q-factor of the HTS coil pair [29]. The magnetic flux coupling is regulated by the adjustment of the matching loop. However, we can utilize the TR switch to couple the HTS coil to a resonant short-circuited transmission line during the probe's load phase to improve the signal bandwidth and response time of irradiation pulses. During NMR signal detection (source phase), most sensitivity improvements provided by the HTS resonators are preserved.

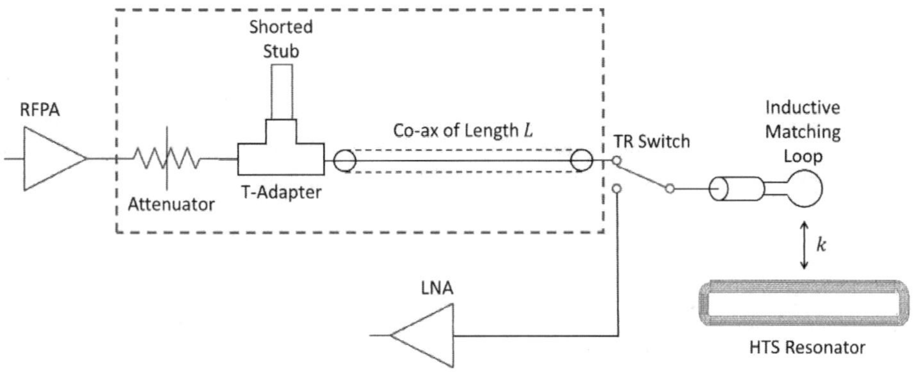

Figure 1. Circuit diagram of an NMR transceiver and coupled HTS resonator. The added resonant short-circuited transmission line elements are contained in the dashed box.

To explain the bandwidth broadening procedure, we first establish a relationship between the bandwidth and the coupling coefficient. As with most RF measurement equipment, the impedances at the ports of an NMR probe are transformed to match the 50 Ω characteristic impedance of a coaxial cable. The maximum transfer of power from the matching loop to the sample coil occurs when impedance presented at the terminals of the matching loop is equal to 50 Ω real at the NMR frequency [20]. Thus, for a particular amount of flux coupling a minimum in the reflection coefficient (S_{11}) is observed. Off resonance, the non-zero reactance presented by the sample coil causes a decrease in inductive power transfer. The bandwidth is calculated from the points at which the power transferred to the sample coil are half of its peak value or 3 dB less than maximum:

$$Q_L = f_c/\Delta f \tag{1}$$

where Q_L, f_c, and Δf are the loaded Q-factor, resonance frequency and 3 dB bandwidth of the loaded sample coil. Equation (1) states that an increase in system bandwidth corresponds to an increase in loss. Thus, the system bandwidth can be increased by increasing the energy transfer between the RF coil and the 50 Ω impedance at the terminals of the matching loop. Therefore, the first degree of freedom (DOF) available to increase the system bandwidth is the amount of flux coupling between the matching loop and resonator. An increase in flux coupling beyond the matched condition (over-coupling) continues to load the resonator. Since this adjustment occurs on the probe side of the TR switch, it increases system bandwidth regardless of the position of the TR switch. It has been previously shown that an inductive coupling loop can be over-coupled to an NMR coil to broaden bandwidth with only modest S/N degradation [20,29]. For high-Q HTS resonators within a multi-resonance NMR probe, the loss in detection sensitivity sustained by an increase in flux coupling depends on the details of the LNA and the level of over-coupling used. For a modest amount of over-coupling and a cryogenic LNA, a significant S/N reduction is not expected.

To further increase the energy exchange without additional degradation of the detection sensitivity, a second DOF can be established on the RFPA side of the TR switch. A section of low coaxial cable is introduced between the RFPA and the TR switch as shown in Figure 1 [20,30]. A coaxial cable with series resistance (R), series inductance (L), shunt conductance (G), and shunt capacitance (C) has the impedance transforming properties of a finite low loss ($R \ll \omega L, G \ll \omega C$) transmission line. For the low loss transmission line of length l that we are inserting, as shown in the dotted box in Figure 1, the input impedance is given by [31]

$$Z_i = Z_0 \frac{Z_L + Z_0 \tanh(\gamma l)}{Z_0 + Z_L \tanh(\gamma l)} \tag{2}$$

where Z_L is the load impedance on the cable, Z_0 is the characteristic impedance of the cable (50 Ω), and γ is the propagation constant defined by

$$\gamma = \alpha + j\beta \cong \frac{1}{2}\left(R\sqrt{C/L} + G\sqrt{L/C}\right) + j\omega\sqrt{LC} \tag{3}$$

If the RFPA is matched to 50 Ω ($Z_L = Z_0$), equation (2) shows that changing the coaxial length l does not affect the matching condition of the probe ($Z_i = Z_0$). However, if there is an impedance mismatch between the RFPA and the probe impedance, the length of the low loss coaxial cable between the RFPA and the probe affects the system bandwidth [30]. We create an impedance mismatch by introducing a shorted stub between the RFPA and the transmission line. Thus, the load impedance of the transmission line is transformed to 50 Ω in parallel with a small impedance determined by the length of the stub. For the purposes of this discussion, we can consider $Z_L \cong 0$ Thus, equation (2) is reduced to

$$Z_{i,shorted} = Z_0 \tanh(\gamma l) \tag{4}$$

and is dependent on the cable length. When the length l is an integer multiple of $\lambda/4$, equation (3) gives

$$\alpha l = \frac{n\pi}{4}\left(\frac{R}{\omega L} + \frac{G}{\omega C}\right) \ll 1, \qquad \beta = n\pi/2 \tag{5}$$

For even n, $\sin(\beta l) = 0$ gives $Z_{i,shorted} \cong Z_0(\alpha l)$: a series-resonant circuit. For odd n, $\cos(\beta l) = 0$ gives $Z_{i,shorted} \cong Z_0/(\alpha l)$: a parallel-resonant circuit. In either case, the reactance is zero. However, only when n is odd ($Z_{i,shorted}$ is large) does the circuit couple strongly with the sample coil. Thus, the probe-to-RFPA cable functions as a quarter-wave resonator with a resonance frequency

$$f_{0m} = n/\left(2l_m\sqrt{LC}\right), \qquad n = \text{odd} \tag{6}$$

where the relevant length l_m is the length from the RFPA to the matching loop. During an irradiation pulse, this resonator is inductively coupled to the sample coil by the matching loop. f_{0m} is tuned by modifying the length of the cable, which in turn changes the system bandwidth and Q-factor. Experimentally, the optimal length may be determined by monitoring the system bandwidth as the cable length is changed. In this way, the system bandwidth during sample irradiation is increased beyond what is achievable with only additional flux coupling.

In summary, the energy exchange between the RFPA and an inductively coupled RF coil during an irradiation pulse is made dependent on two adjustable parameters: the length of the cable between the TR switch and the RFPA and the position of the matching loop with respect to the sample coil [20]. While this procedure was tested for individual isolated HTS coils in [15], its effect on the bandwidth and sensitivity of a built-out HTS probe had yet to be examined. It needed to be shown that the decrease in sensitivity incurred by a desired increase in bandwidth does not overwhelm the benefits produced by our higher Q-factor HTS resonators. If it doesn't, the increase in bandwidth and decrease in the rise/fall times of ^{13}C irradiation pulses is expected to benefit the utility of ^{13}C NMR and HTS NMR probes.

3. Results and Discussion

All recorded measurements were conducted on a ^{13}C-optimized 1.5-mm all-HTS NMR probe in a nominally 600 MHz/54 mm Agilent/Varian VNMRS spectrometer. The probe has a sample fill volume of about 35 microliters and active sample volume of 20 µl. The probe features ^{13}C and ^{1}H detection channels. An additional ^{2}H lock system ensures the spectrometer operates at a constant net magnetic field [32]. At the magnet's static field of ~14.1 T, isotopes ^{13}C, ^{1}H, and ^{2}H have Larmor frequencies of 150.750 MHz, 599.48 MHz, and 92.023 MHz, respectively. The three channels each utilize a pair of planar HTS coils arranged in a Helmholtz-like configuration and tightly nested around the sample (Figure 2). Each of the six individual coils used in the probe was patterned by Star Cryoelectronics (Santa Fe, NM, USA) from ~300 nm-thick superconducting films of $Y_1Ba_2Cu_3O_{7-\delta}$ (YBCO) coated epitaxially onto 430 µm-thick sapphire by Ceraco, GmbH (Ismaning, Germany).

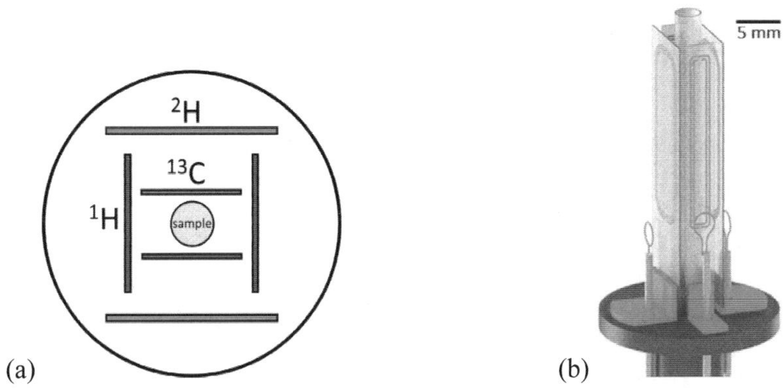

Figure 2. Design of the ^{13}C-optimized all-HTS NMR probe. Some dimensions and details are simplified and/or distorted for illustrative purposes and are not exact reproductions of the design. (a) Cross-sectional schematic of the coil arrangement. Three Helmholtz-like pairs of planar HTS coils are cooled to 20 K and surround the 1.5-mm sample tube. (b) Diagram of the NMR probe-head illustrating the arrangement of two pairs of HTS coils and the corresponding normal metal impedance-matching and tuning loops.

YBCO is used for its highly anisotropic crystal structure. The substrate-plane of each coil is aligned approximately parallel to the static field to preserve resonator Q-factor. To minimize the effect of the diamagnetism of the superconductor on the static magnetic field and improve lineshape, each conductor is patterned as several thin parallel wires. A standard Agilent Gifford-McMahon cooling system is used to generate cold helium gas which cools the coils to ~20K, well below the superconducting transition temperature of YBCO (93 K). To tune the resonance frequency of each coil to the needed probe frequencies the coils were laser trimmed after fabrication. Since each HTS coil's resonance frequency was appreciably shifted by the presence of the other resonators, the relevant probe resonances were the coupled resonances of all local resonators, not just the Helmholtz-like HTS pairs. A magnetic coupling model based on microwave filter theory was used to predict the individual coil resonance required to obtain the desired set of coupled modes and thereby, accurately tune the probe [12].

The ^{13}C coils were placed closest to the sample to maximize sensitivity by optimizing filling factor. The coil design utilized a double-sided counter-wound spiral resonant structure consisting of four turns per side. The full details of the new ^{13}C coil upgrades over the previous coils described in [2] are recorded in [13]. In short, in the previous design a gold overlayer was used to suppress the spurious modes created by the coil's thin parallel wires. In this probe the geometric parameters of the coils are redesigned using IE3D simulations to empirically predict the modes splitting and eliminate the need for the gold overlayer. The ^{1}H and ^{2}H coils employed the same design as in our previous probe of similar design described in [2]. Briefly, a single-sided interdigital-capacitor based design, known as a racetrack design, with 24 fingers and four gaps was used for the ^{1}H detection coils [22]. The ^{2}H coils employed a single-sided spiral resonant structure with 10 turns. Three movable impedance-matching loops were inductively coupled to the HTS coils and were adjusted to modify flux coupling with the nearest HTS resonator. Three movable shorted inductive tuning loops were employed to finely tune the coupled resonances of the probe during routine probe use. The tuning and matching loops were fashioned from normal-metal susceptibility-compensated wire. The entire coil and loop space was evacuated to provide thermal insulation from the sample, which was placed inside a sample tube fabricated from RF transparent material. The sample was maintained near room temperature using a standard Agilent heater/thermocouple unit, which regulates the temperature of the airflow through the sample tube. The impedance-matched Q-factors of the fully loaded and tuned probe in no external magnetic field at ~35 K and 0 dBm incident power were 6000, 3900, and 2100 for ^{13}C, ^{1}H, and ^{2}H, respectively [13].

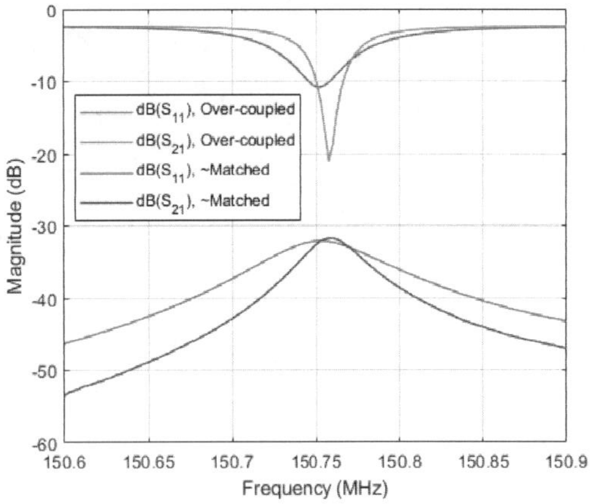

Figure 3. Measured S_{21} and S_{11} for ^{13}C channel

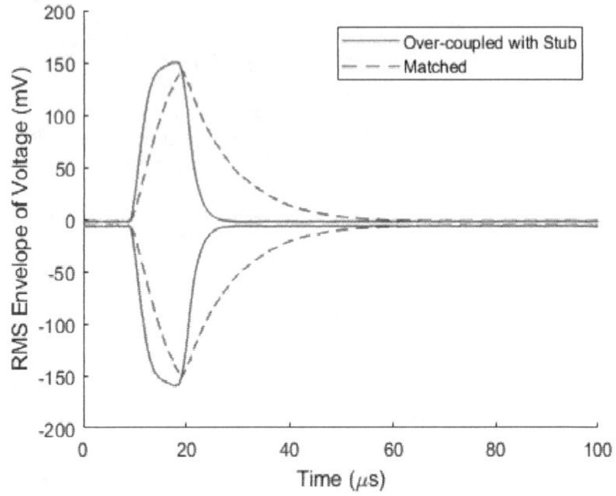

Figure 4. Measured ^{13}C pulse decay time

3.1. RF Specifications

At 14.1 T, the resonance frequencies and impedance-matched 3 dB Q-factors and bandwidths of the HTS probe were measured at ~20 K and 0 dBm incident power and are listed in the table below:

Table 1. RF measurements for the ^{13}C-optimized 1.5 mm HTS NMR Probe.

Nucleus	f_c (MHz)	$Q_{\text{matched, zero-field}}$	$Q_{\text{matched, in-field}}$	$\Delta f_{\text{matched, in-field}}$
^{13}C	150.75	6000	4300	35 kHz (200 ppm)
^{1}H	599.48	3900	1600	375 kHz
^{2}H	92.02	2100	2100	45 kHz

3.1.1. Carbon-13. The new ^{13}C HTS coil design preserved 71% of its zero-field Q-factor and exhibited 3.3 times the in-field Q-factor of 1300 observed in the previous probe. This decrease in loss represented a potentially large increase in ^{13}C detection sensitivity. However, the Q-factor was essentially equivalent to the zero-field Q-factor of the HTS coil tested in [15]. Therefore, distorted pulse shapes and increased NMR dead time were expected in the regular impedance-matched condition. In addition, the measured 3 dB bandwidth was 200 ppm. The 3 dB bandwidth should be several times the ~200 ppm ^{13}C spectral range in order to avoid suppressing peaks at the edges of the NMR spectrum. Thus, the bandwidth broadening procedure described above was applied to increase the detection bandwidth and reduce the rise/fall time of the ^{13}C irradiation pulses. First, the flux coupling between the matching loop and the ^{13}C coil was increased to an over-coupled condition of about 10 dB return loss. Figure 3 shows that the 3 dB bandwidth was increased from about 35 kHz to 70 kHz. To further reduce the rise/fall time, a shorted transmission line stub was added. The stub was made from a M-M and F-F adapter and a shorting plug. We varied the length of the coaxial cable attached to the stub to get the most bandwidth. Here we found that a cable length of 45 cm was optimal. We then applied a series of 9.45 μs square pulses at roughly 25 dBm to quantify the decay time constant (τ) and compare it with the matched condition. Figure 4 shows a reduction in τ from 8.9 μs to 1.3 μs. In total, the bandwidth of the irradiation pulse was increased from 200 ppm (Q_L = 4300) to 1660 ppm (Q_L = 600) by additional flux and stub coupling. For detection, the Q-factor was decreased from 4300 to an over-coupled Q-factor of 2100 by additional flux coupling.

3.1.2. ^{1}H and ^{2}H. The ^{2}H HTS coils used for the lock system demonstrated exceptional Q-factor retention at field. We measured approximately the same Q-factor as in zero-field. This may be due to particularly accurate coil-field alignment working in tandem with the tuning/matching loop optimization schemes laid out in [13]. It may also be due to enhanced HTS film quality. The ^{1}H detection coils, however, preserved only 42% of the zero-field Q-factor. Conversely to the ^{2}H coils, we attribute this to either weak coil-field alignment or, more likely, poor HTS film quality on one or both coils in the ^{1}H pair. This low power Q-factor was still higher than that of the previous probe. However, the RF dissipation of HTS thin films increases rapidly near critical current density [33], limiting the 90° pulse length to a significantly longer value that obtained in the previous version of the probe.

3.2. NMR Specifications

Table 2. NMR sensitivity measurements for the ^{13}C-optimized 1.5 mm HTS NMR probe.

	^{13}C	^{1}H
90° Pulse Length	19 μs	58 μs
S/N	526[a]	193[b]

[a] 40% Dioxane in C_6D_6 ASTM standard. Mean of four measurements with two per duplicate sample.
[b] 0.1% Ethylbenzene in $CDCl_3$ standard. Mean of four measurements with two per duplicate sample.

The ^{13}C signal-to-noise ratio (S/N) was measured to be 526:1 using a sealed ASTM standard of 40% dioxane in C_6D_6 (table 2). ^{1}H S/N was measured to be 193:1 using a sealed ASTM standard of 0.1% ethylbenzene in $CDCl_3$. The minimum 90° pulse lengths tested were 19 μs and 58 μs for ^{13}C and ^{1}H, respectively. While it may have been possible to slightly reduce pulse length by increasing power, an educated decision was made to stop increasing power to avoid damaging the coils when the relationship between power and pulse length became sufficiently nonlinear. The ^{13}C sensitivity was increased by 33% compared to the previous probe (396:1). Therefore, the new ^{13}C coils facilitate a significant increase in measured ^{13}C detection sensitivity. The ^{1}H S/N was 40% of the previous probe (490:1). While the ^{1}H pulse length was sufficient to effectively decouple the ^{1}H signal during ^{13}C measurements, the ^{1}H S/N is not high enough to perform cutting edge ^{1}H based NMR. We expect to be able to correct this issue by replacing these resonators.

NMR focused data such as lineshape and B_1 homogeneity for ^{13}C and ^{1}H will be discussed in a future publication. However, to vindicate the use of the transmission line stub, a ^{13}C 1D spectrum was collected for Ibuprofen dissolved in DMSO-d6 (Figure 5). Ibuprofen has signals that show up over an almost 200 ppm bandwidth. The measured ibuprofen spectrum showcases clear quantization of peak heights and no reduction in S/N for edge signals [34]. Therefore, our new HTS probe is capable of greatly improving sensitivity without sacrificing ^{13}C irradiation bandwidth.

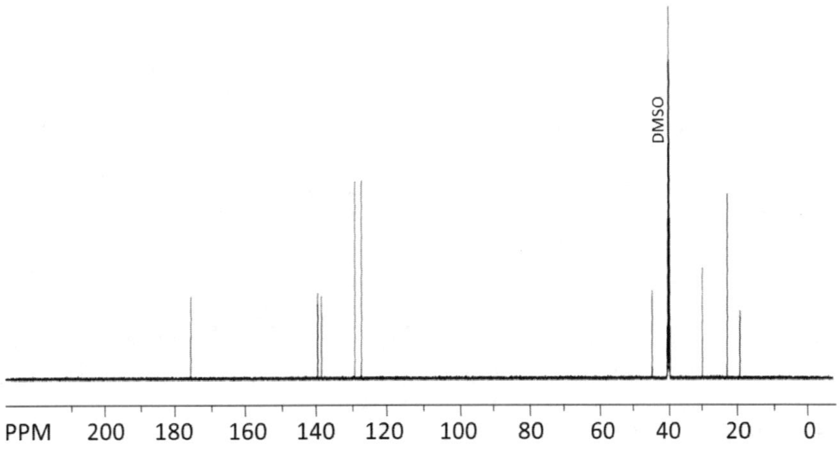

Figure 5. ^{13}C 1D NMR spectrum of Ibuprofen dissolved in DMSO-d6, 0.186 Molar (38.6 μg/μl)

4. Conclusions

We have reported a 330% increase in Q-factor and 33% increase in ^{13}C sensitivity for our 1.5 mm ^{13}C-optimized HTS NMR probe over our previous state of the art HTS probe design. We have also demonstrated that it is possible to utilize HTS coils to further increase ^{13}C NMR sensitivity while preserving a wide irradiation pulse bandwidth and avoiding distorted pulse shapes. We suspect that ^{13}C sensitivity can still be practically improved beyond what has been achieved here by reducing the resonator losses generated by the normal-metal tuning loops. Despite the outstanding improvement in ^{13}C sensitivity performance, the ^{1}H sensitivity was comparatively poor. In the future we will investigate the reason for the observed reduction in ^{1}H coil Q-factor in field and determine if the coils must be replaced. The exceptional ^{13}C sensitivity offered by our probe suggests that it can be used for cutting edge ^{13}C experiments requiring previously unavailable sensitivity. Therefore, we anticipate that this probe will have broad application in metabolomics and natural products research. Future NMR probe designs utilizing high-Q resonators may implement the ascribed over-coupling techniques to broaden bandwidth and reduce the rise/fall time of irradiation pulses.

Acknowledgements

This work was supported by NIH/NIGMS Grant P41 GM122698. It was carried out in part at the National High Magnetic Field Laboratory which is supported by the NSF through Cooperative under Agreement DMR-1644779 and by the State of Florida. The authors would like to thank Richard S. Withers for introducing to us the stub broadening concept. The authors would also like to thank Steve Ranner and Jason Kitchen for valuable technical support.

References

[1] Lee J H, Okuno Y and Cavagnero S 2014 Sensitivity enhancement in solution NMR: Emerging ideas and new frontiers *J. Magn. Reson.* **241** 18–31
[2] Ramaswamy V, Hooker J W, Withers R S, Nast R E, Brey W W and Edison A S 2013 Development of a C-13-optimized 1.5-mm high temperature superconducting NMR probe *J. Magn. Reson.* **235** 58–65
[3] Yamada T, Saito A, Oikawa S, Koshita K, Takahashi M, Maeda H and Ohshima S 2015 Electromagnetic Evaluation of HTS RF Coils for Nuclear Magnetic Resonance *IEEE Trans. Appl. Supercond.* **25** 1–4
[4] Anderson W A, *et al.* 1995 High-sensitivity NMR spectroscopy probe using superconductive coils *Bull. Magn. Reson.* **17** 98–102
[5] Oikawa S, Tanaka Y, Yamada T, Kanamaru A, Takahashi M, Saito A and Ohshima S 2014 Evaluation of superconducting pickup coils with high Q for 700 MHz NMR *J. Phys. Conf. Ser.* **507** 042028
[6] Hill H D W 1997 Improved sensitivity of NMR spectroscopy probes by use of high-temperature superconductive detection coils *Ieee Trans. Appl. Supercond.* **7** 3750–5
[7] Clendinen C S, Lee-McMullen B, Williams C M, Stupp G S, Vandenborne K, Hahn D A, Walter G A and Edison A S 2014 13C NMR Metabolomics: Applications at Natural Abundance *Anal. Chem.* **86** 9242–50
[8] Clendinen C S, Stupp G S, Ajredini R, Lee-McMullen B, Beecher C and Edison A S 2015 An overview of methods using 13C for improved compound identification in metabolomics and natural products *Front. Plant Sci.* **6**
[9] Clendinen C S, Pasquel C, Ajredini R and Edison A S 2015 13C NMR Metabolomics: INADEQUATE Network Analysis *Anal. Chem.* **87** 5698–706
[10] Patterson R E, *et al.* 2016 Lipotoxicity in steatohepatitis occurs despite an increase in tricarboxylic acid cycle activity *Am. J. Physiol. Endocrinol. Metab.* **310** E484-494
[11] Frelin O, *et al.* 2015 A directed-overflow and damage-control N-glycosidase in riboflavin biosynthesis *Biochem. J.* **466** 137–45

[12] Thomas J N, Ramaswamy V, Johnston T L, Belc D C, Freytag N, Hornak L A, Edison A S and Brey W W 2020 Modeling the Resonance Shifts Due to Coupling Between HTS Coils in NMR Probes *J. Phys. Conf. Ser.* **1559** 012022
[13] Thomas J N, Ramaswamy V, Litvak I M, Johnston T L, Edison A S and Brey W W 2021 Progress Towards a Higher Sensitivity 13C-Optimized 1.5 mm HTS NMR Probe *IEEE Trans. Appl. Supercond.* **31** 1–4
[14] Agarwal A and Lang J 2005 *Foundations of Analog and Digital Electronic Circuits* (Morgan Kaufmann)
[15] Amouzandeh G, Ramaswamy V, Freytag N, Edison A S, Hornak L A and Brey W W 2019 Time and Frequency Domain Response of HTS Resonators for Use as NMR Transmit Coils *IEEE Trans. Appl. Supercond.* **29** 1–5
[16] Estill I 1955 *The Story of Q*
[17] Brey W W, Edison A S, Nast R E, Rocca J R, Saha S and Withers R S 2006 Design, construction, and validation of a 1-mm triple-resonance high-temperature-superconducting probe for NMR *J. Magn. Reson.* **179** 290–3
[18] Shulman R G and Rothman D L 2005 *Brain Energetics and Neuronal Activity: Applications to fMRI and Medicine* (John Wiley & Sons)
[19] Ramaswamy V, Edison A S and Brey W W 2017 Inductively-Coupled Frequency Tuning and Impedance Matching in HTS-Based NMR Probes *IEEE Trans. Appl. Supercond.* **27** 1–5
[20] Suddarth S 1998 A method for matching high-temperature superconductor resonators used for NMR signal pickup at 400 MHz *IEEE Trans. Biomed. Eng.* **45** 1061–6
[21] Withers R S 1996 Inductively coupled superconducting coil assembly
[22] Ramaswamy V, Hooker J W, Withers R S, Nast R E, Edison A S and Brey W W 2013 Microsample Cryogenic Probes: Technology and Applications *eMagRes* **2** 215–28
[23] Decorps M, Blondet P, Reutenauer H, Albrand J P and Remy C 1985 An inductively coupled, series-tuned NMR probe *J. Magn. Reson. 1969* **65** 100–9
[24] McNichols R J, Wright S M, Wasser J S and Coté G L 1999 An inductively coupled, doubly tuned resonator for in vivo nuclear magnetic resonance spectroscopy *Rev. Sci. Instrum.* **70** 3454
[25] Froncisz W, Jesmanowicz A and Hyde J S 1986 Inductive (flux linkage) coupling to local coils in magnetic resonance imaging and spectroscopy *J. Magn. Reson.* **66** 135–43
[26] Jacquinot J-F and Sakellariou D 2011 NMR signal detection using inductive coupling: Applications to rotating microcoils *Concepts Magn. Reson. Part A* **38A** 33–51
[27] Taber B and Zens A 2015 Using Magnetic Coupling to Improve the 1H/2H Double Tuned Circuit *J. Magn. Reson.* **259** 114–20
[28] Hoult D I and Tomanek B 2002 Use of mutually inductive coupling in probe design *Concepts Magn. Reson.* **15** 262–85
[29] Raad A and Darrasse L 1992 Optimization of NMR Receiver Bandwidth by Inductive Coupling *Magn. Reson. Imaging* **10** 55–65
[30] Jakobsen H J, Bildsøe H, Gan Z and Brey W W 2011 Experimental aspects in acquisition of wide bandwidth solid-state MAS NMR spectra of low-γ nuclei with different opportunities on two commercial NMR spectrometers *J. Magn. Reson.* **211** 195–206
[31] Cheng D 1989 *Field and Wave Electromagnetics* (Tsinghua University Press)
[32] Hoult D I, Richards R E and Styles P 1978 A Novel Field-Frequency Lock for a Superconducting Spectrometer *J. Magn. Reson. 1969* **30** 351–65
[33] Black R D, Early T A and Johnson G A 1995 Performance of a High-Temperature Superconducting Resonator for High-Field Imaging *J. Magn. Reson. A* **113** 74–80
[34] Spectral Database for Organic Compounds (SDBS); mass spectrum; SDBS No.: 10570; CDS 11-007; http://riodb01.ibase.aist.go.jp/sdbs/ (accessed December 2, 2021).

Design and fabrication of microwave transmitting antenna using HTS thick disk for wireless power transfer

A Saito[1], F Shimada[1], T Sato[1], K Kiyooka[1], M Shibata[2], S Ono[3], M Takeda[4], Y Narita[1], K Nakajima[1]

[1] Graduate School of Science and Engineering, Yamagata University, 4-3-16, Jonan, Yonezawa, 992-8510 Japan.
[2] Fuji Electric Co., LTD., 1-11-2, Osaki, Shinagawa, 141-0032 Japan.
[3] The University of Electro-Communications, 1-5-1, Chofu, 182-8585 Japan.
[4] Graduate School of Engineering, Shizuoka University, Hamamatsu, 432-8011 Japan.

atsu@yz.yamagata-u.ac.jp

Abstract. We designed and fabricated a microwave transmitting antenna that uses an HTS thick disk as a patch antenna with a resonant frequency of around 5.0 GHz. The reflection property S_{11}, electromagnetic fields, current distributions, and far-field of several resonant modes in the antenna were analyzed using the CST studio suite. We compared the properties in several resonance modes and selected the TM11 mode, which has the smallest resonator size and the highest directivity. The antenna was assembled using a GdBCO thick disk, sapphire substrates, Cu housing, and SMA connector. We measured the S_{11} using a vector network analyzer and pulse tube refrigerator and found that it had a resonant frequency of 4.996 GHz and a Q_u of 405 at the temperature of 48 K. These measured properties were in good agreement with the simulated results.

1. Introduction

Wireless power transmission (WPT) is classified into radiative and non-radiative transmission. Non-radiative transmission can achieve high-efficiency transmission over short distances, whereas in radiative transmission, while there is the possibility of long-distance transmission, it is very difficult to improve the transmission efficiencies [1–8]. Yang has reported that the transmission efficiency of 46.9% with a distance of 100 cm can be obtained by using 8×8 array antennas for both transmitting and receiving [8]. In order to improve the transmission efficiency further, it is necessary to dramatically improve both the directivity and the losses of the antenna system. For this purpose, superconducting antenna arrays made of high-temperature superconductor (HTS) thin films have attracted interest because their directivity increases linearly with an increasing number of elements. However, the power-handling capability (PHC) of the antennas is not strong enough for the practical application of WPT [9, 10]. We previously studied microwave bandpass filters (BPFs) using HTS bulk resonators and observed a 100-W class of PHC in five-pole Chebyshev-type BPFs using $GdBa_2Cu_3O_y$ (GdBCO) superconducting thick disk resonators [11]. This result suggests that a high-quality superconducting bulk is a promising material for superconducting antennas intended for high-power transmission, but as yet there has been no investigation into this possibility. Therefore, in the current study, we developed a microwave transmitting antenna that uses an HTS thick disk as a patch antenna and clarified its suitability.

2. Design of microwave antenna using thick disk resonator

2.1. Antenna design

We designed a microwave antenna intended to have a resonant frequency of around 5.0 GHz. The frequency responses, electromagnetic fields, current distributions, and far-field of the antenna were simulated using a three-dimensional electromagnetic simulator (CST studio suite). Figure 1 shows a cross-section of the schematic design from the (a) top and (b) side of the antenna with Cu housing, sapphire substrate, and feed port, the sizes of which were optimized for the desired target frequency response. First, to obtain the resonant frequency of around 5.0 GHz and coupling coefficient (β) of around 1.0, the radius of the HTS disk resonators was optimized (ϕ = 7.50 mm). The distance between the resonator and the center pin of the coaxial connector, gap distance (g), was also optimized (g = 3.81 mm). The thickness of the disk resonator was 0.5 mm, which is the same as the diameter of the center pin of the coaxial connector of the feed port. The dimensions of the Cu housing and the inner sapphire substrate are listed in Table 1. To simulate the frequency responses, we used an effective dielectric constant of 8.35 and a loss tangent of 10^{-4} for the sapphire substrates. Intrinsic value of 5.80×10^{7} S/m was used for the conductivity of the Cu material. The loss for the HTS material could not be precisely defined, so we assumed the conductivity of the HTS material of 5.80×10^{9} S/m that was two orders of magnitude larger than that of Cu material [12].

2.2. Electromagnetic simulations

Figure 2 shows the simulated results of the S_{11} reflections, current distributions, and far-field patterns of the optimized HTS disk antenna. The resonant modes at 5.008 GHz and 7.466 GHz were confirmed to be TM11 and TM21 modes from each current distribution. In our analysis of each resonant mode, the gap distance (g) was adjusted so that the coupling coefficient (β) approached 1.0. The β is expressed by the following equation.

$$\beta = \left(1 \pm \left|S_{11}(f_0)\right|\right) / \left(1 \mp \left|S_{11}(f_0)\right|\right) \quad (1)$$

From the Smith chart of S_{11}, we used $\beta > 1$ for overcoupled and $\beta < 1$ for undercoupled. Here $S_{11}(f_0)$ is a dimensionless S_{11} value at a resonant frequency and can be calculated by the following formula.

$$S_{11}(f_0) = 10^{S_{11}(f_0)[\mathrm{dB}]/20} \quad (2)$$

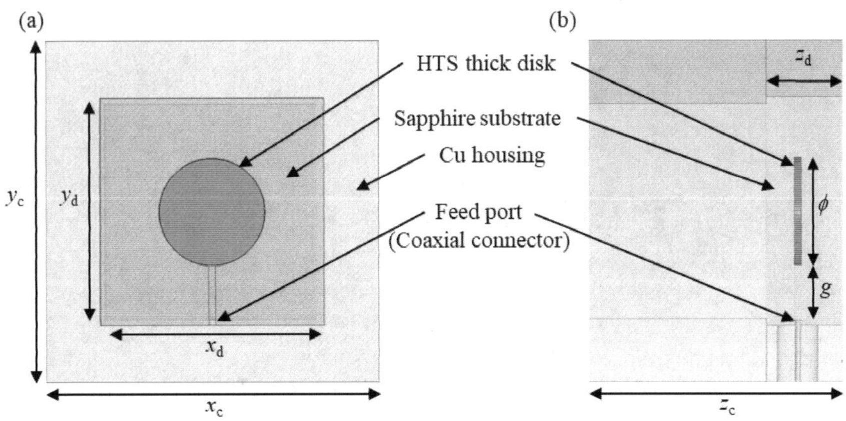

Figure 1. Simulation model of HTS disk antenna. (a) Top view. (b) Side view.

Table 1. Dimensions of HTS disk antenna.

	x [mm]	y [mm]	z [mm]
Cu housing	$x_c = 24.0$	$y_c = 24.0$	$z_c = 18.5$
Sapphire substrate	$x_d = 16.0$	$x_d = 16.0$	$x_d = 6.50$

These simulated values of $S_{11}(f_0)$, β, unloaded quality factor (Q_u), and directivities (D) are listed in Table 2. We compared these properties in several resonance modes and selected the TM11 mode due to its smallest resonator size and highest directivity. For comparison, the Q_u and D of the TM11 mode for the Cu antenna calculated with the resonator conductivity of 5.80×10^7 S/m were 1094 and 8.289 dBi, respectively.

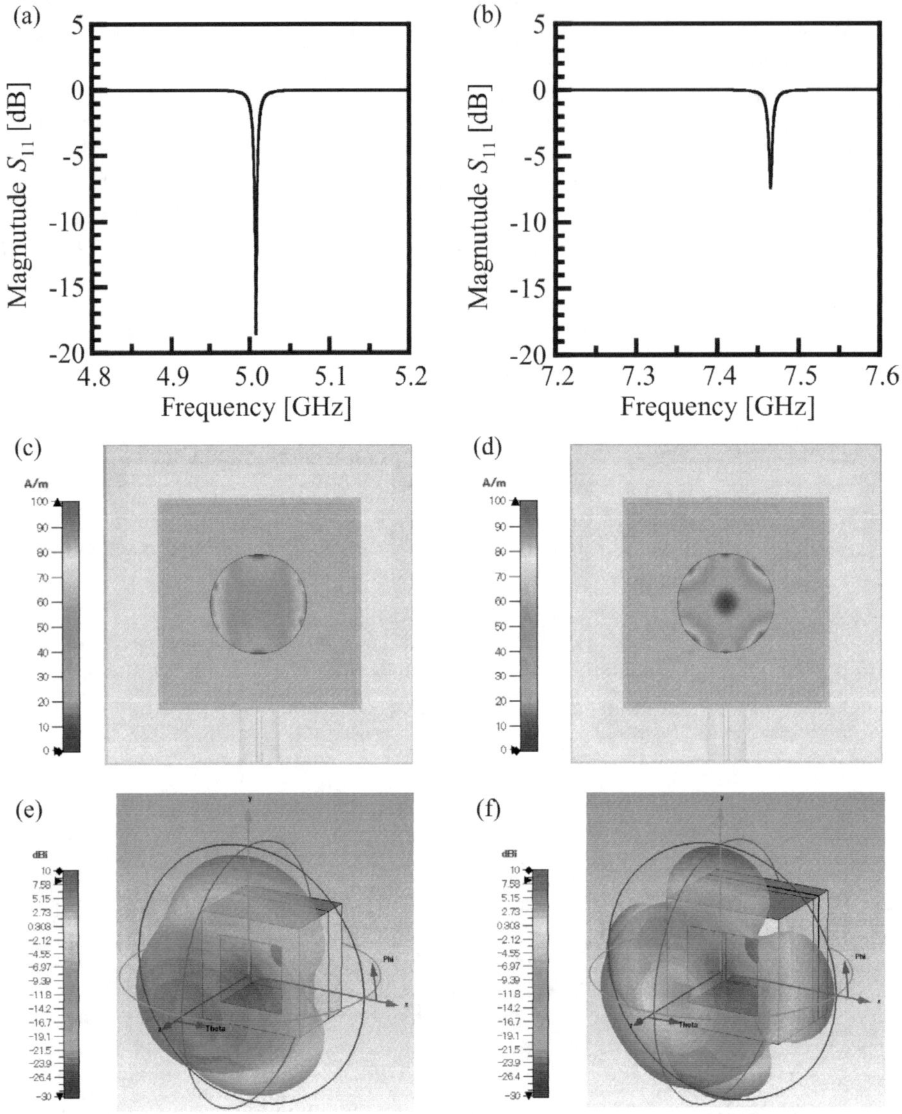

Figure 2. Simulated results of HTS disk antenna. Frequency responses: (a) TM11, (b) TM21. Current distributions: (c) TM11, (d) TM21. Far-field patterns: (e) TM11, (f) TM21.

Table 2. Simulated results of HTS disk antenna.

Resonant mode	Resonant frequency f_0 [GHz]	Gap distance g [mm]	$S_{11}(f_0)$	Coupling coefficient β	Unloaded quality factor Q_u	Directivity D [dBi]
TM11	5.008	3.81	1.266 (-18.6 dB)	1.27	1419	8.289
TM21	7.466	4.00	2.470 (-7.46 dB)	2.41	2879	8.234

Journal of Physics: Conference Series **2323** (2022) 012031 doi:10.1088/1742-6596/2323/1/012031

3. Fabrication of antenna using GdBCO disk resonator

A GdBCO superconducting bulk with AgO_2 was fabricated using a modified QMG process [13]. The typical critical temperature T_c was 93 K. The critical current density J_c was more than 10 kA/cm^2 at 77 K and 1 T. Details of the fabrication of the thick disk resonators were reported previously [14]. Figure 3 shows the assembly process of the antenna. After polishing the top and bottom faces of two c-sapphire substrates (3.0- and 3.5-mm-thick), we excavated the 3.5-mm-thick substrate to a depth of 0.5 mm to implant the disk resonator and feed port. The GdBCO thick disk was placed on the bottom substrate (Fig. 3(a)). The upper substrate was placed on top of the bottom one and the combined substrates, excluding the feed port, were wrapped with aluminum tape (Fig. 3(b)). The combined substrates and SMA connector were then set into the bottom of the Cu housing (Fig. 3(c)). Figure 3(d) shows the completed antenna with the upper Cu housing.

4. Measurement of antenna characteristics

After the antenna was put into a vacuum chamber and cooled to around 50 K by a small pulse tube refrigerator (Fuji Electric Co., Ltd.), the S_{11} reflection of the antenna was measured by a vector network analyzer (HP 8720D). The S_{11} characteristics of the antenna are shown in Fig. 4, where the

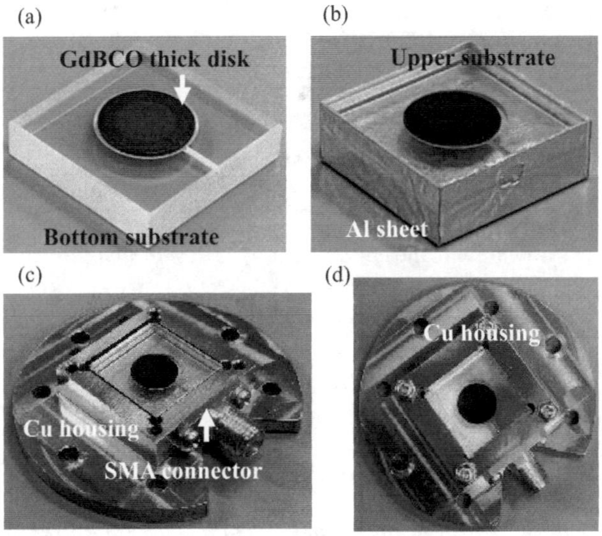

Figure 3. Photograph of assembly process. (a) GdBCO thick disk placed on bottom substrate made by c–sapphire. (b) Fixed by Al sheet after placing the upper substrate. (c) The substrates and SMA connector was set to bottom Cu housing. (d) Completed antenna with upper Cu housing.

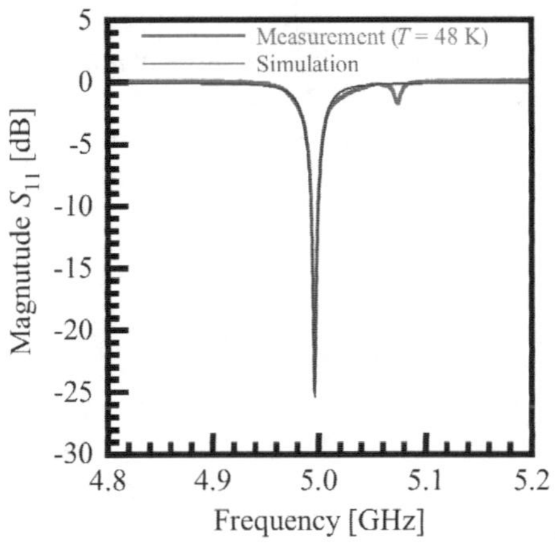

Figure 4. S_{11} characteristics of the antenna. Red line indicates the measured characteristics at 48 K. Blue line indicates the simulated data.

red and blue lines show the simulated and measured responses at 48 K, respectively. The measured S_{11} property showed a resonant frequency of 4.996 GHz, S_{11}(4.996 GHz) of -25.1 dB, and a Q_u of 405 at the temperature of 48 K and was in good agreement with the simulated one, which was used the relative permittivity of 8.428 and the conductivity of the Cu housing of 5.80×10^5 S/m. These results demonstrate that we can achieve a simple single-resonant antenna, and that the superconducting bulk resonator antenna is suitable for use as WPT. The reason for the low Q_u is that the conductivity of the Cu housing was smaller compared with the simulation and/or the radiation loss in the experiments was higher than that of the simulations. The improvement by Au coating of the Cu housing and the transmission experiments will be reported in the next paper.

5. Conclusion

We designed and tested a microwave antenna that uses an HTS thick disk resonator. Simulations to determine the optimum antenna model showed that TM11 mode was the best, demonstrating the directivity of 8.289 dBi. We then fabricated the antenna using a GdBCO thick disk and measured the S_{11} reflection at 48 K. The results demonstrate that we can achieve the designed antenna resonance using the HTS bulk disk antenna.

Acknowledgements

This work was supported by JSPS KAKENHI Grant Number JP20K04615 and Fuji Electric CO., LED. Some of this work was performed in the clean room at Yamagata University.

References

[1] Dharani M and Ravi K 2021 *Turkish Journal of Computer and Mathematics Education* **12** pp 4400-4406
[2] Yi X, Chen X, Zhou L and Hao S 2020 *AIP Advances* **10** 115111
[3] Tomura T, Hirokawa J, Furukawa M, Fujiware T and Shinohara N 2021 *Antennas and Propagation* **2** 170
[4] Wang H, Deng L, Luo H, Du J, Zhou D and Huang S 2021 *Energies* **14** 415
[5] Nguyen D M, Au N D and Seo C 2021 *IEEE Access* **9** 134163
[6] Malisuwan S, Tiamnara N and Suriyakrai N 2017 *J. Clean Energy Tech.* **5** 42
[7] Shinohara N 2020 *IEEE Journal of Microwave* **1** 218
[8] Yang X, Geyi W and Sun H 2017 *IEEE Antennas and Wireless Propagation Letters* **16** pp 1824-1827
[9] Ehata K, Ali M I, Kusunoki M and Ohshima S 1999 *IEEE Trans. Appl. Superdond.* **9** 3081
[10] Ehata K, Kusunoki M and Ohshima S 1999 *Electronics and Communication in Japan* **82** pp 915-923
[11] Saito A, Tsurui R, Kato T, Nakajima K, Teshima H and Ohshima S 2015 *APEX* **8** 043101
[12] Saito A, Teshima H, Ono S, Hirano H, Ohshima S 2007 *Physica C* **463-465** pp 1064-1067
[13] Morita M, Sawamura M, Takebayashi S, Kimura K, Teshima H, Tanaka M, Miyamoto K and Hashimoto M 1994 *Physica* C **445-448** 330
[14] Saito A, Lee J H, Teshima H, Ono S and Ohshima S 2009 *IEEE Trans. Appl. Superdond.* **19** 899

Compact Stochastic Computing Circuits Using the Latching Function of RSFQ Circuits for Computing Polynomials

Koki Wada and Nobutaka Kito

Graduate School of Engineering, Chukyo University, Aichi, Japan

E-mail: t32106m@m.chukyo-u.ac.jp

Abstract. Stochastic computing (SC) is a method of expressing a numerical value by the appearance ratio of 1 in a bitstream to perform an operation. Rapid single flux quantum (RSFQ) circuits can handle SC naturally and operate at ultra-high speed. RSFQ SC circuits are suitable for applications tolerating error. This paper presents a design method of RSFQ SC circuits to calculate polynomials. Using the proposed design method, an internal circuit is designed to feed bitstreams of coefficients from bitstreams of constant to reduce large stochastic number generators. The method inserts storage elements to reduce correlations of signals, and thereby reduce the calculation error of the circuit. Additionally, it minimizes the number of inserted D flip-flops utilizing the latching function of gate inputs of RSFQ logic gates.

1. Introduction

Rapid single flux quantum (RSFQ) circuits [1] have garnered attention as ultra-high-speed and low-power computing devices, which use pulse logic. In particular, these circuits express logical values with voltage pulses. The operation of a gate changes according to the order of pulse arrivals to the input terminals of the gate. Each gate input can obtain a function as a storage element by tuning the order of pulse arrivals. In this paper, we refer to the functionality as the latching function.

Various calculation circuits for RSFQ circuits have been proposed. Particularly, studies have proposed parallel high-throughput calculation circuits [2] [3]. Because RSFQ circuits use splitter elements for wiring distribution, parallel computing circuits consume a large area. Thus, area-efficient circuits such as bit-serial arithmetic circuits have been proposed [4] [5].

Recently, stochastic computing (SC) has been actively studied in the research community of semiconductor integrated circuit designs for applications tolerating error, such as neural network processing. In SC, the stochastic number (SN) is expressed by a bitstream consisting of 1 and 0 on a single wire. The value corresponding to an SN is the probability of 1 in it. An AND gate is used to perform multiplication in SC. Therefore, SC uses a small number of wires and gates. Consequently, we can design compact superconducting circuits using SC. Previously, superconducting SC-based circuits for deep-learning processing [6] using Adiabatic Quantum-Flux-Parametron (AQFP) [7] and SC-like RSFQ multiplication and matrix multiplication circuits [8] [9] have been proposed. In this paper, we utilize the latching function of RSFQ circuits for reduction of gates in SC circuit design.

Here, we propose a design method for RSFQ SC circuit to compute polynomials. The inputs of the SC circuit include the SN of an argument of a polynomial and one or more SNs of arbitrary constant

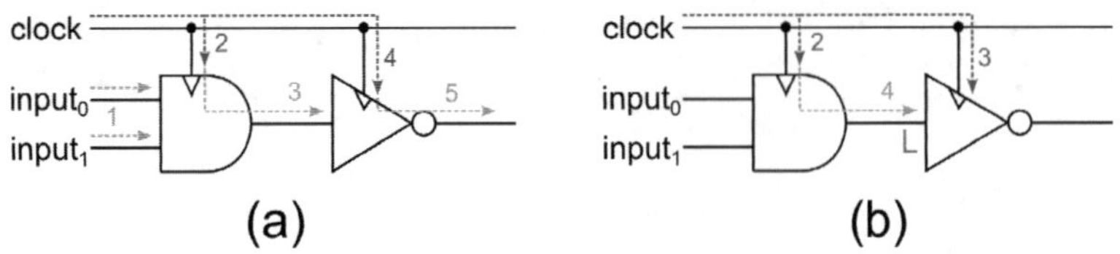

Figure 1. Clocking scheme used in this paper. (a) Circuit using clock-follow-data clocking. The gate input of the inverter in (b) acquires the latching function. “L” denotes the enabled latch-function and the numbers denote the order of pulse arrivals.

values. The output of the SC circuit is the SN of the calculation result. Furthermore, a polynomial to be computed and a set of arbitrary constant values supplied as SNs are input to the design method.

The method consists of three steps. First, it generates an SC circuit from a transformed formula, which contains several coefficients, of the given polynomial. Next, the method prepares an internal circuit to input the SNs of coefficients. Preparing an internal circuit instead of large SN generators saves circuit area. In addition, correlations between SNs compromise the accuracy of the calculation in SC. These correlations can be reduced by inserting storage elements in data lines and shifting arrivals of bits. Finally, the method inserts D flip-flops (DFFs) or enables the latching function to reduce the calculation error caused by the correlations. The method minimizes the number of DFF insertions by using the latching function of gate inputs.

We designed circuits for several polynomials to validate the proposed design method. The gate counts of the circuits were at most 30 including DFFs and were comparable with those of four full adders (FAs) while an ordinary non-SC multiplier uses many FAs. Thus, the circuit area was small. The absolute error of the circuits was not large. Namely, the number of pulses differed around three in average from the correct number in the evaluations of 100 clock cycles.

The remainder of this paper is organized as follows. Section 2 describes RSFQ circuits and SC. Section 3 presents the design of RSFQ SC circuits to calculate polynomials and a design method. Section 4 presents the evaluation results. Section 5 concludes this paper.

2. Preliminaries

2.1. RSFQ Circuits

RSFQ circuits express logical values by voltage pulses. As each logic gate of an RSFQ circuit has a clock terminal, SNs, which are time-divided bitstreams, can be handled efficiently. In RSFQ circuits, the gate operation changes according to the arrival order of pulses to the input terminals of the gate.

A data input that arrives at a gate input terminal before its corresponding clock pulse is used for calculation once the clock pulse arrives. Figure 1(a) shows an example of this process. Once a clock pulse arrives at a gate, the calculation result of the corresponding data inputs arriving before the clock pulse is produced as output. This clocking scheme is known as clock-follow-data clocking.

Additionally, a data pulse arriving after its corresponding clock pulse, as shown in Figure 1(b), is held at the gate input. This data pulse is used for the calculation triggered by the next clock pulse. Therefore, the input works as if it contains a storage element. The functionality of the gate input is called the latching function. Each gate input can acquire the latch function according to the arrival order of pulses.

In this study, we design an entire circuit with clock-follow-data clocking. Then, we selectively enable the latch function by adjusting the order of pulse arrivals for gates.

2.2. Stochastic Computing

SC is a calculation method that uses probabilities to perform calculations [10, 11], and is suitable for applications tolerating calculation error, such as neural networks. A numerical value is expressed

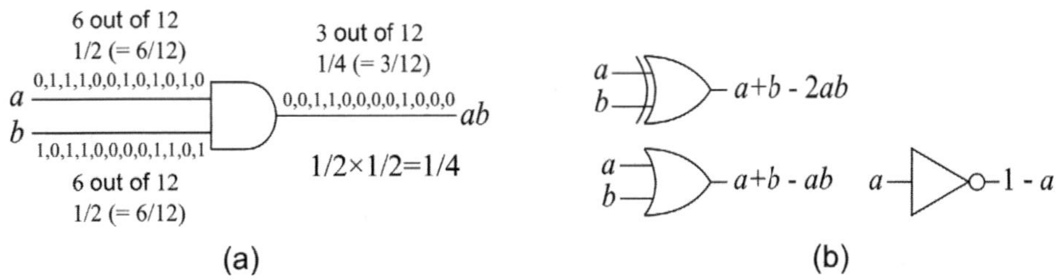

Figure 2. Calculation in SC circuits. (a) Example of multiplication with an AND gate. (b) Calculations of other logic gates.

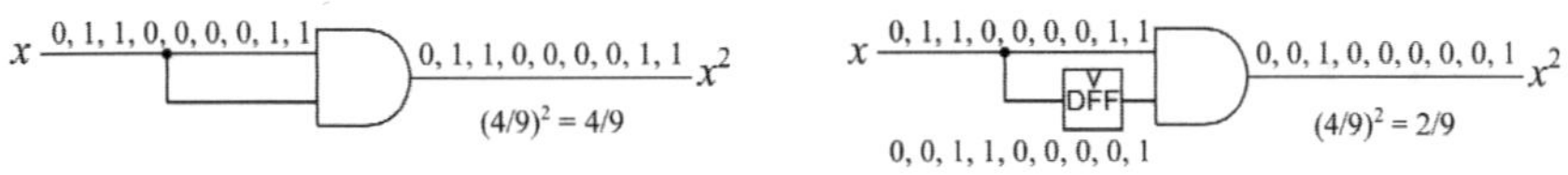

Figure 3. Example of calculation error by a correlation. Both circuits calculate x^2 for $4/9$. The correct result is $16/81 (\approx 1.77/9)$. The correlation worsens the error in the left circuit. The DFF removes the correlation, and the error is small in the right circuit.

by the probability of 1 in a bitstream consisting of 0s and 1s. Although multiplication of values in binary representation consumes many gates, in SC, an AND gate is used for multiplication. Figure 2(a) demonstrates multiplication with an AND gate. In the figure, logic values of 12 time-frames are shown for each line. A value corresponding to a line is the ratio of 1s in 12 time-frames. We can use other logic gates in SC. Figure 2(b) presents calculations of other gates.

An SC circuit produces calculation error inherently, and correlations among gate inputs worsen the calculation error. Thus, feeding bits depending on the same bit to inputs of a gate worsen the error. To eliminate correlations, we insert storage elements, such as DFFs, into signal lines, and thereby avoid feeding values similar to the input bits of the gate. Figure 3 presents an example, wherein the left circuit receives the same input bits, resulting in a significant calculation error. To avoid feeding the same bits, we insert a DFF to shift the input timing of the gate input of the AND gate.

3. Compact RSFQ SC circuits for computing polynomials and a design method

We propose a design of RSFQ SC circuits to calculate polynomials and a design method for the circuits. The inputs of the RSFQ SC circuits are an SN of an argument in a polynomial and one or more SNs of constant values. The SC circuit outputs an SN of the calculation result. We explain the targeted polynomials in Section 3.1 and show the structure of the circuits in Section 3.2.

Section 3.2 presents the design method. A polynomial and a set of arbitrary constant values supplied as SNs for the designed circuit are inputs for the design method.

The design method consists of three steps. First, an initial SC circuit is derived from the given polynomial. Then, a circuit is attached to the initial circuit for feeding SNs of coefficients. However, inputs of many gates are correlated in the circuit. Therefore, in the final step, storage elements, such as DFFs or gate inputs with the latching function, are inserted to remove correlations. We minimize the number of DFFs to be inserted by using the latching function.

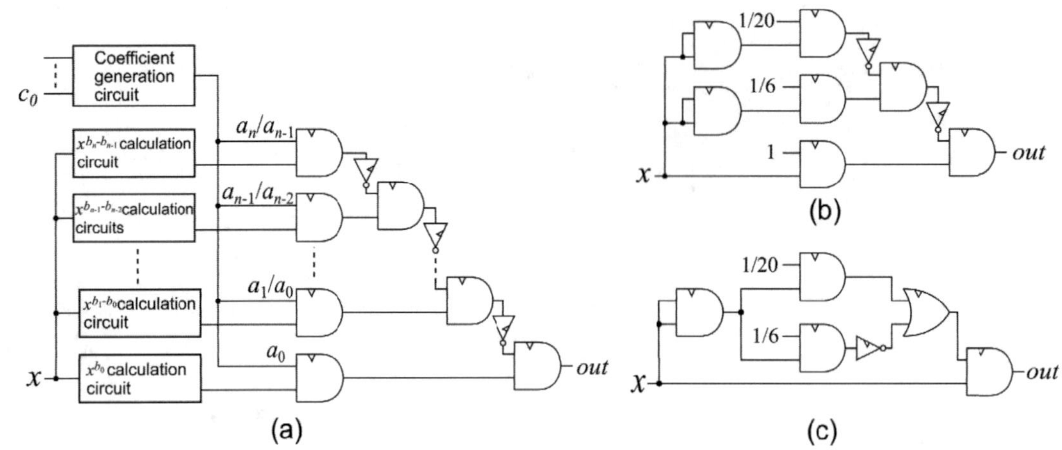

Figure 4. Polynomial calculation circuits. (a) Structure of the polynomial calculation circuits. (b) and (c) Examples for $x - \frac{x^3}{6} + \frac{x^5}{120}$ without coefficient generation circuits. A duplicated gate and a useless gate are removed in (c).

3.1. Targeted polynomials

We consider polynomials of the following structure in this paper:

$$\sum_{i=0}^{n}(-1)^i a_i x^{b_i} = a_0 x^{b_0} - a_1 x^{b_1} + \cdots + (-1)^n a_n x^{b_n}$$

where x is the argument of the polynomial, $0 < a_j \leq a_i \leq 1$, and $0 < b_i < b_j (i < j)$. An example of polynomials of this form is $x - \frac{x^3}{6} + \frac{x^5}{120}$, which includes the first three terms of the Maclaurin expansion of $\sin(x)$.

3.2. Structure of the RSFQ SC circuits

We can transform a polynomial into the following form

$$\sum_{i=0}^{n}(-1)^i a_i x^{b_i} = a_0 x^{b_0}\left(1 - \frac{a_1}{a_0}x^{b_1-b_0}\left(1 - \frac{a_2}{a_1}x^{b_2-b_1}\cdots\left(1 - \frac{a_{n-1}}{a_{n-2}}x^{b_{n-1}-b_{n-2}}\left(1 - \frac{a_n}{a_{n-1}}x^{b_n-b_{n-1}}\right)\right)\cdots\right)\right).$$

The subtractions from 1 can be realized using NOT gates and multiplications, such as $\frac{a_i}{a_{i-1}}x^{b_i-b_{i-1}}$, using AND gates. Figure 4(a) demonstrates the structure of the circuit used to calculate the polynomial. $\frac{a_{i-1}}{a_{i-2}}$ denotes constant coefficients. A coefficient generating circuit generates SNs corresponding to the coefficients from SNs of arbitrary constants provided through $c_0, \ldots$. Moreover, generating the SNs of the coefficients with the circuit instead of using large SN generators can save circuit area. In addition, we can simplify the structure by sharing the same internal circuits, although there are n calculation circuits for $x^{b_i-b_{i-1}}$ in the figure. The circuit contains correlations, which must be removed (Section 3.4) to obtain the circuit for calculation.

As an example, we transform $x - \frac{x^3}{6} + \frac{x^5}{120}$ as follows:

$$x - \frac{x^3}{6} + \frac{x^5}{120} = x\left(1 - \frac{x^2}{6} + \frac{x^4}{120}\right) = x\left(1 - \frac{x^2}{6}\left(1 - \frac{x^2}{20}\right)\right).$$

Figure 4(b) demonstrates a circuit obtained directly from the formula. We omit a coefficient generation circuit here. As shown in the figure, a gate for calculating x^2 is duplicated, and the bottom AND gate

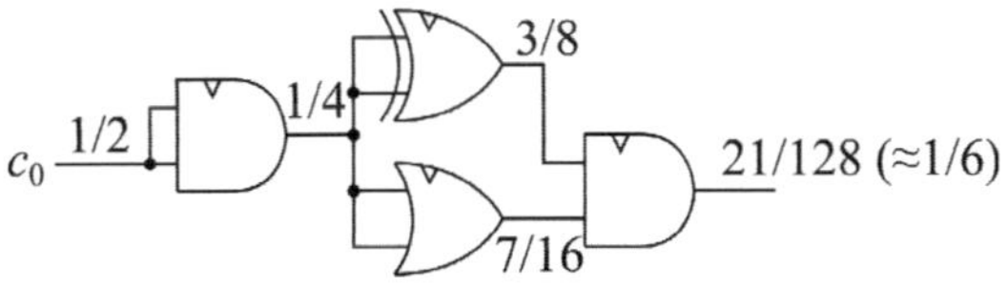

Figure 5. Derivation of an SN for coefficient 1/6 from a SN of 1/2. The removal of correlations (Section 3.4) is necessary for an accurate calculation.

is not used. Therefore, we eliminate such gates, as well as a gate with DeMorgan's law, as depicted in Figure 4(c).

3.3. Design of a coefficient generation circuit

We propose a design method of the coefficient generation circuit. Using this method, we generate SNs corresponding to coefficients in the transformed formula from one or more SNs of arbitrary constants.

Algorithm 1 shows the algorithm to design the circuit. G contains pairs of a constant and its derivation. It gradually expands G by using the calculation of logic gates shown in Figure 2. There are many ways to generate the same constant. We save a circuit with the lowest depth for a constant. The algorithm generates circuits each of which consisting of a single gate at first and then generates larger circuits by combining circuits generated previously. Lines 6, 9, 12, and 16 avoid adding deeper circuits for the values obtained with circuits with lower depth. $const(G)$ represents a set of constants which circuits in G can generate. Once we insert a constant and a circuit deriving the constant into G_d which is merged into G, any circuit found for the constant after the insertion is not smaller than this one. Those lines prevent to insert such new circuits to G_d. For example, we can derive $3/4$ from $1/2$ as "$OR(1/2, 1/2)$" and "$NOT(AND(1/2, 1/2))$." The algorithm finds the former one at the first iteration of Lines 2-21. Then it finds the latter one in the second iteration. It adds the former one into G_d and avoids inserting the latter one according to Line 16. Finally, we choose the nearest values from G for generating coefficients.

For example, we show a circuit for coefficient $1/6$ from an SN of $1/2$ for input c_0. The exploration depth d_{max} is set to be 3. The algorithm outputs the nearest value $21/128$ of $1/6$ and its derivation "$AND(XOR(AND(c_0, c_0), AND(c_0, c_0)), OR(AND(c_0, c_0), AND(c_0, c_0)))$." Figure 5 demonstrates the circuit derived from the result, which contains correlations between gate inputs. The method for removing these correlations and consequently reducing calculation error is described in the next section.

3.4. Insertion of storage elements for removing correlations

Correlations at gate inputs worsen the calculation error, as mentioned in Section 2.2. Therefore, we eliminate correlations by inserting storage elements in signal lines and shifting the data arrivals. We insert both DFFs and the latch function and minimize the number of DFFs to be inserted. We formulate the insertion as an integer linear programming (ILP) problem, which identifies assignments of integers for variables to minimize the linear objective function under linear constraint formulas.

To avoid correlations, we take care of the range of time frames of a circuit input that a gate input value depends on. We insert DFFs such that the range for an input of a gate and the range for the other input of the gate are not overlapped. Therefore, we insert DFFs so that each path from a circuit input to a gate goes through a different number of storage elements with each other. For example, at time frames t, both inputs of the AND gate in the left circuit shown in Figure 3 depend on the circuit input of time frame t. The range of the two inputs overlaps. The two paths to the AND gate go through the same number of storage elements, i.e., zero in this circuit, resulting in correlation in the circuit. In the right circuit, the time frame of the gate input without DFF is t and that of the other with DFF is $t-1$, i.e., the value of the previous time frame. Thus, the ranges do not overlap; one path to the AND gate goes through no storage element, and the other goes through one storage element, resulting in no correlation. In this case, the gate output depends on the circuit input of time frames from $t-1$ to t.

We treat time frames as a pair of integers in the ILP formulation. We represent the time frames of input i of gate $g2$ as $(min_{g,i}, max_{g,i})$ with variables $min_{g,i}$ and $max_{g,i}$. We represent the time frames of

Algorithm 1 Design of a coefficient generation circuit

Input: Set of coefficients to generate T, set of pairs of a given constant and its corresponding input G_0, and exploration depth d_{max}
Output: Generated coefficients and derivations for them

$G = G_0$ ▷ G is initialized as pairs of a given constant and its corresponding input
for d from 1 to d_{max} **do**
  $G_d = \Phi$
  for $(i_c, i_p) \in G$ **do**
    for $(j_c, j_p) \in G$ **do** ▷ i_c, j_c are constant values and i_p,j_p are their derivations
      if $i_c j_c \notin const(G)$ **then**
        $G_d = G_d \cup \{(i_c j_c, AND(i_p, j_p))\}$
      end if
      if $i_c + j_c - i_c j_c \notin const(G)$ **then**
        $G_d = G_d \cup \{(i_c + j_c - i_c j_c, OR(i_p, j_p))\}$
      end if
      if $i_c + j_c - 2i_c j_c \notin const(G)$ **then**
        $G_d = G_d \cup \{(i_c + j_c - 2i_c j_c, XOR(i_p, j_p))\}$
      end if
    end for
    if $1 - i_c \notin const(G)$ **then**
      $G_d = G_d \cup \{(1 - i_c, NOT(i_p))\}$
    end if
  end for
  $G = G \cup G_d$
end for
for $target \in T$ **do**
  $result = \phi$
  $error = 1$
  for $(i_c, i_p) \in G$ **do**
    if $|i_c - target| < error$ **then**
      $result = i_p$
      $error = |i_c - target|$
    end if
  end for
  print "Result for $target$ is $result$"
end for

gate g as (min_g, max_g) with variables min_g and max_g. In addition, we describe the relations between time frames as linear equations. We prepare variable $d_{g1,g2,i}$ to express the number of DFFs between the output of gate $g1$ and input i of gate $g2$, and prepare variable $l_{g,i}$ to indicate whether the latching function of input i of gate g is enabled. We describe the range of time frames of gate input i of gate g with those variables as follows:

$$(min_{g,i},\ max_{g,i}) = (min_{src(i)} + d_{src(i),g,i} + l_{g,i},\ max_{src(i)} + d_{src(i),g,i} + l_{g,i})$$

where $src(i)$ represents the source gate of gate input i. The range of gate output is described as $(min_g, max_g) = (\min_i(min_{g,i}), \max_i(max_{g,i}))$. We also prepare linear equations to avoid overlaps of frames. The objective function is the sum of variables $d_{g1,g2,i}$. The problem can be solved using commercial ILP solvers such as CPLEX and Gurobi or open-source solvers such as lpsolve.

Figure 6 illustrates the circuit after inserting storage elements to the circuit in Figure 4(c). The circuit

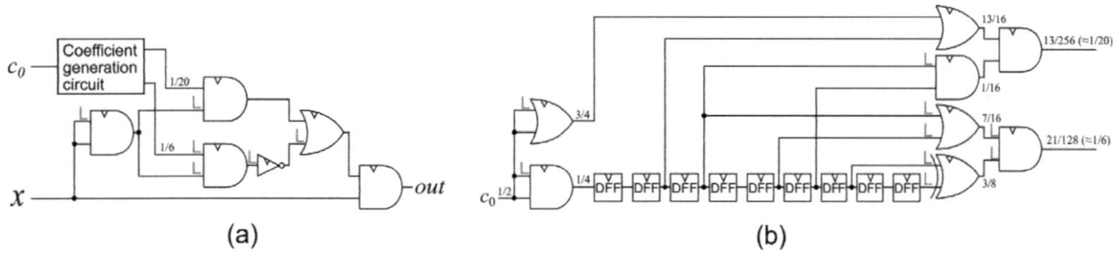

Figure 6. Designed RSFQ SC circuit for $x - \frac{x^3}{6} + \frac{x^5}{120}$. (b) is the coefficient generation circuit in (a).

consists of 14 logic gates and 9 DFFs. The latching function reduces the number of DFFs. There are two circuit inputs. In the circuit, for each pair of a gate and a circuit input, each path from the circuit input to the gate goes through a different number of storage elements.

4. Evaluation results

We implemented the proposed design method and used Python for implementing the program for designing the coefficient generation circuits. We employed lpsolve as the solver of the ILP problem for circuit design. To validate the proposed design method, we designed circuits for several polynomials. We used a 2 GHz quad-core Intel Core i5 processor for the evaluations.

The following three polynomials were used for the evaluations:

$$x - \frac{x^3}{6} + \frac{x^5}{120}, \quad x - \frac{x^3}{3} + \frac{2x^5}{15}, \text{ and } \quad \frac{x}{4} - \frac{x^4}{48} + \frac{x^5}{480}.$$

We evaluated the execution time of the exploration time in the design of the coefficient generation circuit and the execution time of the ILP solver for the insertions of storage elements, both of which are time-consuming processes of the design method. Table 1 lists the execution time of these two steps. As shown in the table, the execution time of the ILP solver is short. The execution time of the exploration was also short, even though programs written in Python tend to be slow. Table 1 also shows the gate counts of designed circuits. They were composed of at most 30 gates including DFFs and comparable with three or four full adders each of which consists of seven gates, i.e., five logic gates and two DFFs for path-balancing. They were smaller than RSFQ parallel adders and RSFQ multipliers such as [4]. As shown in Table 1, the latching function was utilized at many gate inputs. The seventh column of Table 1 shows the number of DFFs in circuits designed without the latching function. It indicates that the number of gates including DFFs reduced by about 30% with the latching function. For example, the circuit for the third polynomial consisted of 23 gates including 16 logic gates and 7 DFFs while it needed 33 gates including 17 DFFs if the latching function was not used. Thus, the number of gates for the polynomial reduced by 30.3%(= 10/33). We can conclude that the design algorithm for coefficient generation circuits and the reduction of DFFs utilizing the latching function work effectively.

We evaluated the absolute calculation error and the effectiveness of removing correlations by performing simulation of circuits. We used bitstreams of 100-bits for the evaluation. We varied the input x of the polynomial from 0 to 1 with 0.01 steps and ran 100 times. Table 2 presents the results, which demonstrate that the average error of the circuits with the correlation removal is smaller than that of circuits without the removal. The unit of the calculation result is 0.01 (= 1/100) here because bitstreams consisting of 100 bits were used. The average error is not large considering the unit.

5. Conclusion

We proposed a design of RSFQ SC circuits to calculate polynomials and a design method for them. The design method generates coefficient SNs from given constant SNs to reduce SN generators and save

Table 1. Execution time and size of the designed circuits

Polynomial	Design of the coefficient generation circuit [s]	lpsolve for removing correlations [s]	Logic gates	DFFs	Latching function	DFFs without latching function
$x - \frac{x^3}{6} + \frac{x^5}{120}$	0.013	0.79	14	9	15	17
$x - \frac{x^3}{3} + \frac{2x^5}{15}$	0.014	0.09	13	6	15	14
$\frac{x}{4} - \frac{x^4}{48} + \frac{x^5}{480}$	0.014	0.24	16	7	19	17

Table 2. Absolute calculation error (Average of 100 runs)

Polynomial	Obtained circuits		Without removing correlations	
	Maximum error	Average error	Maximum error	Average error
$x - \frac{x^3}{6} + \frac{x^5}{120}$	0.130	0.039	0.373	0.209
$x - \frac{x^3}{3} + \frac{2x^5}{15}$	0.114	0.032	0.213	0.074
$\frac{x}{4} - \frac{x^4}{48} + \frac{x^5}{480}$	0.083	0.028	0.280	0.133

circuit area. The proposed method eliminates correlations that worsen the calculation error by inserting storage elements. It utilizes the latching function for minimizing the number of DFFs to eliminate the correlations. The evaluated results show that the calculation error of the designed circuits is not large, and the circuit size is small.

References

[1] Likharev K and Semenov V 1991 *IEEE Transactions on Applied Superconductivity* **1** pp 3–28
[2] Nagaoka I, Ishida K, Tanaka M, Sano K, Yamashita T, Ono T, Inoue K and Fujimaki A 2021 *IEEE Transactions on Applied Superconductivity* **31** pp 1–5
[3] Nagaoka I, Tanaka M, Inoue K and Fujimaki A 2019 *IEEE International Solid-State Circuits Conference (ISSCC)* pp 460–462
[4] Hara H, Obata K, Park H, Yamanashi Y, Taketomi K, Yoshikawa N, Tanaka M, Fujimaki A, Takagi N, Takagi K and Nagasawa S 2009 *IEEE Transactions on Applied Superconductivity* **19** pp 657–660
[5] Tanaka M, Akaike H, Fujimaki A, Yamanashi Y, Yoshikawa N, Nagasawa S, Takagi K and Takagi N 2011 *IEEE Transactions on Applied Superconductivity* **21** pp 792–796
[6] Cai R, Ren A, Chen O, Liu N, Ding C, Qian X, Han J, Luo W, Yoshikawa N and Wang Y 2019 *46th International Symposium on Computer Architecture (ISCA '19)* pp 567–578
[7] Takeuchi N, Yamae T, Ayala C L, Suzuki H and Yoshikawa N 2022 *IEICE Transactions on Electronics* **E105.C**
[8] Kito N, Odaka R and Takagi K 2019 *IEICE Transactions on Electronics* **E102.C** pp 607–611
[9] Kito N and Takagi K 2021 *Journal of Physics: Conference Series* **1975** 012025
[10] Gaines B R 1967 *Spring Joint Computer Conference* pp 149–156
[11] Alaghi A and Hayes J P 2013 *ACM Trans. Embed. Comput. Syst.* **12** ISSN 1539-9087

Operation of a 4-bit RSFQ digital-to-analog converter based on a binary split-confluence configuration

Naonori Sega, Yuta Somei, Hiroshi Shimada and Yoshinao Mizugaki

Department of Engineering Science, The University of Electro-Communications (UEC Tokyo), 1-5-1 Chofugaoka, Chofu, Tokyo, 182-8585, Japan

E-mail: sega@w8-7f.ee.uec.ac.jp, y.mizugaki@uec.ac.jp

Abstract. One of the unique applications of single-flux-quantum (SFQ) circuitry is a digital-to-analog converter (DAC) of which the output voltage is defined by the Josephson effect. Recently, we have designed and operated rapid single-flux-quantum (RSFQ) DACs with frequency modulation of SFQ pulse sequences. In this paper, we report our design and test results of a 4-bit RSFQ-DAC based on the configuration of the "sum of selected bits sequence (Σ-SBS)". By adjusting the timing margins, in addition to the fabrication using the Nb integration process with the critical current density as high as 10 kA/cm^2, the maximum output voltage up to 260 μV was demonstrated.

1. Introduction

Rapid single-flux-quantum (RSFQ) logic has been developed as a next-generation circuit technology because of its high speed and low power consumption [1]. Besides digital signal processing, high-precision voltage generation based on the Josephson effect is a unique application of RSFQ circuitry for the realization of AC voltage standards [2]–[5]. Among several types of RSFQ-DACs, we have developed circuit configurations of single-flux-quantum (SFQ) pulse-frequency modulation (PFM) [6]. Our SFQ-PFM DACs are composed of a DC/SFQ converter, a variable pulse number multiplier (VPNM), and a voltage multiplier (VM). The final output voltage V_{out} is given by the equation of $V_{out} = \Phi_0 m(t) N f_{\mathrm{in}}$, where Φ_0, $m(t)$, N, and f_{in} are the flux quantity of an SFQ, the variable multiplication factor of the VPNM, the fixed multiplication factor of the VM, and the repetition frequency of SFQ pulses synchronized to the microwave reference. By changing a digital code for determination of $m(t)$, a variable voltage output, of which the accuracy depends on that of f_{in}, is generated. For realization of a wide dynamic range of SFQ-PFM DACs, operation over a wide bandwidth is necessary.

We have demonstrated SFQ-PFM DACs of a "counter-forwarding type" [7],[8] and of a "sum of selected bits sequence (Σ-SBS) type" [9]. Their Nb chips were fabricated using the standard process 2 (STP2, the critical current density of 2.5 kA/cm^2) of the National Institute of Advanced Industrial Science and Technology (AIST) [10]. The maximum output voltage of our previous SFQ-PFM DACs was 2.54 mV, where the configuration of the counter-forwarding type was employed with a built-in 100-fold VM [8]. Its internal maximum SFQ pulse frequency and the corresponding average voltage were 12.3 GHz and 25.4 μV, which were mainly limited by the feedback delay in the counter unit. On the other hand, the SFQ-PFM DAC of the Σ-SBS type did not have an internal feedback structure. Instead, the output voltage was determined by selecting and summing binary sequences of SFQ pulses according to the digital data stored in non-destructive read-out memory cells (NDROs). We demonstrated the operation of a 6-bit SFQ-PFM DAC of the Σ-SBS type [9]. The maximum output voltage was 1.7 μV, whereas the numerical simulation indicated the maximum voltage of 23 μV. A possible reason for such a small voltage value was that SFQ pulses of high repetition frequency were partly annihilated due to collisions at confluence nodes. Thus, timing design of pulses propagating toward the output terminal seems to be a key to realize larger resolution and higher repetition frequency.

In this paper, we report improved performance of RSFQ-PFM DAC of the Σ-SBS type. The operation frequency and the bias margins were numerically improved by adjusting the arrival timing of SFQ pulses at confluence buffers (CBs). Test circuit of a 4-bit SFQ-PFM DAC of the Σ-SBS type was fabricated using the AIST high-speed standard process (HSTP) of which the critical current density was 10 kA/cm^2 [11]. The maximum output voltage of 244 μV was confirmed, which was consistent with the numerical result.

2. Circuit design

Figure 1 shows the block diagram of the 4-bit SFQ-PFM DAC of the Σ-SBS type that we designed in this work. The circuit includes 3 kinds of components. The circuit on the left-hand side consists of 4-cascaded Toggle Flip Flops (TFFs), TFF$_1$–TFF$_4$. The input SFQ pulses of the repetition frequency f_{in} are divided into four frequencies ($f_{in}/2$, $f_{in}/4$, $f_{in}/8$, and $f_{in}/16$) through the 4-cascaded TFFs. The center part is composed of 4 NDROs, which allow or prohibit SFQ pulse transmission of each bit according to the external digital code. Finally, the transmitted SFQ pulse sequences are merged at CBs on the right-hand side. Consequently, the frequency f_{out} of the output SFQ pulses is expressed as $f_{\mathrm{out}} = f_{\mathrm{in}}(\sum_{i=1}^{4} D_i 2^{i-1})/16$, where the data D_i is the digital datum for the NDRO$_i$, where $i = 1 - 4$ This means that the maximum output frequency becomes $15/16\, f_{\mathrm{in}}$ for $\{D_1, D_2, D_3, D_4\} = \{1, 1, 1, 1\}$.

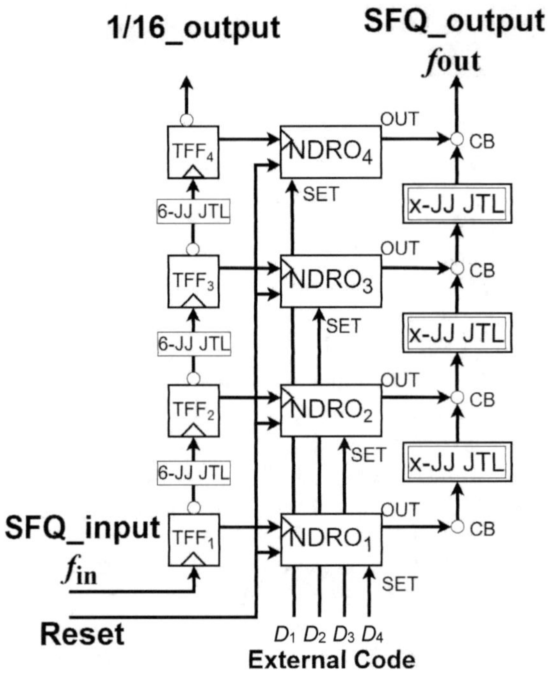

Figure 1. Block diagram of a 4-bit SFQ-PFM DAC of the Σ-SBS type. The label of "x -JJ JTL" denotes a JTL composed of x Josephson junctions (JJs).

The advantage of the Σ-SBS type to the counter-forwarding type is that there is no SFQ feedback in the circuit. In addition, it is possible to update NDRO data asynchronously with the DAC operation.

Journal of Physics: Conference Series **2323** (2022) 012033 doi:10.1088/1742-6596/2323/1/012033

Furthermore, there is no need to initialize the NDRO data before changing the digital code. One of the weaknesses of the Σ-SBS type, on the other hand, is that SFQ pulses collide and disappear when they are merged within the prohibited timing duration of the CB. Therefore, the bandwidth of the SFQ pulse sequence is mainly limited by the CBs.

In this study, we adjusted the timings of SFQ splitting at TFFs and SFQ merging at CBs by tuning the number x of Josephson junctions (JJs) included in Josephson transmission lines (JTLs). (We used the jsim program for analog simulation [12].) As shown in figure 1, we used a 6-JJ JTL between TFFs, wheras the number x of JJs in a JTL between CBs were vaired to change the propagation delays between CBs. We should note that there might be room to optimize the number of JJs in the JTL between TFFs. (That is, the 6-JJ JTL might not be the best.) Finer timing adjustment may be possible if we use a longer JTL including JJs more than 6, for which we have to pay higher hardware cost.

Figure 2 shows the simulation results of operation regions on the pulse-frequency vs. the bias voltage plane for $\{D_1, D_2, D_3, D_4\} = \{1, 1, 1, 1\}$, where 15 SFQ pulses appear at the SFQ output terminal for 16 input SFQ pulses. The parameters are the number x of JJs in a JTL between CBs. It is found that the circuit structures using "4-JJ JTL" and "5-JJ JTL" (JTLs composed of 4 and 5 JJs) have the largest bias voltage margin for the input SFQ frequencies lower than 50 GHz with a range of −32% to +32% of the nominal bias voltage of 2.5 mV. On the other hand, the circuit structure using "6-JJ JTL" has the widest bandwidth, 159.9 GHz at 3.3 mV.

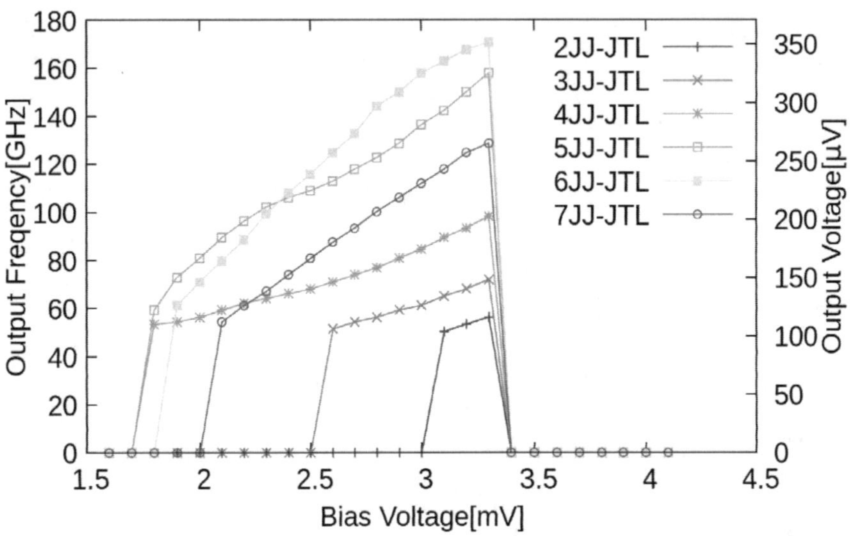

Figure 2. Numerical operation regions on the pulse-frequency vs. the bias voltage plane for various JTLs composed of 2−7 JJs. The nominal bias voltage is 2.5 mV. In simulation, the maximum operation frequency for the bias voltages between 1.6 and 4.2 mV was evaluated with a 0.1 mV step.

3. Experimental results and discussion

For experiments using Nb integrated circuits, we designed a 4-bit SFQ-PFM DAC of the Σ-SBS type with "5JJ-JTL" configuration in order to emphasize operation stability with wider bias margins. We used the digital cell library referred to as the "CONNECT" [14] updated for the HSTP of the AIST. Test chips were fabricated by using the HSTP of the AIST. Figure 3 shows an optical microphotograph of a fabricated circuit.

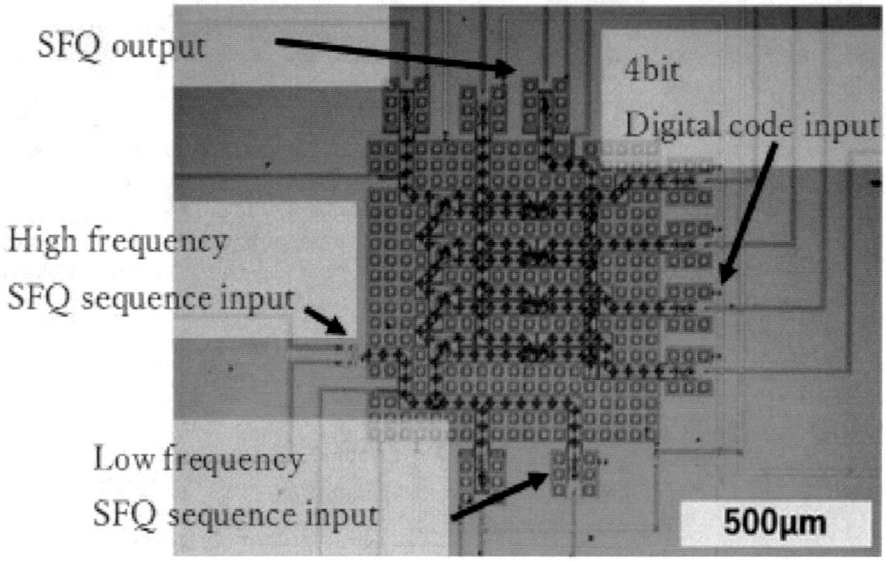

Figure 3. Optical microphotograph of a fabricated 4-bit SFQ-PFM DAC of the Σ-SBS type.

The low-frequency measurement setup is shown in figure 4. A 16 kHz square wave from a function generator was fed to the DC/SFQ converter that converted the square wave into SFQ pulse sequence of 16 kHz. The input SFQ pulse sequences were monitored via the SFQ/DC converter and a 40-dB preamplifier. Two signals from the SFQ-PFM DAC of the Σ-SBS type, "output" and "1/16 output" were also monitored via the SFQ/DC converters and 40-dB preamplifiers. The 4-bit digital code was fed from a digital signal generator at room temperature via the DC/SFQ converter.

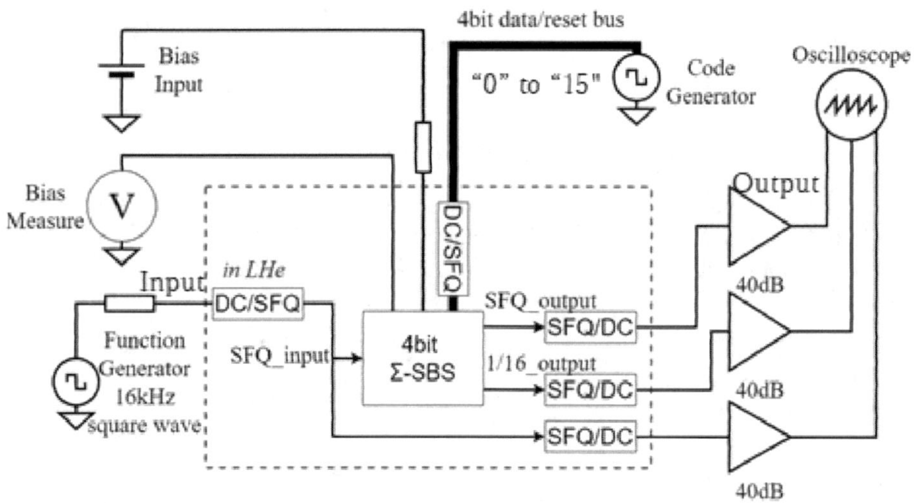

Figure 4. Low-speed measurement

Figure 5 shows low-frequency measurement results for the digital code of 1, 2, 4, and 8. Correct operations were confirmed, that is, the number of outputs SFQ pulses was changed to 1, 2, 4, and 8 for 16 input SFQ pulses. We also confirmed correct operations for all digital codes from 0 to 15.

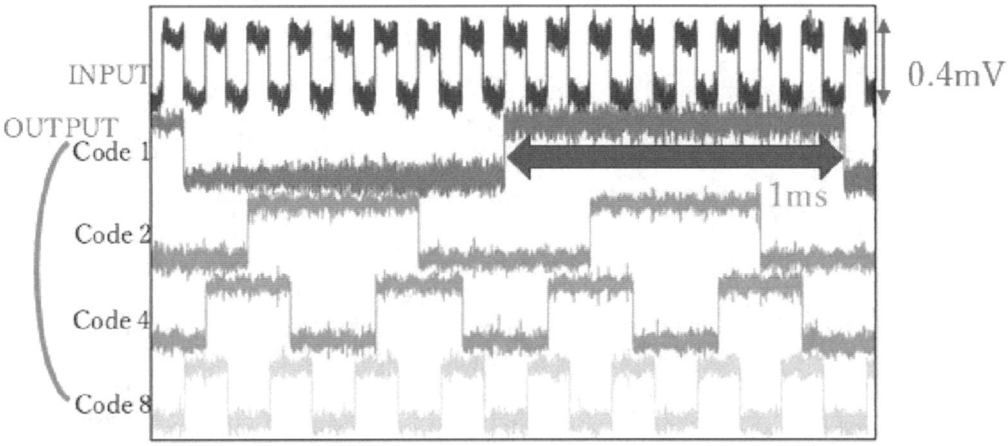

Figure 5. Experimental results of low-speed measurements for a 16 kHz input signal. Waveforms of the SFQ/DC converter for the digital code of 1, 2, 4, 8 are presented.

Next, high-frequency measurement setup is shown in figure 6. An over-biasing method was used for feeding SFQ pulses on the order of 100 GHz. The input and output average voltages (V_{in}, and V_{out}) at each terminal were measured and converted to the Josephson frequencies by calculating V_{in}/Φ_0 and V_{out}/Φ_0.

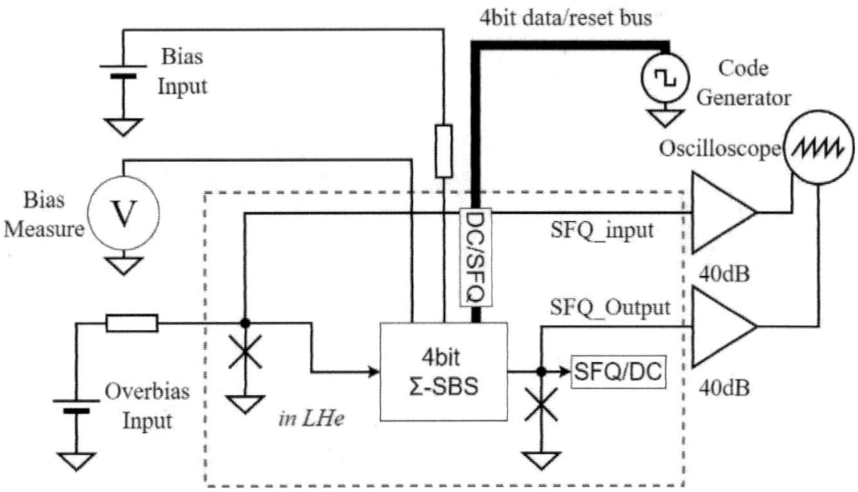

Figure 6. High-frequency measurement setup.

Journal of Physics: Conference Series **2323** (2022) 012033 doi:10.1088/1742-6596/2323/1/012033

Figure 7 shows the measurement results of the relationship between V_{in} and V_{out} in high-frequency measurement. The results for the digital code of 15 and the bias votage of 2.9 mV are presented, for which the ideal relationship becomes $V_{out} = 15V_{in}/16$. The maximum output voltage was determined as the point where the relative error between V_{out} and $15V_{in}/16$ exceeded the range of $\pm 1\%$. The maximum output voltage V_{out} was 244 μV, of which the corresponding Josephson frequency was 244 μV/Φ_0 = 118 GHz.

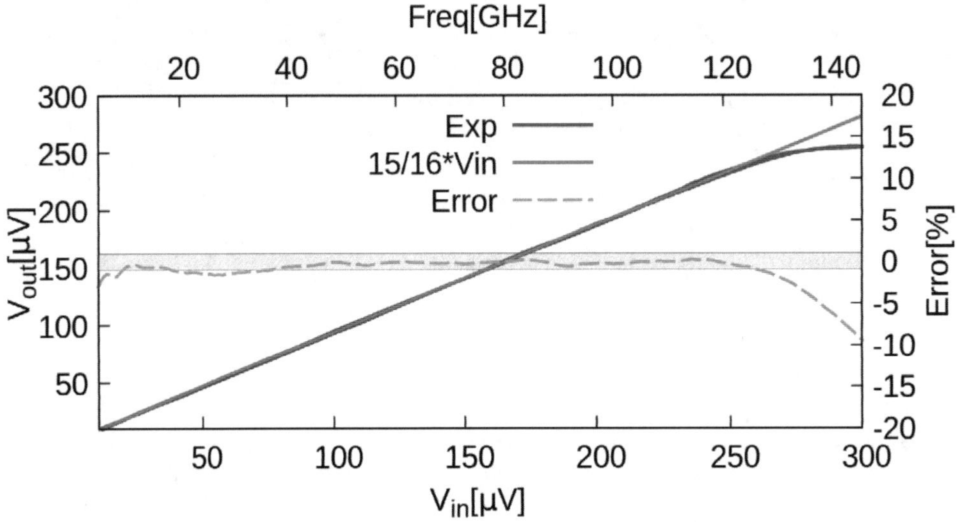

Figure 7. High-speed measurement results for the digital code of 15. The bias voltage was set at 2.9 mV. The yellow band shows the area where the error is within $\pm 1\%$.

Figure 8 shows the maximum operation frequencies for various bias voltages. The digital code was set at 15. For the bias voltages of 2.7, 2.8 and 2.9 mV, the measured maximum frequencies were in good agreement with the numerical results. It should be noted that we could not confirm correct operation for the bias voltage of 2.5 mV, where the relative errors were flat at −5% for 30–88 GHz range. It is not clear the reason for this failure at this moment.

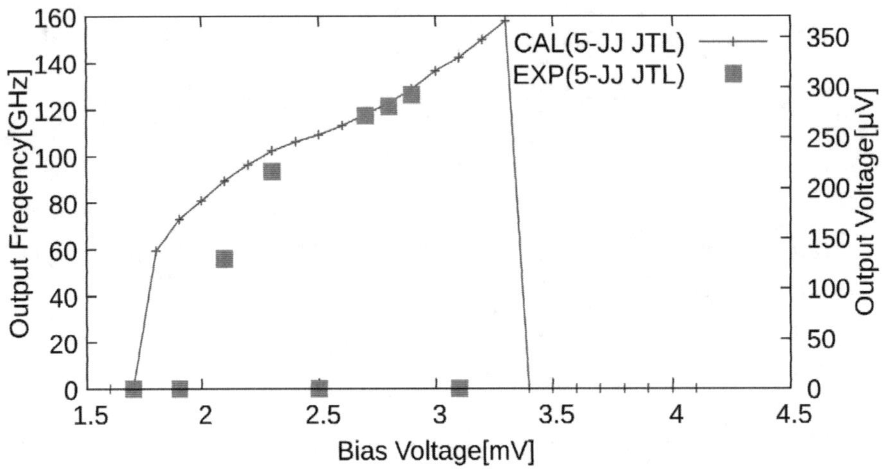

Figure 8. Maximum operation frequencies for various bias voltages. The digital code was 15.

4. Conclusion

We designed and demonstrated a 4-bit SFQ-PFM DAC of the Σ-SBS type. We optimized the timing margins of JTLs between CBs, which had been pointed out as a crucial part in our previous work. By optimization, the operation frequencies were improved up to 159.9 GHz. Test circuits were fabricated using the Nb HSTP (10 kA/cm^2) of the AIST. The maximum output voltage was 244 μV, of which the corresponding Josephson frequency was 118 GHz. This result is 143 times greater than our previous result, where the maximum output voltage was 1.7 μV.

Acknowledgement

The authors thank the lab members in UEC Tokyo for their fruitful discussion and technical support. The circuits used in this work were fabricated using the Nb HSTP in the CRAVITY, AIST, Japan. This work was partly supported by JSPS Grant-in-Aid for Scientific Research JP17K04979 & JP20H02201, and by VLSI Design and Education Center (VDEC), the University of Tokyo in collaboration with Cadence Design Systems. The stable supply of liquid helium from the Coordinated Center for UEC Research Facilities, UEC Tokyo, Japan, is also acknowledged.

References

[1] Likharev K K and Semenov V K 1991 RSFQ Logic/Memory Family: A New Josephson-Junction Technology for Sub-Terahertz-Clock-Frequency Digital Systems *IEEE Trans. Appl. Supercond.* **1** 3
[2] Hamilton C A1992 Josephson voltage standard based on single-flux-quantum voltage multiplier *IEEE Trans. Appl. Supercond.* **2** 139
[3] Semenov V K 1993 Digital to analog conversion based on processing of the SFQ pulses *IEEE Trans. Appl. Supercond.* **3** 2637
[4] Sasaki H, Kiryu S, Hirayama F, Kikuchi T, Maezawa M, Shoji A and S V Polonsky 1999 RSFQ-based D/A converter for AC voltage standard *IEEE Trans. Appl. Supercond.* **9** 3561
[5] Maezawa M, Hirayama F and Suzuki M 2005 Rapid single flux quantum digital-to-analog converter for AC voltage standard *Physica* C **426** 1674
[6] Saito J, Tanaka T, Moriya M, Kobayashi T, Mizugaki Y and Maezawa M 2011 Design and functional tests of variable SFQ pulse number multiplier *Physica* C **471** 1270
[7] Mizugaki Y, Takahashi Y, Shimada H and Maezawa M 2011 9-bit superconductive single-flux-quantum digital-to-analog converter *Electronics Letters* **50** 1637
[8] Takahashi Y, Shimada H, Maezawa M and Mizugaki Y 2014 Design and operation of 6-bit, 0.25-mV$_{pp}$ quasi-sine voltage waveform generator based on SFQ pulse-frequency modulation *Physics Procedia* **58** 220
[9] Mizugaki Y, Saito J, Moriya M and Maezawa M 2011 Design and operation of 64-fold variable single-flux-quantum pulse number multiplier *IEEE Trans. Appl. Supercond.* **21** 3604
[10] Nagasawa S, Hashimoto S, Numata Y and Tahara S 1995 A 380 ps, 9.5 mW Josephson 4-Kbit RAM operated at a high bit yield *IEEE Trans. Appl. Supercond.* **5** 2447
[11] Takeuchi N, Nagasawa S, China F, Ando T, Hidaka M, Yamanashi Y and Yoshikawa N 2017 Adiabatic quantum-flux-parametron cell library designed using a 10kAcm^{-2} niobium fabrication process *Supercond. Sci. Technol.* **30** 035002
[12] Fang E S and Van Duzer T 1989 A Josephson integrated circuit simulator (JSIM) for superconductive electronics application *Ext. Abstr. Int. Supercond. Electronics Conf.* (The Japan Society of Appl. Phys) p 407
[13] Yorozu S, Kameda Y, Terai H, Fujimaki A, Yamada T and Tahara S 2002 A single flux quantum standard logic cell library *Physica* C **378** 1471

High-fidelity circuit simulation of AQFP circuits through compact models extracted from layout

Coenrad J Fourie[1] **and Kyle Jackman**[1]

[1]Department of Electrical and Electronic Engineering, Stellenbosch University, Banghoek Road, Stellenbosch 7600, South Africa

E-mail: coenrad@sun.ac.za

Abstract. Adiabatic Quantum Flux Parametron (AQFP) superconductor logic circuits rely on magnetic coupling between the gate and clock and control lines to function. Circuit designers start by designing a circuit netlist, selecting parameters to fit design objectives, optimizing the netlist in simulation and then progressing to integrated circuit layout. Hand-designed netlists mostly do not contain all the mutual inductances between inductors. The lack of a complete set of mutual inductances can limit the accuracy of the designed netlist. For a circuit netlist to be an exact representation of the layout, the number of inductors should be equal to the number of fundamental cycles in the netlist graph and all inductors should be coupled. In this paper, we show that full-circuit inductance extraction of AQFP layout where self and mutual inductances are specified by the designer produces small but possibly significant errors due to mismatch between the design schematic and the layout. With compact simulation model extraction, which we added to the InductEx tool chain, a much more accurate simulation model that includes all mutual inductances can be obtained to verify circuit performance after layout. We propose compact model extraction as the final step in cell library characterisation.

1. Introduction

Although significant development effort has been invested in expanding the capabilities and scope of electronic design automation tools for superconductor electronics recently [1, 2], the standard design flow for developing superconductor digital electronic circuits still relies heavily on verifying circuit fidelity through circuit simulations based on schematics drawn by designers.

Usually, a superconductor logic cell designer composes a circuit schematic to obtain a desired circuit functionality, then adds the most obvious parasitic components and inductive coupling to the schematic. A simulation netlist is generated from this schematic, and verified electrically with a SPICE-like simulation engine. Optimisation procedures also rely on the circuit schematic and the electrical simulator. An integrated circuit layout is then done to match the optimised circuit netlist. Finally, parameter extraction is performed on the layout, with extracted component values back-annotated to the circuit netlist for final electrical verification. This back-annotated circuit netlist is then added to a cell library for use in the simulation of larger systems.

We show that this design flow needs an extra step for superconductor circuits with dense coupling, such as Adiabatic Quantum Flux Parametron (AQFP) circuits. A compact model extracted from the layout provides a better fit between the circuit netlist and the layout structure, and should be included in a cell library for accurate simulation of systems.

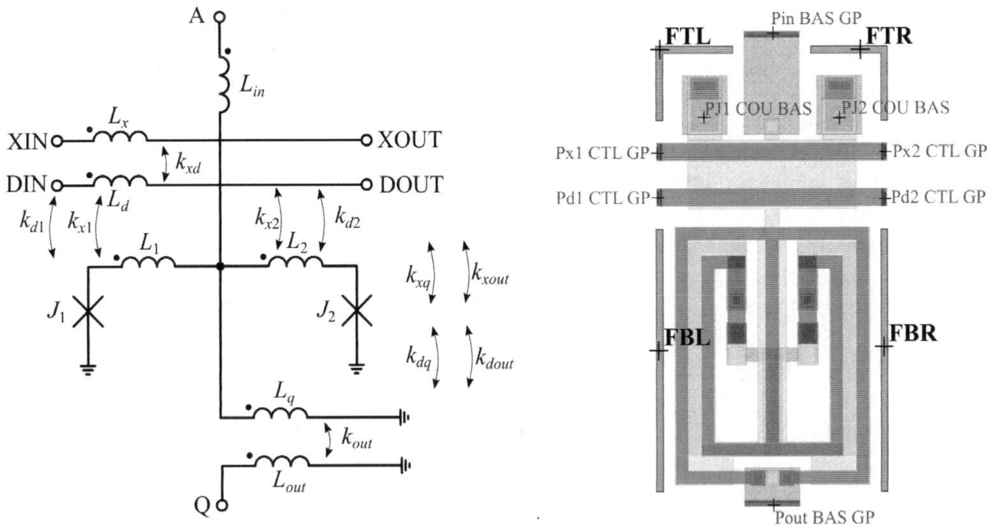

(a) (b)

Figure 1: AQFP buffer cell (a) schematic and (b) layout with InductEx ports.

2. Adiabatic Quantum Flux Parametron test circuit

2.1. AQFP buffer

We have previously shown the evaluation of moat position on AQFP cell performance in the presence of trapped flux [3]. The same AQFP buffer cell [4] is used here, and the designer's circuit schematic and the integrated circuit layout are shown in Figure 1 (a) and (b) respectively. The AQFP buffer cell layout is taken from the cell library for the 10 kA cm^{-2} high-speed standard process provided by the National Institute of Advanced Industrial Science and Technology (AIST) in Japan, as presented in [5].

Simulation of the cell is done in a cascaded test bench, as detailed in [3].

3. Effect of circuit model on extraction results

3.1. Calculation with InductEx

Self and mutual inductance values for the AQFP buffer cell layout are extracted with the three-dimensional parameter extraction tool InductEx [6].

InductEx uses a magnetoquasistatic analysis to obtain the current in every circuit branch when any one port is excited with a known voltage at a defined frequency while all other ports are zeroed. From the branch currents obtained when every port has been excited, a vector of voltages, **b**, and a rectangular $m \times n$ branch current matrix **A** are obtained. The voltage around any loop in the circuit is determined by branch currents and branch impedances around the loop, leading to a set of linear equations of the form $\mathbf{A}\mathbf{x} = \mathbf{b}$, where **x** is the vector of unknown impedances. Singular value decomposition (SVD) is used to obtain the least squares solution when the system is overdetermined ($m > n$) and the minimum norm solution when the system is underdetermined ($m < n$).

The self and mutual inductances are obtained by dividing frequency out of **x**.

3.2. Errors

Errors in calculated inductance values are introduced by perturbations to **A** and **b** that arise from computer precision and from neglected coupling. The use of double precision floating point numbers internally by InductEx and the TetraHenry [7] engine keeps errors from computer precision small. However, it is very easy for a circuit designer to unintentionally neglect a mutual inductance in a circuit netlist and thereby introduce significant errors into calculated inductance results.

Circuits that have multiple mutual inductors in a tight layout, such as typical AQFP cells, are especially vulnerable to errors introduced by neglected coupling.

There are a few quick ways to determine how significant the errors in solution are.

Firstly, if **A** is rank deficient, the minimum norm solution is often mathematically correct but physically impossible, so that negative inductances or coupling factors larger than 1 can be obtained. In general the solution from a rank deficient system should be disregarded, and the circuit model revisited to provide more ports.

Secondly, the condition number of **A**, which provides the ratio between its largest and smallest singular values, gives an indication of the largest error in precision. For every factor of ten increase in the condition number, one digit of precision in the solution is compromised. In general, the solution from a system with a condition number that exceeds 1×10^6 should be distrusted. More ports in the circuit netlist can often improve the calculation result.

Thirdly, the *relative error* $\|\mathbf{Ax} - \mathbf{b}\|/\|\mathbf{b}\|$ can be calculated after solution of $\mathbf{x}$ and evaluated to determine if the solution is reliable. InductEx calculates this relative error automatically and prints it as a percentage. There is no clear threshold as to what constitutes a sufficiently precise solution, but in general a worst-case error of more than 5 % is indicative of considerable magnetic coupling in the circuit layout that has not been accounted for by the circuit netlist.

3.3. AQFP buffer cell with no coupling

If all magnetic coupling in the AQFP cell of Figure 1(a) is ignored in the extraction circuit netlist, the system is full rank with a condition number of 12.2. The warning flag is the relative error of 34.4 %. The extracted results (listed in Table 2), which only include self inductance, do not accurately reflect the circuit parameters of the circuit as it is laid out.

3.4. AQFP buffer cell with basic coupling

The AQFP buffer cell relies on magnetic coupling for its operation. It is obvious that at least the mutual inductance between the clock and bias lines and the inductors L_1 and L_2, as well as the mutual inductance of the output stage between L_q and L_{out} must be calculated, as these are the couplings that give the circuit its functionality.

Extraction results with InductEx are listed in Table 2. The system is full rank with a condition number of 24.0. The results appear reasonable, but since there is actually coupling between more pairs of inductors than specified in the circuit netlist, the layout does not match the circuit schematic very well. We see this from the relative error of 16.7 %.

3.5. AQFP buffer cell with circuit designer schematic

The circuit designer usually elects to include the most obvious mutual inductances, as well as some coupling factors that need to be minimized. With the addition of the mutual inductance between L_d and L_x,the mutual inductances between L_d and both L_q and L_{out} as well as the mutual inductances between L_x and both L_q and L_{out} (thus the complete circuit shown in Figure 1(a)), the system remains full rank with a condition number of 24.1. However, the relative error reduces to 6.41 %.

3.6. AQFP buffer cell with dense coupling

For inductors galvanically connected to the same subnet, such as L_1, L_2, L_{in} and L_q, and where the number of inductors in the subnet is larger than the number of fundamental cycles, the current distribution in the inductors when there is coupling between any of the inductors can also be obtained by a different set of inductor values without any coupling. A dense coupling circuit schematic would thus include all coupling, except coupling between inductors in the same netlist. It also excludes coupling from any inductor outside of a multi-inductor subnet to one of the inductors in the multi-inductor subnet. If we exclude L_{in} from any coupling, but add coupling between L_1 and L_{out} as well as between L_2 and L_{out}, the condition number is 24.1 and the relative error is 6.39 %.

This is the best accuracy that can be obtained from the user-defined circuit schematic that is intuitive to interpret for layout purposes.

3.7. Full coupling

It may seem intuitive to add all possible coupling to the circuit schematic (so that every inductor is coupled to every other inductor), but in a circuit netlist with more than one inductor connected galvanically to the same subnet the result is almost always an underdetermined system. For this AQFP circuit, if all 21 mutual inductances are included in the circuit netlist, the system is rank deficient. The minimum norm SVD solution returns negative inductances for L_{in}, L_2 and L_q, which makes calculation of the mutual inductances that includes these inductors impossible.

4. Compact model

4.1. Fundamental inductor set

For a circuit netlist to be an exact representation of the superconductor layout, the number of inductors should be equal to the number of *fundamental cycles* in the netlist graph. We will refer to this set of inductors as the *fundamental inductor set.* The *fundamental inductor set* represents the exact number of inductors required to extract a full inductance matrix, containing all the mutual inductances.

The *fundamental cycles* can be obtained by first constructing a *spanning tree* of the netlist graph and then identifying the *chords* of the *spanning tree* [8]. The *chords* are the edges (branches) of the graph that do not form part of the *spanning tree* [8]. Each *fundamental cycle* consists of a *chord* together with the path in the *spanning tree* connecting the endpoints of the *chord.* There are exactly $m-n+c$ *fundamental cycles,* where m is the number of edges, n is the number of vertices, and c is the number of connected components [8]. The *fundamental cycles* are linearly independent from the remaining cycles, because each *fundamental cycle* contains a unique *chord.* The *fundamental inductor set* can therefore be obtained by placing inductors only in the *chord* branches of the circuit netlist. This will ensure that each *fundamental cycle* contains only one inductor.

4.2. Compact model of the AQFP buffer

To obtain the *fundamental inductor set* of the AQFP buffer cell, we first convert the schematic to a graph, as shown in Figure 2(a). The edges with label "P*" correspond to the InductEx ports defined on the layout in Figure 1(b). Next, we contract all the inductor edges and keep the excitation ports, as shown in Figure 2(b). The *chords,* shown as red edges in Figure 2(c), can then be identified using the method discussed in Section 4.1. The *fundamental inductor set* is obtained by placing an inductor in series with each *chord.* Figure 3(a) shows the final graph of the AQFP buffer cell with the *fundamental inductor set* and Figure 3(b) shows the corresponding schematic.

Each *fundamental cycle* must contain at least one unique excitation port in order to calculate the mutual inductance between all the *fundamental inductors.* This can be accomplished by placing an excitation port in series with each *chord.* The *fundamental cycle* in Figure 3(a) all contain at least one unique excitation port, which provide enough linear equations to calculate all the mutual inductances between the six *fundamental inductors* (L_{c1} to L_{c6}).

The *fundamental inductor set* is mostly not intuitive. In the compact model here, for instance, there is no obvious inductor between L_{c4} and L_{c5} and ground to couple to the output inductor L_{c6}. The output stage coupling is fully modelled by the coupling between L_{c6} and *all the other inductors*. It is difficult to design and adjust the layout from the compact model, which is why compact model extraction is the final step before the simulation netlist is committed to a library.

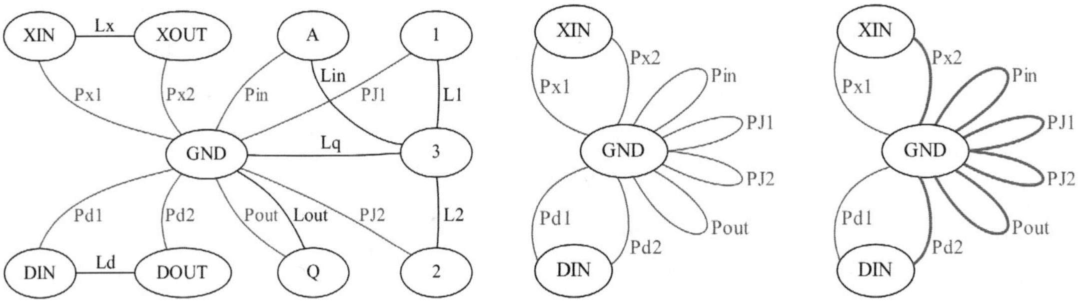

Figure 2: (a) Graph of AQFP buffer cell. (b) Graph with inductor edges contracted. (c) The *chords* of the graph represented by the red edges.

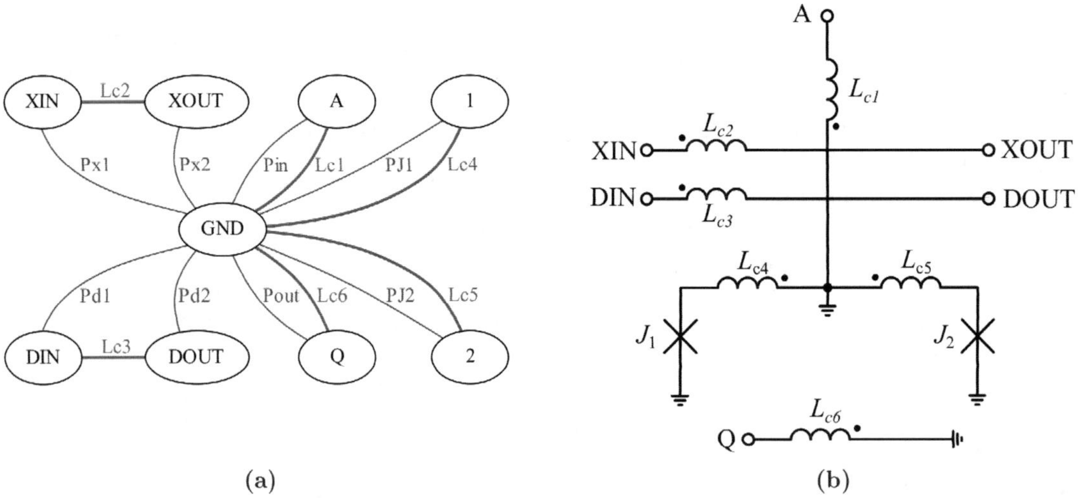

Figure 3: (a) Graph of AQFP buffer cell with red edges representing the *fundamental inductors*. (b) Schematic of the AQFP buffer cell with the *fundamental inductor set* (L_{c1} to L_{c6}).

The extraction results for the compact model of the AQFP buffer are listed in Table 1. The relative error $||\mathbf{Ax} - \mathbf{b}||/||\mathbf{b}|| = 0.0075$ %.

5. Simulation results

All simulations were performed with 5 GHz clock signals.

5.1. *AQFP buffer excitation margins*

The AQFP buffer cell is tested in simulation with JoSIM [9] for the different coupling strategies. The dc bias current is set to provide an offset flux of $0.5\Phi_0$ to the gate [5], and the amplitude of

Table 1: InductEx extraction results for the compact model of the AQFP buffer.

Component	Value	Component	Value	Component	Value
L_{c1}	9.33 pH	k_{c1-c3}	-3.64×10^{-5}	k_{c2-c6}	6.58×10^{-7}
L_{c2}	7.27 pH	k_{c1-c4}	0.852	k_{c3-c4}	6.16×10^{-2}
L_{c3}	7.27 pH	k_{c1-c5}	0.852	k_{c3-c5}	-6.17×10^{-2}
L_{c4}	9.21 pH	k_{c1-c6}	0.476	k_{c3-c6}	-8.81×10^{-6}
L_{c5}	9.21 pH	k_{c2-c3}	0.257	k_{c4-c5}	0.832
L_{c6}	30.4 pH	k_{c2-c4}	0.86×10^{-2}	k_{c4-c6}	0.481
k_{c1-c2}	-9.74×10^{-6}	k_{c2-c5}	-0.86×10^{-2}	k_{c5-c6}	0.481

the ac excitation currents are varied to determine margins. The margins are listed in dBm for a 50 Ω system. The results are shown in Table 3. The circuit does not work correctly when there is no coupling, as is expected. The margins for the designer schematic and the densest coupling possible with the designer schematic are almost exactly similar. The circuit model with basic coupling requires significantly more excitation current than the designer, dense coupling and compact model circuits to work, and the margins are consequently so far off that this circuit model is clearly inadequate. Finally, the margins for compact model are about 0.5 dB below those of the designer schematic.

5.2. Response to flux trapping

Coupling from the moats to the circuit components can be calculated with InductEx, as discussed in [3]. We extracted the circuit models for both the designer schematic and the compact model, with coupling from the four moats (FTL, FTR, FBL and FBR) shown in Figure 1(b). Figure 4(a) shows the graph of the compact model with the four moats and Figure 4(b) shows the corresponding schematic. The *fundamental inductors*, shown as red edges in Figure 4(a), are identified using the method discussed in Section 4.1.

We simulated each circuit model with one fluxon trapped in moat FBR and observed the excitation margins. Those are listed in Table 4.

6. Conclusion

We have shown that for superconductor digital circuits that rely heavily on magnetic coupling, the circuit designer has to carefully develop a circuit model that accounts for all magnetic coupling in the circuit to enable high-fidelity parameter extraction and accurate simulation of the circuit response. Inclusion of only the most obvious coupling leads to large errors in extracted parameters and under or overestimation of operating margins.

Even with careful circuit schematic compilation, a circuit that has more inductors than fundamental cycles in the netlist graph cannot take into account all coupling without resulting in an underdetermined system of linear equations.

Parameter extraction of the highest precision and circuit simulation models of the highest fidelity can be achieved by using a compact model extraction technique to find the fundamental set of inductors in a circuit, and to couple every inductor pair. We showed that the precision of extraction of a compact model is vastly superior to that of a good design schematic (0.0075 % compared to 6.4 % for the AQFP buffer cell analysed here), and that there is about a 0.5 dB change in excitation current operating margins for the compact model circuit compared to the best designer schematic in electrical simulation.

Table 2: InductEx extraction results for AQFP buffer with different coupling modelling strategies.

Component	No coupling	Basic	Designer	Dense coupling
L_{in}	1.42 pH	1.43 pH	1.43 pH	1.43 pH
L_1	1.27 pH	1.37 pH	1.42 pH	1.42 pH
L_2	1.27 pH	1.37 pH	1.42 pH	1.42 pH
L_q	4.62 pH	7.74 pH	7.72 pH	7.71 pH
L_{out}	17.1 pH	30.5 pH	30.3 pH	30.3 pH
L_x	5.40 pH	6.50 pH	7.21 pH	7.21 pH
L_d	6.16 pH	6.54 pH	7.24 pH	7.24 pH
k_{out}	–	-0.522	-0.523	-0.521
k_{x1}	–	+0.170	+0.213	+0.213
k_{x2}	–	+0.170	+0.213	+0.213
k_{d1}	–	+0.092	+0.153	+0.153
k_{d2}	–	+0.092	+0.153	+0.153
k_{xd}	–	–	+0.225	+0.225
k_{xq}	–	–	$+2.8 \times 10^{-5}$	$+2.7 \times 10^{-5}$
k_{dq}	–	–	$+4.8 \times 10^{-5}$	$+1.5 \times 10^{-5}$
k_{xout}	–	–	$+3.2 \times 10^{-6}$	$+1.1 \times 10^{-5}$
k_{dout}	–	–	$+2.0 \times 10^{-5}$	-2.8×10^{-5}
k_{1out}	–	–	–	-3.2×10^{-13}
k_{2out}	–	–	–	$+3.2 \times 10^{-3}$

Table 3: Simulated excitation margins in dBm for AQFP buffer with different coupling modelling strategies.

Excitation current	Basic	Designer	Dense coupling	Compact model
I_{x1}	-14.5 – -10.2	-23.3 – -14.8	-23.3 – -14.8	-23.9 – -15.3
I_{x2}	-14.4 – -10.2	-23.0 – -14.9	-23.0 – -14.9	-23.6 – -15.4

Table 4: Simulated excitation margins in dBm for AQFP buffer with a fluxon in moat FBR.

Excitation current	Basic coupling	Designer schematic	Compact model
I_{x1}	-13.6 – -10.5	-22.1 – -15.3	-22.7 – -15.8
I_{x2}	-13.7 – -10.4	-22.1 – -15.2	-22.7 – -15.7

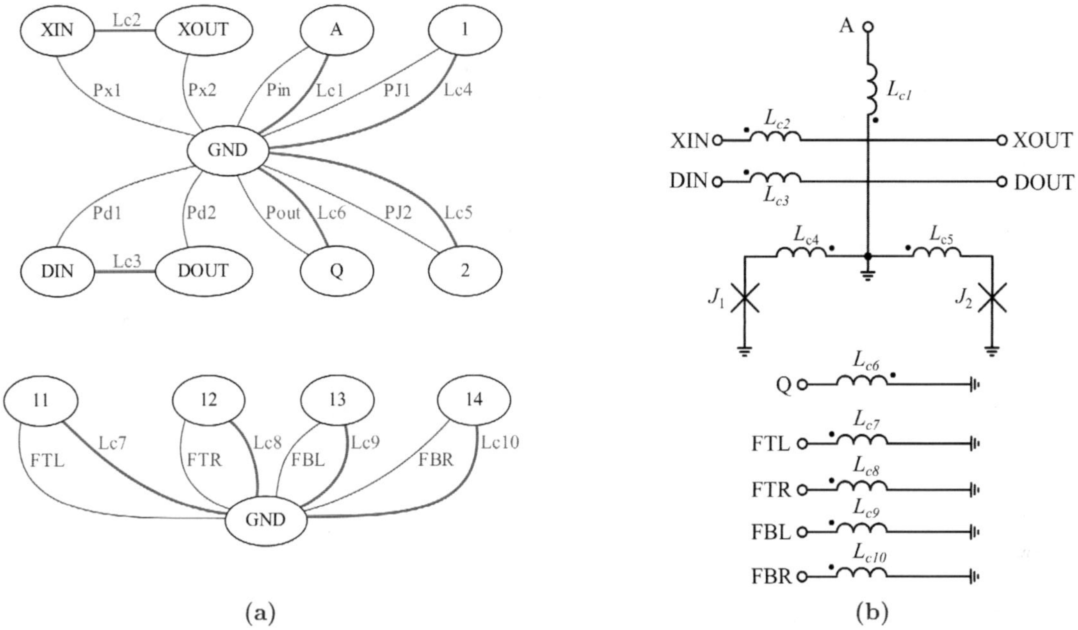

Figure 4: (a) Graph of AQFP buffer cell with four moats (FTL, FTR, FBL and FBR). The red edges represent the *fundamental inductors*. (b) Schematic of the AQFP buffer cell with the *fundamental inductor set* (L_{c1} to L_{c10}).

It is thus advisable to do a compact model extraction of a circuit with mutual inductance after layout to nullify any omissions by the circuit designer, and to add this compact model to the cell library for the highest fidelity simulation of cell in a larger system.

Acknowledgments
The research is based upon work supported by the Office of the Director of National Intelligence (ODNI), Intelligence Advanced Research Projects Activity (IARPA), via the U.S. Army Research Office grant W911NF-17-1-0120.

References

[1] Fourie C J 2018 *IEEE Trans. Appl. Supercond.* **28** 1300412

[2] Fourie C J, Jackman K, Botha M M, Razmkhah S, Febvre P, Ayala C L, Xu Q, Yoshikawa N, Patrick E, Law M, Wang Y, Annavaram M, Beerel P, Gupta S, Nazarian S and Pedram M 2019 *IEEE Trans. Appl. Supercond.* **29** 1300407

[3] Fourie C J, Takeuchi N, Jackman K and Yoshikawa N 2021 *Journal of Physics: Conference Series* **1975** 012027

[4] Takeuchi N, Suzuki H, Fourie C J and Yoshikawa N 2021 *IEEE Transactions on Applied Superconductivity* **31** 1300605

[5] Takeuchi N, Nagasawa S, China F, Ando T, Hidaka M, Yamanashi Y and Yoshikawa N 2017 *Superconductor Science and Technology* **30** 035002

[6] Fourie C, Xizhu Peng, Takahashi A and Yoshikawa N 2013 *IEEE 14th Int. Superconductive Electronics Conf. (ISEC)* pp 1–3

[7] Jackman K and Fourie C J 2016 *IEEE Trans. Appl. Supercond.* **26** 0602305

[8] Deo N 1974 *Graph theory with applications to engineering and computer science* (Prentice Hall)

[9] Delport J A, Jackman K, l Roux P and Fourie C J 2019 *IEEE Trans. Appl. Supercond.* **29** 1300905

Braiding Operations for a Topological Josephson Junction Array using SFQ Current Pulses

Takuro Funazawa, Masato Naruse, and Hiroaki Myoren

Graduate School of Science and Engineering, Saitama University, 255 Shimo-Okubo, Sakura-ku, Saitama 338-8570, JAPAN

E-mail: (t_funazawa, myoren)@super.ees.saitama-u.ac.jp

Abstract. Majorana bound states (MBSs) can occur in topological Josephson junction (JJ) arrays with the presence of a fractional vortex with 2π-phase winding. It has been theoretically shown that the braiding operation can be achieved by using current pulses, compatible with single flux quantum (SFQ) technology. Therefore, we designed an SFQ logic circuit for the braiding operations of the topological JJ array using single flux quanta in a conventional JJ array. Braiding operations could be realized by an SFQ logic circuit with double flux quantum driver. To obtain positive and negative current pulses using the SFQ logic circuit, we used SFQ logic circuits with ac biasing for simulation. Finally, we confirmed the CNOT operation using an array-type SFQ circuit by means of simulation and estimated the CNOT operation speed.

1. Introduction

A topological quantum computer (TQC) based on Majorana bound states (MBSs) has attracted great interest because of its excellent computational robustness[1, 2]. These properties imply a smaller quantum error correction circuit compared to any other quantum computers based on the two-energy level system. Many solid-state material platforms have experimentally shown the ability to host Majorana signatures[3, 4]. Some examples are topological semimetals in combination with superconductivity[5]. When superconductivity is induced at the surface of a topological material using the proximity effect, an MBS can occur either at an Abrikosov vortex core[6] or at a planar Josephson junction[7]. Since a two-dimensional surface state is realized without applying a strong magnetic field, single flux quantum (SFQ) logic circuits as classical computing parts can be placed near a topological quantum computing part. This can reduce heat load through many coaxial cables for microwave manipulation of conventional superconducting qubits. Therefore, TQC based on MBS with SFQ logic circuits can lead us to the $10^6 - 10^8$-scale quantum computers.

Recently, manipulation methods with current pulses for MBS were proposed for topological Josephson junction (JJ) arrays[8]. Via configuration with multiple line junctions meeting at a point, this is equivalent to a schematic of a superconducting loop with multiple junctions, where the criteria for the MBS to exist coincides with the presence of a fractional Josephson vortex with 2π phase winding (2π vortex). The existence and dynamic stability of 2π and 4π vortices were determined in triangular and square JJ arrays because 2π vortices were required for MBS. The resistively shunted junction (RSJ) model with a current-phase relationship $I_{\rm cp}(\theta)$

Journal of Physics: Conference Series **2323** (2022) 012035 doi:10.1088/1742-6596/2323/1/012035

with mixed 2π (fractional) and 4π-periodic components is expressed as follows:

$$I_{\rm cp}(\theta)/I_{\rm c0} = (1-\alpha)\sin(\theta) + \alpha\sin(\theta/2),$$

where α is the percentage of 4π periodic current components and $I_{\rm c0}$ is the critical current of the JJ. In this study, it is assumed that all topological JJs have the same series inductance L and that mutual inductance between junctions is neglected in the topological JJ array. The existence criterion of 2π vortices depends strongly on the inductance parameter $\beta_L = 2\pi L I_{\rm c0}/\Phi_0$ and the percentage of 4π periodic current component α, where Φ_0 is the SFQ. Typical values are reported to be $\beta_L = 3.77$ and $\alpha = 0.2$[8].

In this study, we proposed SFQ logic circuits for braiding operations in conventional JJ arrays using the RSJ model with $\beta_L = 2.12$ and $\alpha = 0$. The proposed SFQ logic circuits will be extended to topological JJ arrays with $\alpha \neq 0$.

2. Braiding operations using a double flux driver gate

First, we confirmed braiding operations of SFQs in the square JJ arrays. Assuming SFQ logic circuits with critical current density of 10 kA/cm^2 Nb process, a double flux quantum (DFQ) driver gate[9] was required to manipulate 2π vortices (SFQs). Fig. 1(a) shows a 4×4 topological JJ array and four MBSs bound to 2π-vortex in loops labeled 1, 2, 3, and 4 with three current injection points labeled A, B, and C. In this study, we focused on manipulating SFQs in the conventional square JJ array using the SFQ logic circuit. A DFQ driver is connected to a current

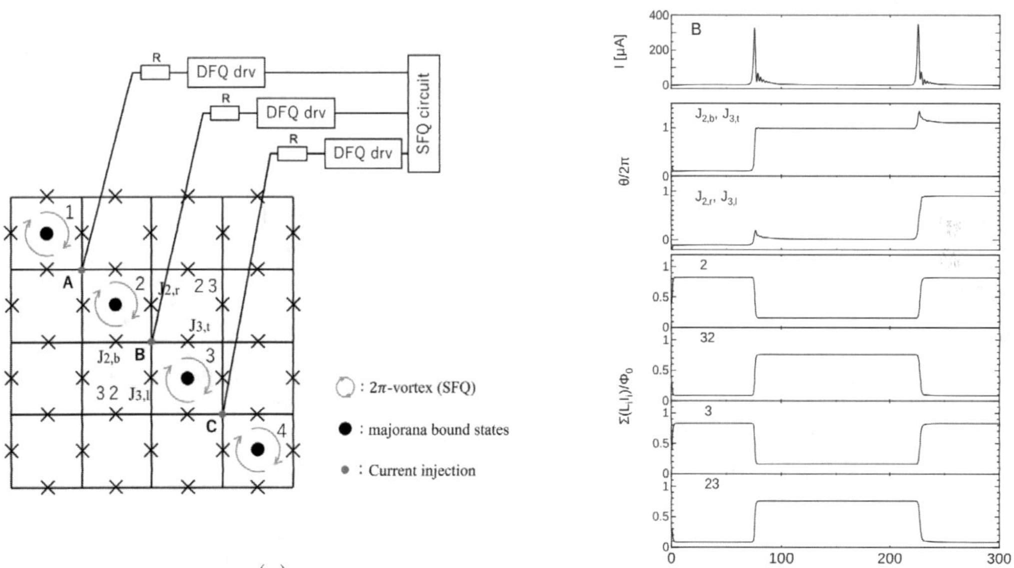

(a) (b)

Figure 1. (a) 4×4 topological Josephson junction array and four MBSs bound to 2π-vortex with three current injection points, labeled A, B, and C. Each point is connected to a DFQ driver through a resistance R. (b) Simulation results of a braiding operation around the current injection point B. Parameters are as follows: R=0.8 Ω, L=14 pH, and $I_{\rm c0}$=50 μA. Phase difference θ is defined from B to the outer branches of JJs. Current pulses induce 2π-leaps of JJs and movement of SFQs. Magnetic flux $\varphi_{\rm loop} = \Sigma(L_i I_i)$ indicates locations of SFQs.

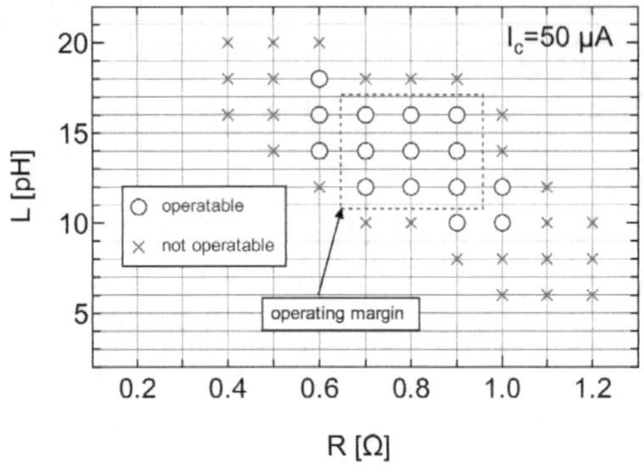

Figure 2. Operating margin of square JJ array at $I_c = 50\ \mu$ A.

injection point in the JJ array through a resistance R, and the current shape is dominantly determined by L and R values. Fig. 1(b) shows the simulation result of a braiding operation around point B using WRSpice[10]. Phase differences θ were defined from B to the outer branches of JJs. The first current pulse induces 2π-leaps at $J_{2,\mathrm{b}}$ and $J_{3,\mathrm{t}}$ and then the SFQ at loop 2 moves to loop 32 and the SFQ at loop 3 moves to loop 23. The second current pulse induced 2π-leaps at $J_{3,\mathrm{l}}$ and $J_{2,\mathrm{r}}$, and then the SFQ at loop 32 moves to loop 3 and the SFQ at loop 23 moves to loop 2. So, a braiding operation around point B is achieved by two consecutive current pulses. Locations of SFQs are determined by magnetic flux $\varphi_{\mathrm{loop}} = \Sigma(L_i I_i)$, as shown in Fig. 1(b).

Fig. 2 shows the operating margin of the square JJ array at $I_c = 50\ \mu$ A when changing the parameters of R and L. Correct braiding operations were confirmed for the square JJ array with $\beta_L = 1.82 - 2.43$, $I_c = 50\ \mu$A.

3. Quantum operations using SFQ logic array

Two consecutive pulses are required to exchange SFQs, labeled 2 and 3, as shown in Fig. 1(b). Positive current pulses and negative current pulses into an injection point induce counterclockwise and clockwise rotation, respectively. When we operate braiding with counterclockwise rotation of SFQs labeled 1 and 2, it is described by σ_A^n. Superscript n indicates the number of braiding operations and subscript A indicates the current injection point label. Braiding operations with clockwise rotation are described by σ_A^{-n}. For quantum gate operations both counterclockwise and clockwise braiding operations are needed.

Next, we examined a CNOT operation for a square JJ array with $\alpha = 0$, $\beta_L = 2.12$ and $I_c = 50\ \mu$A. The CNOT operation using 4×4 JJ array shown in Fig. 1 is described as follows[11]:

$$\sigma_A^{3}\sigma_B^{-2}\sigma_A^{-4}\sigma_B^{2}\sigma_A^{4}\sigma_B^{2}\sigma_A^{-2}\sigma_B^{-2}\sigma_A^{-4}\sigma_B^{-4}\sigma_A^{-2}\sigma_B^{4}\sigma_A^{2}\sigma_B^{-2}\sigma_A^{2}\sigma_B^{2}\sigma_A^{-2}\sigma_B^{3}$$
$$\sigma_C^{-2}\sigma_B^{-4}\sigma_C^{4}\sigma_B^{-2}\sigma_C^{2}\sigma_B^{2}\sigma_C^{-2}\sigma_B^{4}\sigma_C^{2}\sigma_B^{-4}\sigma_C^{-2}\sigma_B^{-2}\sigma_C^{2}\sigma_B^{-2}\sigma_C^{-2}$$
$$\sigma_B^{3}\sigma_A^{-2}\sigma_B^{2}\sigma_A^{2}\sigma_B^{-2}\sigma_A^{2}\sigma_B^{4}\sigma_A^{-2}\sigma_B^{-4}\sigma_A^{-4}\sigma_B^{-2}\sigma_A^{-2}\sigma_B^{2}\sigma_A^{4}\sigma_B^{2}\sigma_A^{-4}\sigma_B^{-2}\sigma_A^{3}.$$

One hundred forty braiding operations are needed.

To simulate the CNOT operation of the square JJ array, two consecutive and bipolar current pulses were needed. Therefore, we used nondestructive readout (NDRO) gates as a main component for efficient consecutive pulse injections and we applied ac biasing for the SFQ circuit[12] in the simulation. Fig.3(a) shows an SFQ current pulse generator using an NDRO gate, (b) 4×4 pulse generator array composed of SFQ current pulse generators, and (c) shows its pulse sequence in realizing CNOT operation.

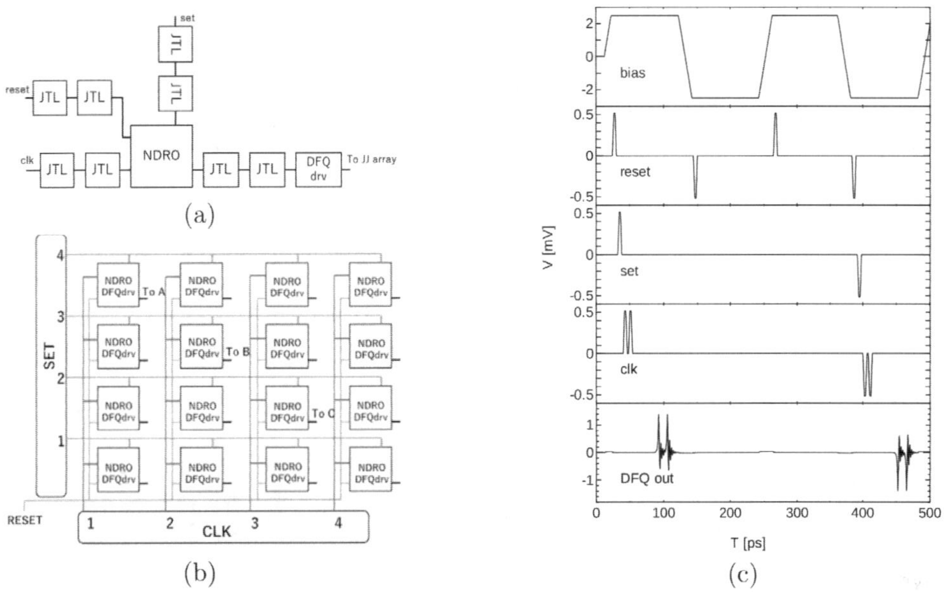

Figure 3. (a) SFQ current pulse generator using an NDRO gate and a DFQ driver. (b) 4 × 4 pulse generator array composed of SFQ current pulse generators as shown (a). (c) Input and output pulse sequence for a SFQ pulse generator with ac bias waveform.

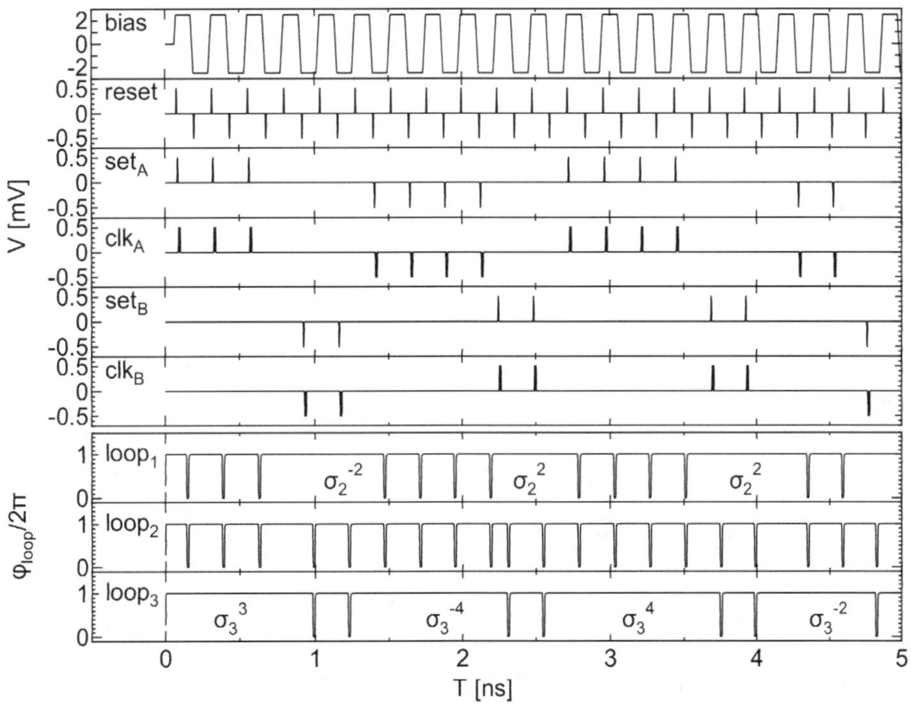

Figure 4. Simulation result for first part of the CNOT operation.

The pulse generator array is connected to reset, a column selector for set, and a row selector for clock signals. An NDRO gate array was placed corresponding to the square JJ array, including a SFQ pulse propagation path of jtl, driver, ptl, receiver, and DFQ gates. Fig. 4 shows the simulation result of the CNOT braiding operation. The upper part in Fig. 4 corresponds to Fig. 3(c) and the bottom part corresponds to Fig. 1(b), which is closely related to braiding operations and indicate locations of SFQs rotating by current pulses. We confirmed 240 ps for the braiding operation and 33.6 ns for the CNOT operation by simulation using WRSpice.

Hadamard gate is one of the quantum gates. Its braiding operation is already known as[13],

$$\sigma_B^4\sigma_A^{-2}\sigma_B^2\sigma_A^{-2}\sigma_B^2\sigma_A^2\sigma_B^{-2}\sigma_A^4\sigma_B^2\sigma_A^{-2}\sigma_B^{-2}\sigma_A^2\sigma_B^2.$$

Thirty braiding operations are needed for the Hadamard operation. 7.2 ns for the Hadamard operation was predicted from the 240 ps braiding operation time.

4. Practical methods to realizing positive and negative current pulses

In the previous section, we introduced an SFQ circuit working with ac biasing to demonstrate the CNOT operation. This method introduced a digital SQUID[12], which is a relatively small system and unsuitable for large systems such as a quantum computer. So, we tried to realize bipolar current pulses using the conventional SFQ circuit.

One possible method would be to realize a virtually negative bias current by applying positive SFQ current simultaneously to around the bias point. Fig. 5(a) shows a 4×4 topological JJ array

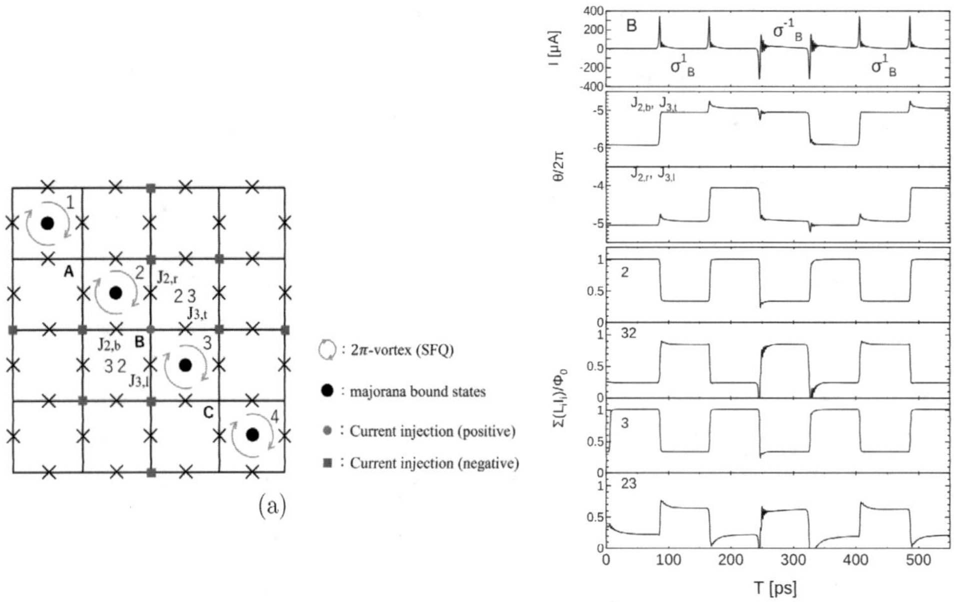

(a) (b)

Figure 5. (a) 4×4 topological Josephson junction array and four MBSs bound to 2π-vortex with three current injection points, labeled A,B and C. Braiding operations with counterclockwise rotation realized by applying SFQ pulses to point B and the operation with clockwise rotation by applying SFQ pulses to ten bias points around point B. (b) Simulation results for the braiding operation of σ_B^1, σ_B^{-1} and σ_B^1.

with the bias point B for a SFQ positive current pulse and 10 bias points (blue square symbols) around B to realize a virtually negative current pulse at point B. Fig. 5(b) shows a simulation result for the braiding operation of σ_B^1, σ_B^{-1}, and σ_B^1. The first and last two positive pulses generated SFQ current pulses applied to point B, and two middle negative current pulses were generated using 10 bias points around B. Though loop flux wave forms are fluctuated by virtual negative SFQ current, both braiding operations were realized using this method.

Another possible method is to use bipolar double flux quantum (DFQ) amplifiers described in ref.[14]. DFQ drivers could be replaced by bipolar DFQ amplifiers, and small circuit designs are needed for high-density circuits for quantum computers.

5. Conclusion

We showed the design of a SFQ logic circuit for braiding operation in square JJ array. By means of simulation, braiding operations of SFQs with SFQ current pulses were confirmed using an ac biased NDRO gate with DFQ driver. Finally, by simulation, we confirmed CNOT operation, which is the most important quantum operation. The proposed SFQ logic circuits will be extended for topological JJ arrays with $\alpha \neq 0$.

Acknowledgments

This work was supported by JSPS KAKENHI Grant Number 21K18708. This study has been partially supported by the VLSI Design and Education Center (VDEC) at the University of Tokyo, in collaboration with Cadence Design Systems, Inc. Circuits were fabricated in the clean room for analog-digital superconductivity (CRAVITY) of the National Institute of Advanced Industrial Science and Technology (AIST) with the Nb 10 kA/cm^2 high-speed standard process (HSTP). The AIST-STP2 process and AIST-HSTP process are based on the Nb circuit fabrication process developed by the International Superconductivity Technology Center (ISTEC).

References

[1] Das Sarma S, Freedman M, and Nayak C 2015 *npj Quantum Inf.* **1** 15001
[2] Kitaev A Y 2003 *Ann. Phys., NY* **303** 2
[3] Laubscher K, and Klinovaja J 2021 *J. Appl. Phys.* **130** 081101
[4] Oreg Y, Refael G, and von Oppen F 2010 *Phys. Rev. Lett.* **105** 177002
[5] Qi X L, and Zhang S C 2011 *Rev. Mod. Phys.* **83** 1057
[6] Rakhmanov A L, Rozhkov A V, and Nori F 2011 *Phys. Rev. B* **84** 075141
[7] Fu L and Kane C L 2008 *Phys. Rev. Lett.* **100** 096407
[8] Lankhorst M, Jansen T, Brinkman A and Golubov A 2021 *Supercond. Sci. Technol.* **34** 035024
[9] Herr Q P, Smith A D, and Wire M S 2002 *Appl. Phys. Lett.* **80** no. 17 3210
[10] Whiteley S R 1991 *IEEE Trans. Mag.* **27** 2902
[11] Nayak C, Saimon S H, Stern A, Freedman M, and Sarma S D 2008 *Rev. Mod. Phys.* **80** 1084
[12] Foroughi F, Bozbey A, and Fardmanesh M 2014 *J. Supercond. Novel Magn.* **27** no. 7 623
[13] Rouabah T, Belaloui N E, and Tounsi A 2021 *J. Phys.: Conf. Series* **1766** 012029
[14] Somei Y, Shimada H and Mizugaki Y 2021 *Jpn. J. Appl. Phys.* **60** 073001

Heat leak variation with the surface temperature of a cryogenic pipe of the superconducting power transmission

Hirofumi Watanabe, Shiori Iizuka, Takaya Kato, Masae Kanda, and Satarou Yamaguchi

Chubu University, Kasugai, Aichi 487-8501, Japan

h_watanabe@isc.chubu.ac.jp

Abstract. The measurements of the heat leak variation with the surface temperature of a cryogenic pipe of the superconducting power transmission were performed in the range of 23.7 °C–36.4 °C. The cryogenic pipe which was an object of the present study was that used in the project of the superconducting DC power transmission conducted in Ishikari, Japan (Ishikari project). This cryogenic pipe has two inner pipes in one outer pipe, with a radiation shield used to conceal the inner pipe installing the cable from the outer pipe, which is at room temperature, to reduce heat leakage. A test pipe with the length of 12 m was used for the measurements. The liquid nitrogen was filled in the test pipe and the evaporated nitrogen gas rate was measured to obtain the heat leak. The heat leak to the inner pipe installing the cable was almost constant at around 0.04 W/m, whereas the heat leak to the other inner pipe used to return liquid nitrogen for circulation was around 1.3 W/m at surface temperatures ranging from 23.7 °C to 36.4 °C. The latter, with previous measurements, was well fitted by a function considering radiative heat transfer and conductive heat transfer. Portions of the radiative heat transfer and the conductive heat transfer were separated with this function. This information can be used to improve the cryogenic pipe in the future.

1. Introduction

Superconducting power transmission is an efficient method of electric power transmission, because there is no loss due to electric resistance when using superconductivity. Alternatively, to use superconductivity, power is required to achieve cryogenic temperature, which results in transmission loss. Therefore, the heat leak to the parts at cryogenic temperatures should be minimized to reduce the capacity of cryocoolers and, thus, the power for cooling. The cryogenic pipe has been used for the thermal insulation. Because the heat leak of the cryogenic pipe dominates all other heat leaks in the transmission system, particularly for long-distance transmissions, accurate information of this heat leak is needed to design the capacity of the cryocoolers and the circulation pumps. In the case of the cryogenic pipe, the heat leak occurs mainly by radiative heat transfer and conductive heat transfer whereas convective heat transfer can be avoided using vacuum insulation. Consequently, the heat leak depends on the cryogenic pipe's surface temperature, which varies with environmental conditions, such as ambient temperature, weather and sunshine duration. Therefore, knowing how the heat leak changes with surface temperature is crucial for designing a superconducting power transmission system.

We have measured the heat leak with respect to the surface temperature of the cryogenic pipes developed for the project of the superconducting DC power transmission performed in Ishikari, Japan

(Ishikari project) [1,2,3]. Two different cryogenic pipes were developed in this project, one of which is the object of this study [4]. Previously, the measurements were performed using an actual transmission line constructed in this project and using a test pipe, with surface temperatures ranging from -3 °C to 17.4 °C [4,5,6]. In this study, higher surface temperature measurements from 23.7 °C to 36.4 °C were performed to consider summer temperature. This information can be used to design future superconducting power transmission systems.

In a previous study, it was shown that a simple function which reflects the temperature dependence of radiative heat transfer and conductive heat transfer could accurately reproduce the heat leak variation with the cryogenic pipe's surface temperature, and the function could be used to extract portions of the radiative heat transfer and the conductive heat transfer from the heat leak [7]. The application of this function to the present result was investigated. The possibility of further improvements to the cryogenic pipe is discussed by extracting the portions of the radiative heat transfer and the conductive heat transfer from the heat leak.

2. Experiment

Figure 1 shows a schematic cross-section of the cryogenic pipe in the present study. This cryogenic pipe was used for the Ishikari project, whose details are provided in a previous paper [4]. This cryogenic pipe has two inner pipes in one outer pipe for the circulation of the liquid nitrogen. One inner pipe installs the cable while the other circulates liquid nitrogen and are referred to as 'cable pipe' and 'return pipe', respectively. Unlike the cryogenic pipes using corrugated pipes commonly used in the superconducting power transmission, smooth pipes are used to minimize circulation pressure loss, although they have a bellow section with a length of a few tens of centimeters per 12 m to compensate for thermal shrinkage. The cable and return pipes are stainless steel pipes with the outer diameters of 76.3 and 60.5 mm, respectively, and the outer pipe is a carbon steel pipe with the outer diameter of 318.5 mm. The cable pipe is surrounded by a radiation shield of aluminum alloy and the return pipe is tightly clamped by the radiation shield. The thermal radiation emitted from the outer pipe, which is at room temperature, is received by the radiation shield and the resultant heat is taken away by the liquid nitrogen flowing in the return pipe. Using this structure, the heat leak to the cable pipe is expected to be significantly reduced and the temperature increase along the cable becomes small even with the extremely slow liquid nitrogen circulation. This results in a reduced circulation pressure loss, making long-distance transmissions possible. The radiation shield is wrapped using a bundle of 21 layers of multilayer insulation (MLI), whereas the cable pipe is wrapped by one layer of MLI. The used MLI was type KFN-9B05 from Kaneka Co.

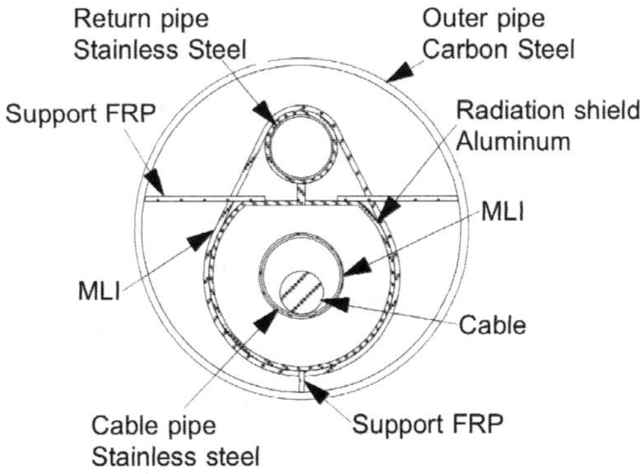

Figure1. The cryogenic pipe used in this study [4].

Figure 2 is a setup of the experiment. This experiment was conducted using a test pipe with the length of 12 m, which was tilted 2° from the ground and connected to reservoirs with liquid nitrogen. The cryogenic pipe used in the Ishikari project had a unit length of 12 m, and the unit pipes were connected at the transmission line's location to create a longer cryogenic pipe. The test pipe used in this study was the same as the unit pipe used in the Ishikari project. After the inner pipes of the test pipe were filled with liquid nitrogen, the temperature in the test pipe was monitored using thermocouples to see if it dropped and stabilized. After refilling the liquid nitrogen, the measurement of the evaporated nitrogen gas rate was conducted with calibrated flow meters at 20 °C and 1 atm from Alicat Scientific inc., depending on the flow rates from the cable and return pipes. To use the data when just the return pipe was full and the reservoir empty, the liquid nitrogen levels were monitored using the thermocouples attached near the return pipe inlet (TC1) and at the bottom of the reservoir for the return pipe (TC2). The data just before increasing the temperature measured by TC1 and after increasing that measured by TC2 were used for the analysis. The surface temperature along the outer pipe was monitored using thermocouples and controlled by heaters based on the temperature measurements. The data of the flow rates and the temperatures were accumulated using a data logger with the vacuum degree in the test pipe. The vacuum degree is maintained below 10^{-3} Pa, where the vacuum insulation is achieved.

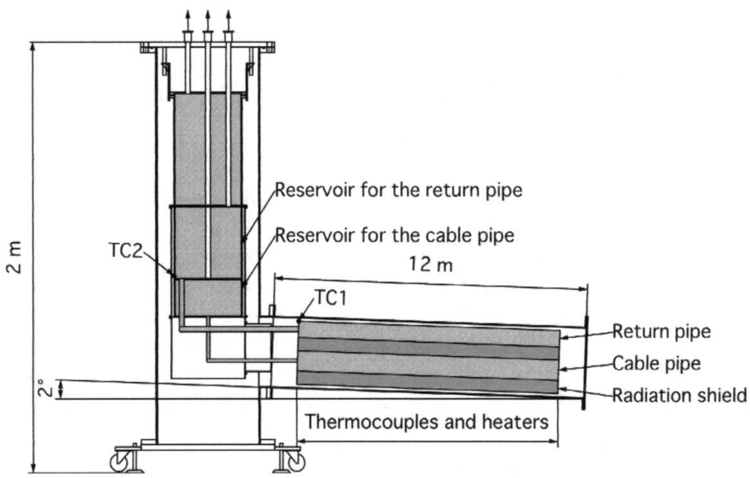

Figure 2. Schematic drawing of the experimental setup. TC1 and TC2 are the thermocouples near the inlet of the return pipe and at the bottom of the return pipe's reservoir, respectively.

3. Results

Figure 3 shows an example of the experimental results at a surface temperature of 23.7 °C. The horizontal axis is the date and time, the left axis is the flow rate measured in L/min, and the right axis is the temperature in the test pipe measured in °C. 'Cable pipe' and 'Return pipe' on the figure are the flow rates from the cable and return pipes, respectively. TC1 and TC2 are the temperatures near the inlet of the return pipe and at the bottom of the return pipe's reservoir, respectively (Figure 2). The return pipe was just full and the reservoir was empty during the period between the vertical lines. The data between the vertical lines were used for the analysis.

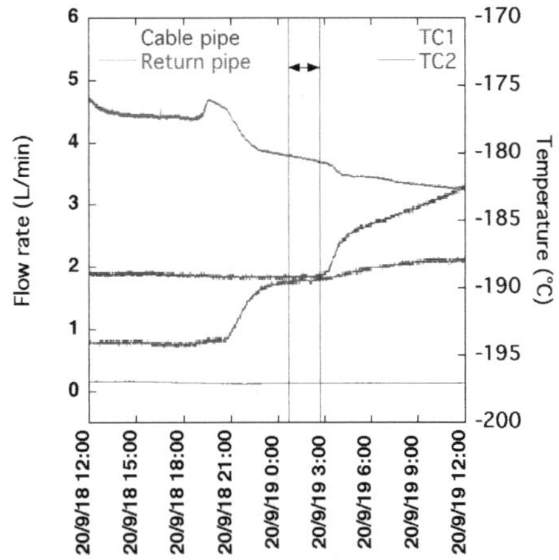

Figure 3. An example of the experimental results at the surface temperature of 23.7 °C. TC1 and TC2 are the temperature near the inlet of the return pipe and at the bottom of the return pipe's reservoir, respectively. The data between the vertical lines were used for the analysis.

In addition to the surface temperature of 23.7 °C, the measurements were performed at 26.4 °C, 29.0 °C and 36.4 °C. Figures 4 and 5 summarize the results of the heat leak for the return and cable pipes, respectively, with the previous results [4,5,6]. The figures show the uncertainties of the results. For the heat leak, the statistical dispersion of the data, and the uncertainties of the flow meters and the data logger were considered. Known heat loads were added to the liquid nitrogen in the inner pipes by heaters to validate the accuracy of the measurements. The differences between the applied and measured heat loads were included in the uncertainties. For the surface temperature, the statistical dispersion of the data and uncertainties of the thermocouples were considered. The previous results were obtained in Ishikari Line 2 and with the test pipe. The results of Ishikari Line 2 were achieved using a different experimental procedure than the current investigation, which involved a temperature rise and the circulating liquid nitrogen flow rate. The details of the analysis can be seen in the previous reports [5,6]. The results of the cable pipe of Ishikari Line 2 have large uncertainties. This is mainly because the temperature rise was small for the small heat leak, and the relative uncertainties of the temperature sensors to the temperature rise became larger. The results of the heat leak of the current measurements were around 0.04 W/m for the cable pipe and around 1.3 W/m for the return pipe. The results of the heat leak for the cable pipe were considerably small as expected and about 3% of those of the return pipe. The heat leak of the return pipe becomes 1.6 times larger by the surface temperature rise from -3 °C to 36.4 °C, which is a fairly large increase. Therefore, the information on the surface temperature is crucial in treating the heat leak. Alternatively, the heat leak of the cable pipe is nearly constant. This is because the cable pipe is hidden by the radiation shield at the liquid nitrogen temperature from the outer pipe at room temperature, and the radiative heat transfer is independent of the surface temperature of the cryogenic pipe. The current results are extrapolated from the previous results. The data obtained using various experimental methods are consistent.

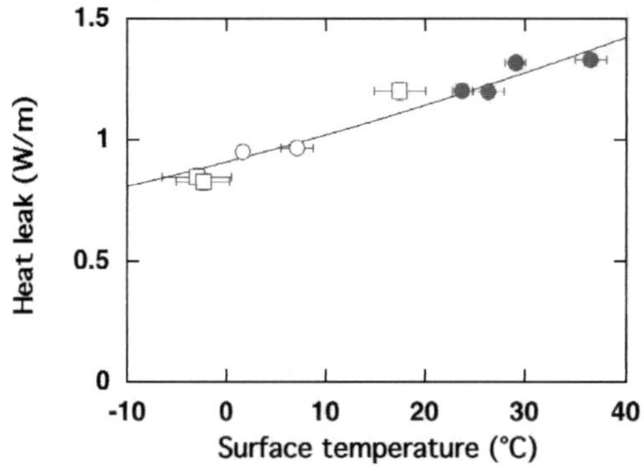

Figure 4. The heat leak variation to the surface temperature of the cryogenic pipe, for the return pipe. The circles are obtained using the test pipe, whereas the squares are obtained using Ishikari Line 2. The solid symbols are the current results, whereas the open symbols are the previous results [4,5,6]. The curve is obtained by fitting a function considering the radiative heat transfer and the conductive heat transfer.

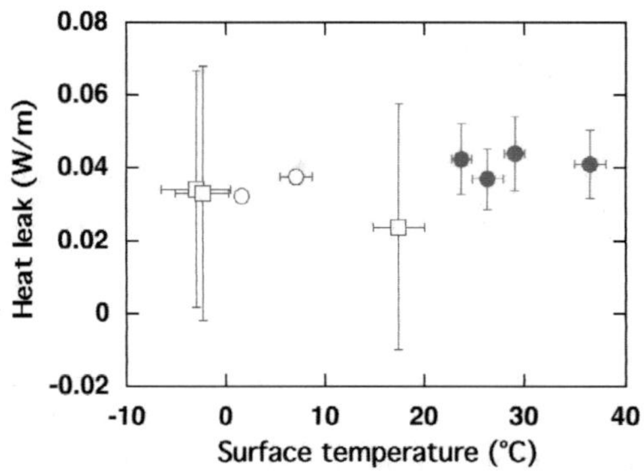

Figure 5. The heat leak variation to the surface temperature of the cryogenic pipe, for the cable pipe. The circles are obtained using the test pipe, whereas the squares are obtained using Ishikari Line 2. The solid symbols are the current results, whereas the open symbols are the previous results [4,5,6].

The curve in figure 4 is obtained by the function proposed in previous study [7] of

$$y = a\{(T + 273.15)^4 - 77.3^4\} + b\{(T + 273.15) - 77.3\}$$

where a and b are the fitting parameters with the units of W/(m K^4) and W/(m K), respectively, and T is the surface temperature with the unit of °C. The first term of the function represents the portion of the radiative heat transfer and the second term represents the conductive heat transfer. This function was fitted to all experimental data, including the previous results [4,5,6], and the values of the fitting parameters were obtained. In the previous study, it was shown that this function could well reproduce experimental results [7]. Figure 4 shows that this function also accurately reproduces the data, though the type of the cryogenic pipe in this study differed from that in the previous study [7]. This function can accurately predict the heat leak variation of the cryogenic pipe with the surface temperature. Table 1 shows the results of the fitting parameters. The uncertainties of the fitting parameters are the propagated values of each experimental result by the fitting procedure. From this result, the portions of the radiative heat transfer and the conductive heat transfer are separately calculated. Table 2 shows the results at the surface temperature of 20 °C. The portions of the radiative heat transfer and the conductive heat transfer are 0.824±0.069 W/m and 0.322±0.069 W/m, respectively. The uncertainties are estimated from those of the fitting parameters. The portion of the conductive heat transfer is 28% of the total heat leak at 20 °C. Though the focus of this cryogenic pipe's construction was to reduce radiative heat transfer, the conductive portion of the total heat leak is not trivial. Further optimization of the support structures and materials for the inner pipes could reduce the heat leak of this cryogenic pipe.

Table 1. The results of the fitting parameters

a ($\times 10^{-10}$ W/(m K^4))	b ($\times 10^{-3}$ W/(m K))
1.121±0.094	1.49±0.32

Table 2. Calculated heat leak at 20 °C.

Radiative heat leak (W/m)	Conductive heat leak (W/m)	Total heat leak (W/m)
0.824±0.069	0.322±0.069	1.146±0.098

4. Summary

The heat leak variation of the cryogenic pipe which equips the radiation shield developed for the Ishikari project was measured against its surface temperature. The heat leak of the cable pipe was almost constant at around 0.04 W/m. Alternatively, the heat leak of the return pipe depended on the surface temperature and averaged around 1.3 W/m at the surface temperatures ranging from 23.7 °C to 36.4 °C. The heat leak of the cable pipe was considerably small, only about 3% of the return pipe. The portions of the radiative heat transfer and the conductive heat transfer were separated by fitting the function considering both the transfers to the return pipe's experimental data. The conductive portion of the total heat leak was found to be quite substantial. Further optimization of the support structures and materials for the inner pipes could reduce the heat leak of the cryogenic pipe used in this study.

Acknowledgement

This work was supported by Chubu University Grant (A).

References

[1] Yamaguchi S, Koshizuka H, Hayashi K and Sawamura T 2015 *IEEE Trans. Appl. Supercond.* **25** 5402504
[2] Yamaguchi S, Ivanov Y, Watanabe H, Chikumoto N, Koshiduka H, Hayashi K and Sawamura T 2016 *Physics Procedia* **81** 182–6
[3] Chikumoto N, Watanabe H, Ivanov V Y, Takano H, Yamaguchi S, Koshizuka H, Hayashi K and Sawamura T 2016 *IEEE Trans. Appl. Supercond.* **26** 5402204
[4] Watanabe H, Ivanov V Y, Hamabe M, Chikumoto N, Takano H and Yamaguchi S 2015 *Physics Procedia* **67** 239–44
[5] Watanabe H *et al.* 2017 *IEEE Trans. Appl. Supercond.* **27** 5400205
[6] Watanabe H, Ivanov V Y, Chikumoto N, Yamaguchi S, Ishiyama K, Oishi Z, Watanabe M and Masuda T 2018 *J. Phys. Conf. Ser.* **1054** 012076
[7] Watanabe H, Takeuchi T, Miyake K and Yamaguchi S 2020 *J. Phys. Conf. Ser.* **1590** 012059

Numerical simulation of a high-power density 10 MW REBCO superconducting synchronous generator cooled by sub-cooled LN_2 for low AC loss

Kaito Noda[1], Hiromasa Sasa[1], Hiroshi Miyazaki[1], Shun Miura[1], Takashi Yoshida[1], Teruyoshi Sasayama[1], Masataka Iwakuma[1], Akifumi Kawagoe[2], Teruo Izumi[3], Masayuki Konno[4], Yuichiro Sasamori[4], Hirokazu Honda[4], Yoshiji Hase[4] and Masao Shutoh[4]

[1]Kyushu University, Fukuoka 819-0395, Japan.
[2]Kagoshima University, Kagoshima 890-0065, Japan.
[3]National Institute of Advanced Industrial Science and Technology, Tsukuba 305-8560, Japan.
[4]Fuji Electric Company Ltd., Ichihara 290-8511, Japan.

E-mail: noda.kaito.531@s.kyushu-u.ac.jp

Abstract. Generators used in electric aircraft require a high-power density, and AC loss is also a significant problem. We designed 10 MW REBCO superconducting synchronous generators at 64 K to achieve a power density of 20 kW/kg and low AC loss. In this study, electromagnetic analyses were performed using finite element method software. Consequently, the thickness of the back yoke is 50 mm or less so that the generator can meet the target power density. The method of decreasing the magnetic field of the armature winding and increasing that of the field winding was used to effectively reduce the AC loss. As a result, the generator achieved a high-power density of 21.0 kW/kg, reducing the AC loss from over 600 kW to 415 kW.

1. Introduction

Electric propulsion systems that use motors and generators have been proposed for a high-efficiency aircraft [1]. High-power density rotating machines are required in the system. Using a superconducting wire can reduce the weight of the machine owing to its high current capacity. Therefore, many research groups have studied superconducting rotating machines with a power density of 20 kW/kg [2] [3] [4]. We previously designed 10 MW superconducting synchronous generators and compared the field windings operating at 20 K and 64 K [5]. The armature winding was operated at 64 K in both cases. We found that the generator in which the field winding was cooled by liquid nitrogen (LN_2) at 64 K was optimal for the following two reasons. First, its structure is simple because the field and armature windings were cooled simultaneously. Second, the dielectric strength of the superconducting transformer cooled by LN_2 can be regulated [6].

However, the power density was slightly less than the target because the weight of the back yoke accounts for most of the total weight of the generator. Therefore, the thickness of the back yoke was set as a parameter in this study to achieve a target power density of 20 kW/kg or less. Furthermore, AC loss is a significant problem for superconducting generators [7], [8]. We also set the number of turns of the armature winding as a parameter because the AC loss of the armature winding is extremely high. The armature winding is a racetrack-type with concentrated winding, which is easy to manufacture. In this study, electromagnetic analyses were performed using finite element method software JMAG and we obtained the specifications of a 10 MW generator operating at 64 K that achieved 20 kW/kg and low AC losses.

The next section discusses the specifications of the generator and the properties of the REBCO superconducting tape used in this study. Section 3 discusses the effects of the thickness of the back yoke and the number of turns of the armature winding on the AC loss and power density.

2. Design of a 10 MW superconducting generator operating at 64 K

2.1. *Specifications of the generator*

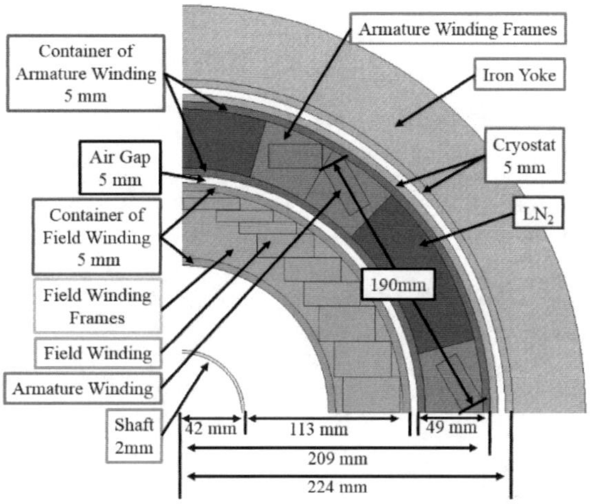

Figure 1. Cross-section of the generator.

Table 1. Specifications of the generator.

Output		10 MW
Voltage		6297–6439 V
Current of	Field	62.5 A
	Armature	894–912 A
Number of Parallel Conductors of	Field	1
	Armature	17
Load Factor		70–80 %
Pole		2
Phase		3
Frequency		100 Hz
Rated Revolution		6000 rpm
Operating Temperature		64 K
Lagging Power Factor		0.9
Effective Length		500 mm
Radius of Rotor		155 mm
Air Gap Length		5 mm

Figure 1 shows the cross-section of the generator used in the analyses and Table 1 lists the specifications of the generator. The analyses were performed using the two-dimensional finite element method to simplify the calculations and only the straight part of the racetrack coil was considered. The armature winding is composed of two double pancakes. The outer diameter of the double pancake was 190 mm. Both the field and armature windings were cooled using subcooled LN_2 at 64 K. The back yoke used was 50JN270 (JFE Steel Corp). The frames for the wire, containers, shafts, and cryostats were made of glass-fiber reinforced plastic (GFRP). The field current used in the analyses was applied from a current source of 62.5 A. We used transposed parallel conductors to apply a large current to the windings [9]. The current capacity can be increased by increasing the number of parallel conductors. The number of parallel conductors and the current were set such that the load factor of the windings is 70–80 %.

2.2. *Properties of REBCO superconducting tape*

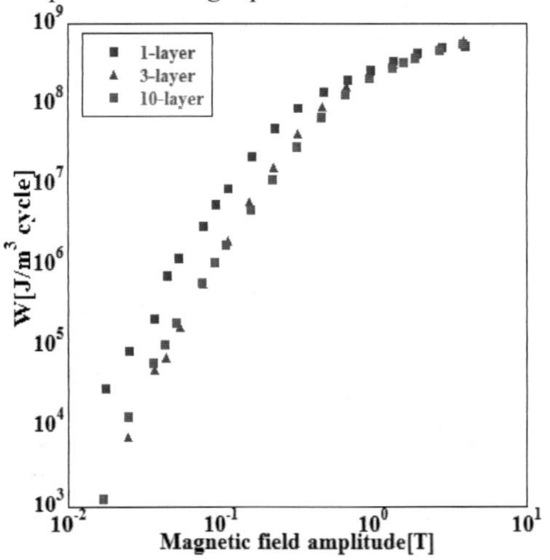

Figure 2. Hysteresis loss dependence on the magnetic field amplitude at 64 K.

In this study, we calculated the AC loss from only the hysteresis loss because the external magnetic field is larger than the self-magnetic field and the hysteresis loss is dominant. The hysteresis loss of the superconducting tape, critical current, width, and thickness of the tape are needed to calculate the AC loss of the generator. Figure 2 shows the dependence of hysteresis loss on the magnetic field amplitude when the number of stacked tapes is 1, 3, and 10. In this study, we used FYSC-SCH04 superconducting tape (Fujikura Ltd). In the characteristics, only the perpendicular component of the applied magnetic field was considered, which is the main contributor of the AC loss. This property was measured using a saddle-shaped pickup coil, and the magnetic field amplitude was calculated from the magnetic field applied vertical to the tape face [10]. We considered a decrease in the hysteresis loss by stacking the tapes [11]. The calculations of the AC loss of the windings are given below.

$$B_m = \frac{(B_{max} - B_{min})}{2}, \tag{1}$$

$$V = S \times l_e \times \frac{t}{t_s}, \tag{2}$$

$$W_{mesh} = B_m \times V \times f \times W_d, \tag{3}$$

$$W_{total} = \sum_{k=1}^{n} W_{mesh}. \tag{4}$$

In the above equations, B_{max}, B_{min}, S, l_e, t, t_s, f, W_d, and n are the maximum and minimum magnetic fields vertical to the tape in one cycle, area of one mesh, the straight part of the racetrack coil, thickness of the tape, thickness of the superconducting layer, frequency of the magnetic field vertical to the winding, hysteresis loss, and number of meshes, respectively. Figure 3 shows the magnetic field fluctuations vertical to the (a) field and (b) armature windings. This is the magnetic field fluctuation at the point with the largest fluctuation.

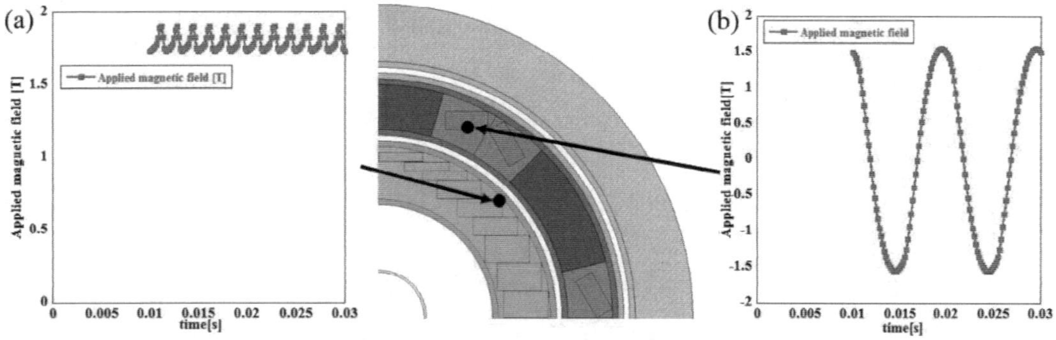

Figure 3. Magnetic field fluctuations vertical to the (a) field winding and (b) armature winding.

The I_c–B characteristics were obtained from reference [12]. We used the fitting curve of the nonlinear current transport characteristics of the critical currents measured at various external magnetic field conditions. In the analyses, the critical current depends only on the external magnetic field, and the self-magnetic field obtained by energizing the parallel conductor was not considered. Table 2 lists the specifications of the superconducting tape.

Table 2. Specifications of the superconducting tape.

Width	4 mm
Thickness	0.18 mm
Substrate	Hastelloy (75 mm)
Superconducting Layer	GdBCO (2 µm)
Stabilizing Layer	Cu plating (20µm)

3. Evaluation of weight and AC loss

3.1. Effect of the thickness of the back yoke

First, we set the thickness of the back yoke as a parameter because it is the heaviest component of the generator. Table 3 lists the parameters of the generator with the thickness of the back yoke as a parameter. The number of turns of one phase of the armature winding was fixed at 60×2. Then, the thickness of the back yoke was set to 30–70 mm. Variations in the parameter affect the output value because the back yoke enhances the magnetic field from the generator. Therefore, the size of the field winding was changed in each case to obtain the desired output. For instance, the field winding turns should be increased if the back yoke is thinner. On the other hand, the increase in the leakage field due to the thinner back yoke was not taken into account because we focused on decreasing the weight.

Table 3. Parameters of the generator with the thickness of the back yoke as a parameter.

Number of Turns of	Armature	60×2 / phase
	Field	12672–15552
Thickness of Back Yoke		30–70 mm
Generator Weight		346–595 kg

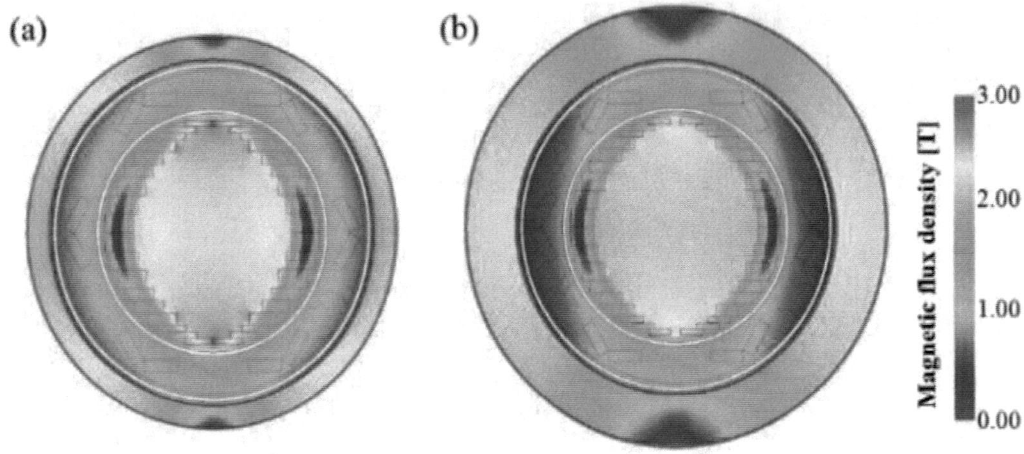

Figure 4. Magnetic flux density distribution in the cross-section for back yoke thicknesses of (a) 30 mm and (b) 70 mm.

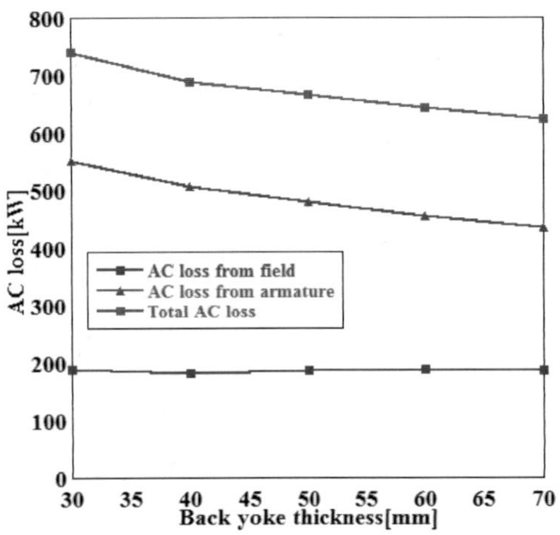

Figure 5. AC loss dependence on the back yoke thickness.

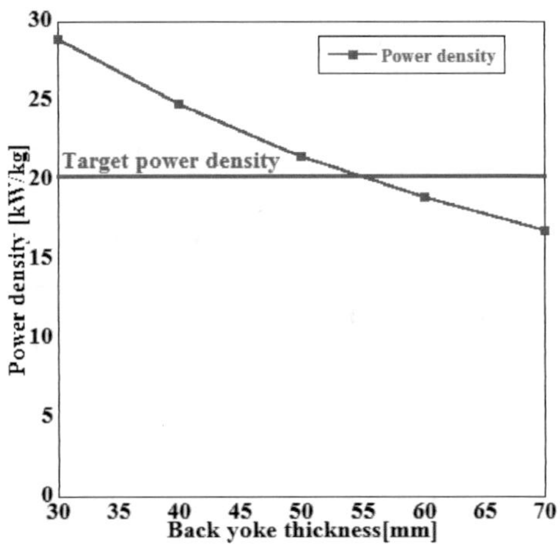

Figure 6. Power density dependence on the back yoke thickness.

Figure 4 shows the magnetic field distributions for back yoke thicknesses of 30 mm and 70 mm. The maximal applied magnetic fields in the armature winding were 1.51 T and 1.39 T, respectively.

Figure 5 shows the dependence of the AC loss on the back yoke thickness. The total AC loss at a back yoke thickness of 70 mm was 624 kW, which is the minimal value. The loss from only the armature winding decreased when the back yoke thickness increased. The AC loss of the armature winding is caused by fluctuations in the magnetic field from the field winding and the self-magnetic field. A shorter field winding was achieved by increasing the back yoke thickness. This decreased the fluctuations in the magnetic field of the field winding as well as the AC loss. On the other hand, the AC loss of the field winding is due to fluctuations in the magnetic field of the armature winding. The AC loss of the field winding was approximately 180 kW in all cases because the length of the armature winding was fixed.

Figure 6 shows the power density dependence on the back yoke thickness. The power density increased drastically at a thinner back yoke because the weight of the back yoke accounts for most of the total weight. The target power density of 20 kW/kg or more was achieved by making the back yoke thickness 50 mm or less.

A back yoke thickness of 50 mm was found to be optimal because it yielded the lowest AC loss for all conditions that satisfied the target power density. Therefore, the thickness was fixed at 50 mm, as described in the next section. However, this is a case study, and the optimal solution of the back yoke thickness is valid only for this generator.

3.2. *Effect of the number of turns of the armature winding*

We set the number of turns of the armature winding as a parameter because it is the primary source of AC loss. Table 4 lists the parameters of the generator with the number of turns of the armature winding as a parameter. The number of turns of one phase of the armature winding was set to 36×2-68×2. The longer the armature winding, the higher the output. The length of the field winding was changed in each case to obtain the desired output, as presented in the previous section. At a shorter armature winding, the length of the field winding increased. The thickness of the yoke was fixed at 50 mm, which is the optimum thickness, as indicated in the previous section.

Table 4. Parameters of the generator with the number of turns of the armature winding as a parameter.

Number of Turns of	Armature	36×2-68×2 / phase
	Field	13248–19904
Thickness of Yoke		50 mm
Generator Weight		464–476 kg

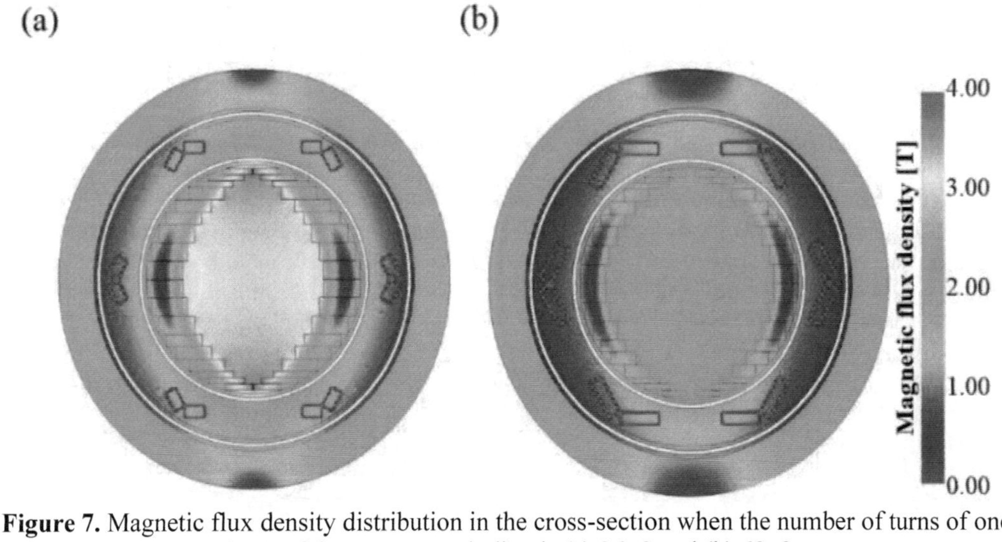

Figure 7. Magnetic flux density distribution in the cross-section when the number of turns of one phase of the armature winding is (a) 36×2 and (b) 68×2.

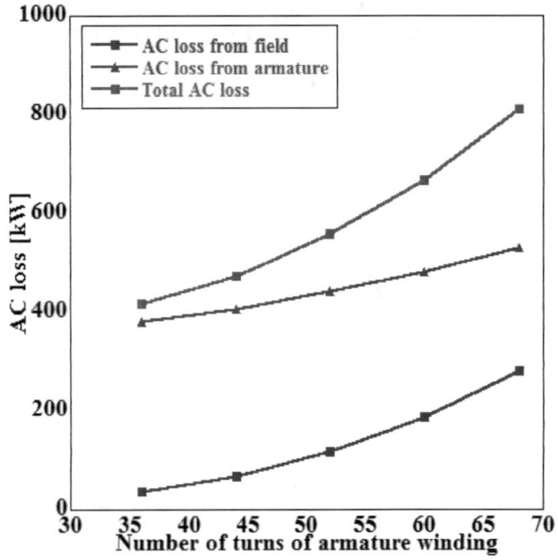

Figure 8. AC loss dependence on the number of turns of the armature winding.

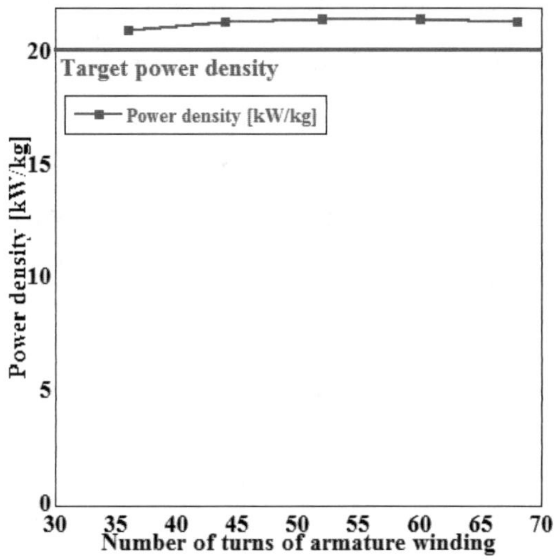

Figure 9. Power density dependence on the number of turns of the armature winding.

Figure 7 shows the magnetic field distributions when the number of turns of one phase of the armature winding is 36×2 and 68×2. The applied maximal magnetic fields in the armature winding are 1.56 T and 1.40 T.

Figure 8 shows the dependence of AC loss on the number of turns of the armature winding. The shorter the armature winding, the lower the loss from both the armature and field windings. The AC loss of the armature winding is proportional to the length of the armature winding. Fluctuations in the magnetic field from the armature winding decreased at a lower number of turns of the armature winding. This also decreased the AC loss of the field winding. When the number of turns of one phase of the armature winding was 36×2 turns, the total AC loss was 415 kW, which is the minimal value. The AC loss can be further reduced through laser scribing of the wire when the heat generated exceeds the capacity of the cooling system [13].

Figure 9 shows the dependence of power density on the number of turns of the armature winding. The power density was approximately 21 kW/kg in all cases because the weight of the wire is very small compared with the total weight.

4. Conclusion

A 10 MW superconducting synchronous generator cooled by LN_2 for an electric aircraft was designed to achieve low weight and AC loss. First, the thickness of the back yoke was set to 50 mm or less to achieve a power density of 20 kW/kg or more. However, the thinner the back yoke, the higher the AC loss. It is possible to realize both low AC loss and high-power density by determining the optimal thickness of the back yoke. Second, the AC loss decreased when the magnetic field from the armature winding decreased and that from the field winding increased. Consequently, the power density was up to 21.0 kW/kg, and the AC loss was reduced from over 600 kW to 415 kW. However, it is not clear whether the heat generated as a result of the AC loss can be cooled. Further reduction in the AC loss can be achieved by combining this design with laser scribing of the superconducting tape or full-pitch winding. Three-dimensional analyses will be performed in future research to consider the entire generator.

Acknowledgments
This research is based on results obtained from a project commissioned by the New Energy and Industrial Technology Development Organization (NEDO), the Japan Science and Technology Agency (JST): Advanced Low Carbon Technology Research and Development Program (JPMJAL1405) and the Japan Society for the Promotion of Science (JSPS): Grant-in-Aid-for Scientific Research (JP18H03783 and JP19K14144).

References

[1] DelRosario R 2014 A future with hybrid electric propulsion systems: A NASA perspective
[2] Dezhin D, Ivanov N, Kovalev K, Kobzeva I and Semenihin V 2018 System Approach of Usability of HTS Electrical Machines in Future Electric Aircraft *IEEE Trans. Appl. Supercond.* **28** 5201905
[3] Filipenko M *et. al*, 2020 Concept design of a high power superconducting generator for future hybrid-electric aircraft *Supercond. Sci. Technol.* **16** 054002
[4] Terao Y, Seta A, Ohsaki H, Oyori H and Morioka N 2019 Lightweight Design of Fully Superconducting Motors for Electrical Aircraft Propulsion Systems *IEEE Trans. Appl. Supercond.* **29** 5202305
[5] Komiya M, Sugouchi R, Sasa H, Miura S, Iwakuma M, Yoshida T, Sasayama T, Yamamoto K, Tomioka A, Konno M and Izumi T 2019 Conceptual Design and Numerical Analysis of 10 MW Fully Superconducting Synchronous Generators Installed with a Novel Casing Structure *IEEE Trans. Appl. Supercond.* **29** 5206607
[6] Iwakuma M, Sakaki K, Tomioka A, Miyayama T, Konno M, Hayashi H, Okamoto H, Gosho Y, Eguchi T, Yoshida S, Suzuki Y, Hirai H, Iijima Y, Saitoh T, Izumi T and Shiohara Y 2015 Development of a 3φ-66/6.9 kV-2 MVA REBCO Superconducting Transformer *IEEE Trans. Appl. Supercond.* **25** 5500206
[7] Ainslie M D, Rodriguez-Zermeno V M, Hong Z, Yuan W, Flack T J and Coombs T A 2011 An improved FEM model for computing transport AC loss in coils made of RABiTS YBCO coated conductors for electric machines *Supercond. Sci. Technol.* **24** 045005
[8] Pardo E, Grilli F, Liu Y, Wolftaedler S and Reis T 2019 AC loss Modeling in Superconducting Coils and Motors With Parallel Tapes as Conductor *IEEE Trans. Appl. Supercond.* **29** 5202505
[9] Iwakuma M, Tanaka H and K. Funaki 2002 Experimental study on ac loss properties of two-strand parallel conductors composed of superconducting multifilamentary strands *Supercond. Sci. Technol.* **15** 1537-1546
[10] Iwakuma M, Nanri M, Fukui M, Fukuda Y, Kajikawa K and Funaki K 2003 Theoretical investigation on the detection ratio of the magnetization in superconducting wires by a saddle-shaped pick up coil *Supercond. Sci. Technol.* **16** 545-556
[11] Sasa H, Miura S, Iwakuma M, Izumi T, Machi T and Ibi A 2021 Estimation method for AC loss of perpendicularly stacked REBa2Cu3Oy superconducting tapes under magnetic field *Physica C: Supercond. Appl.* **580** 1353801
[12] Fujikura Ltd. Fujikura technical document, introduction of FUJIKURA RE-based HTS wire Available:
https://www.fujikura.co.jp/eng/products/newbusiness/superconductors/01/superconductor.pdf
[13] Katayama K, Machi T, Nakamura T, Takagi Y, Nakaoka K, Yoshizumi M, Izumi T and Shiohara Y 2014 Development of low AC loss TFA-MOD coated conductors *Phys. Procedia* **58** 142-145

Improvement of Damping Factor by Optimal Shape and Installation Position of the Damper Coils at Low Velocity Range in Electrodynamic Suspension System

Ryo Yamamoto[1], Ryotaro Betsunoh[1], Shunsuke Ohashi[1]

[1]Department of Electrical and Electronic Engineering, Kansai University, 3-3-35, Yamate-cho, Suita, Osaka, Japan

k680415@kansai-u.ac.jp

Abstract. The electrodynamic suspension system (EDS) realizes stable levitation without the gap control. One of its application is JR Maglev. The EDS system has basically no damping factor against the disturbance. The damping factor is improved by installing the additional coil. However, the damping factor is needed especially at the low velocity. The normal operation velocity in the JR Maglev is the velocity v_x = 120[m/sec]. The optimal shape and installation position of the damper coils at the low velocity range such as v_x < 80[m/sec] is discussed. The type of the damper coil system and the velocity do not affect the optimal height of the coil. The optimal width of the damper coil in the semi-active system is 8.9% smaller than that in the passive one. The optimal installation position is higher at the low velocity than that at the normal operation velocity in both damper system type. By this optimization, the vertical oscillation is up to 13.6% smaller and the pitching oscillation is up to 11.0% smaller.

1. Introduction

High speed magnetically levitation systems are developed for transportation. The electrodynamic suspension system (EDS) is applied to the levitation [1][2]. This system enables stable running without the control at a large levitation gap. This paper discusses about the EDS system used in the JR Maglev. The EDS has basically no damping against the external force. Once the vertical oscillation of the bogie occurs, it does not converge immediately. Thus, the damper coil system is installed[3].

A normal operation velocity in the JR Maglev is the velocity v_x = 120[m/sec]. The damping factor is improved by the damper coil. However, at the low velocity v_x < 60 - 80 [m/sec], the damping factor is small because the time constant becomes large. In the previous study, the improved switching method at low velocity of the semi-active damper system is undertaken[4]. In this paper, the characteristics of the EDS system at low velocity region is studied. The optimal shape and position of the damper coil are discussed in order to increase time derivative of the interlinked flux variation of the damper coil and improve the damping factor. To confirm the effect of the optimization, running simulation with the optimal damper coil at the low velocity is discussed.

2. Analysis method

2.1. EDS system

Figure 1 shows cross section and top view of the EDS system. Table 1 shows the specification of the bogie and coils[5]. JR Maglev is propelled by the linear synchronous motor(LSM) and EDS system is used for levitation and guidance. Superconducting (SC) is used for both field magnet of the LSM and levitation magnet to realize levitation height 100 [mm] and running velocity 150 [m/s][6]. The levitation force is generated by the SC coil on the bogie and the levitation coil. For the connection of the levitation coil, the eight-figure null-flux is used to generate levitation and guidance force[7].

The electric circuit for the EDS is considered to analyse the motion of the bogie in this EDS system. By solving the electric circuit equations, the current of the levitation coil is given. The motion equations of the system are modelled. The transient motion of the bogie is given from this equations. By repeating these procedures, the transient motion of the bogie is given. To solve these equations, Runge-Kutta method is used.

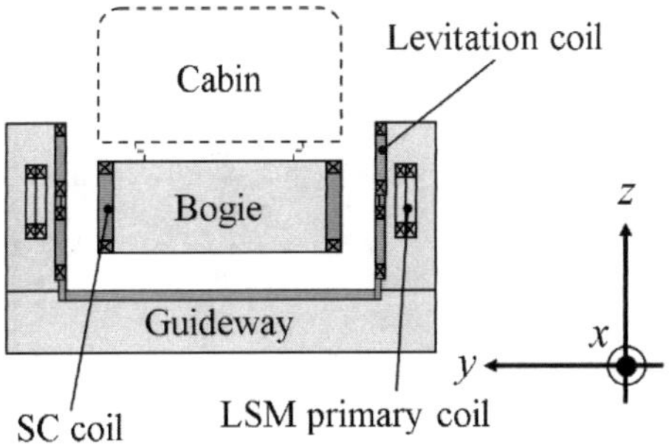

(a) Cross section

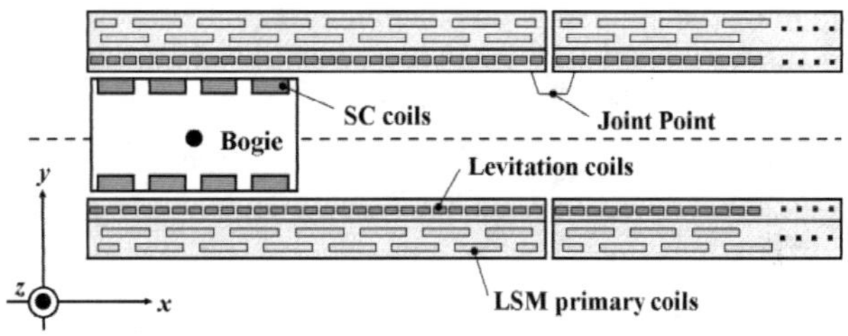

(b) Top view

Figure 1. EDS system (the JR Maglev).

Table 1. Specification of the EDS system.

Bogie	
Mass	2.00×10^4 [kg]
Height	0.90 [m]
Width	5.40 [m]
Depth	2.98 [m]
SC coil	
Electromotive force	700 [kA]
Width	0.963 [m]
Height	0.50 [m]
Pitch	1.35 [m]
Levitation coil	
Width	0.35 [m]
Height	0.34 [m]
Pitch	0.45 [m]

2.2. Damper coil system

The damper coil is installed in front of the SC coil. The damping force is basically generated to reduce oscillation of the bogie. However, the phase of the current of the damper coil is changed by the self-inductance of the damper coil. As a result, damping forces acts to increase the oscillation because of this phase difference. Thus, the semi-active damper system is considered[3]. The switch is added to the damper coils. Switching ON/OFF is decided by the oscillation velocity and levitation force. The damper coils are short circuited by switched ON and opened by switched OFF. To prevent large voltage by the coil inductance, resistance R_s are inserted in parallel. The power in the vertical oscillation is defined as eq(1). F_{zld} is force by the damper coils, and v_z is velocity of the bogie.

$$P_{zld} = F_{zld} \cdot v_z \tag{1}$$

Power in pitching motion is defined as eq(2). T_{pld} is pitching torque by the damper coils, and ω_p is pitching angular velocity of the bogie. Table 2 shows switching condition. The switching control in the semi-active damper system is analysed in the ideal state.

$$P_{pld} = T_{pld} \cdot \omega_p \tag{2}$$

Table 2. Switching condition (semi-active).

Condition	Switch
$P_{zld} + P_{pld} > 0$	OFF
$P_{zld} + P_{pld} < 0$	ON

Figure 2 shows the configuration of the damper coil and Figure 3 shows position and shape at v_x = 120[m/sec]. Table 3 shows the specification of the damper coil optimized at v_x = 120[m/sec].

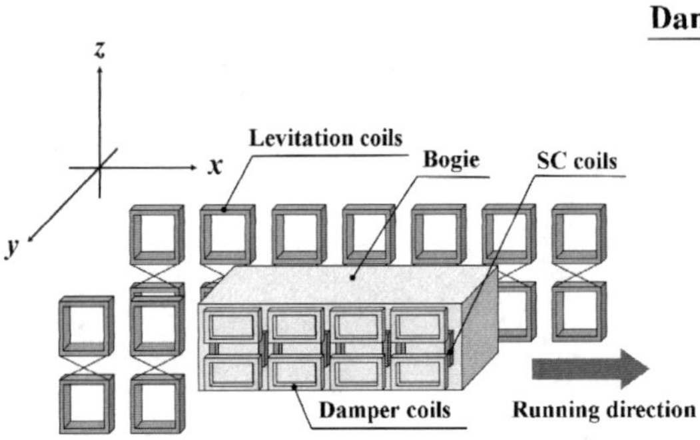

Figure 2. Configuration of the coils.

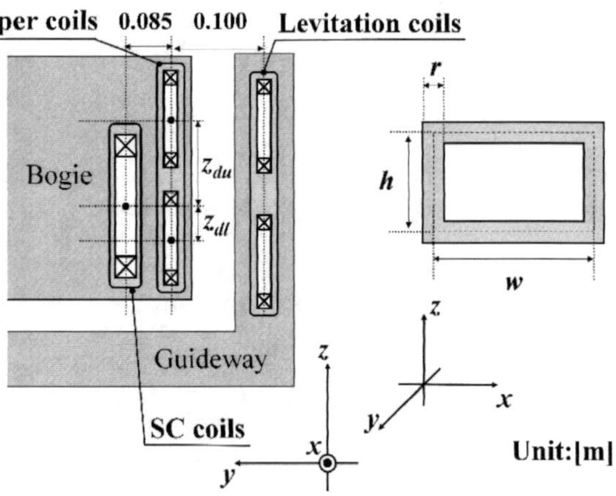

Figure 3. Position and shape of the coil.

Table 3. Specification of the damper coil optimized at the velocity v_x = 120 [m/sec].

	Passive damper system	Semi-active damper system
Mass	10.31 [kg]	10.31[kg]
Turn	2 [turn]	2[turn]
Height h	0.36[m]	0.35[m]
Width w	0.90[m]	0.81[m]
Wire diameter r	0.031[m]	0.032[m]
Resistivity	2.635×10^{-8} [Ω·m]	2.655×10^{-8} [Ω·m]
Upper coil position z_{du}	0.28 [m]	0.27[m]
Lower coil position z_{dl}	-0.13 [m]	-0.14[m]
Distance from SC coil	0.085 [m]	0.085[m]
Pitch	1.35 [m]	1.35[m]

2.3. Method for the evaluation for the position and shape of the damper coil

The shape and installation position of the damper coil are discussed. The bogie runs at the constant velocity v_x = 60 and 80[m/sec]. At same time, the bogie is moved for vertical direction. The damper coils are moved by the external force to keep for one cycle based on sine wave. The period and amplitude of the sin wave is natural frequency of the EDS system. The energy W_t is the energy caused by the EDS system including damper system. And the energy W_0 is the energy caused by the external force added to keep the sine wave oscillation. In this case, the following relational expressions hold[8].

$$W_t = \int F_z \cdot \frac{dz}{dt} dt \tag{3}$$

$$W_o + W_t = 0 \tag{4}$$

The energy W_t working on the bogie by the external force is the indicator of damping. If there is no vertical damping and energy loss for vertical direction, W_t =0. When the energy $W_t > 0$, W_o become negative and it means damper system increases oscillation of the bogie. When the energy $W_t < 0$, damper system increases oscillation of the bogie. And larger the absolute value of W_t shows larger influence on the damping factor. The forced oscillation waveform is obtained by the basic characteristic of the EDS system. Amplitude of the oscillation is 3.99×10^{-2}[m] and 3.86×10^{-2}[m] at v_x = 60 and 80[m/sec]. The natural frequency of EDS system is 2.26[Hz] and 2.31[Hz] at v_x = 60 and 80[m/sec].

3. Results and discussion

3.1. The optimal shape of the damper coil

The weight of the damper coil is 10.31[kg]. The optimal width of the damper coil is around the width w = 0.8[m] because variation of the flux linkage is largest. The width of the damper coils is analyzed at w = 0.70 - 0.90[m]. The height of the damper coil is considered around the height of the levitation coil because the electromagnetic force works between the damper coil and levitation coil. The height of the damper coils is analyzed at h = 0.30 - 0.40[m]. In the semi-active damper system, the resistance inserted into the circuit of the damper coil is R_s = 0.03 - 0.08[Ω]. The wire diameter r depends on the height, width and mass of the damper coil.

Figure 4 and Figure 5 show dependence of energy consumption on the damper coil shape at v_x = 60 and 80[m/sec]. In the passive damper system, the optimal shape is h = 0.36[m] and w = 0.90[m] at v_x= 60[m/sec], and as same as that at the normal operation velocity. The optimal shape of the damper coil at v_x = 80[m/sec] is h = 0.35[m] and w = 0.86[m]. The optimal height of the damper coil at

v_x = 80[m/sec] is 0.01[m] smaller than that at the normal operation velocity. The optimal width of the damper coil at v_x = 80[m/sec] is 0.04[m] smaller than that at the normal operation velocity.

In the semi-active damper system, the optimal shape of the damper coil at v_x = 60 and 80[m/sec] is h = 0.35[m] and w = 0.82[m]. The optimal height of the damper coil at v_x = 60 and 80[m/sec] is as same as that at the normal operation velocity. The optimal width of the damper coil at v_x = 60 and 80[m/sec] is 0.01[m] larger than that at the normal operation velocity. The optimal insertion resistance at v_x = 60 and 80[m/sec] is 0.06[Ω], and as same as that at the normal operation velocity.

The optimal height is around the h = 0.35 [m] near the height of the levitation coil, independent of the damper coil system and the velocity. The optimal width in the semi-active damper is 8.9% smaller than that in the passive one.

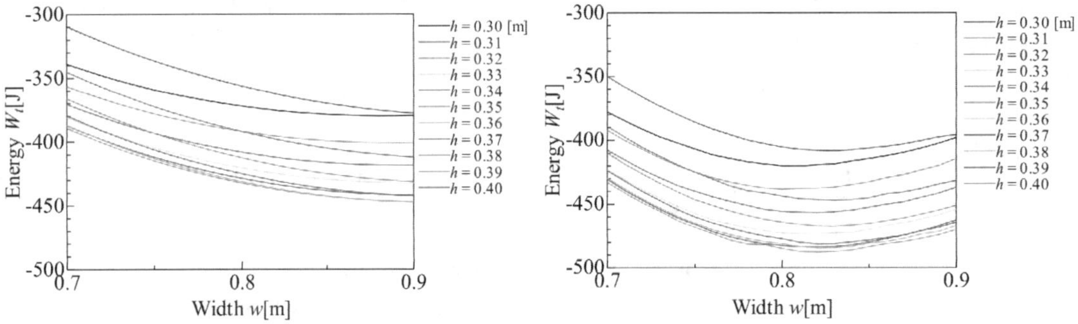

(a) Passive (b) Semi-active

Figure 4. Dependence of energy consumption on the damper coil shape (v_x = 60[m/sec]).

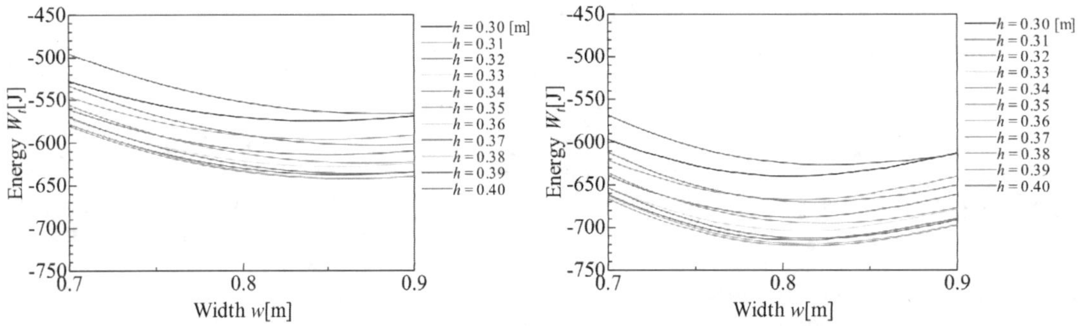

(a) Passive (b) Semi-active

Figure 5. Dependence of energy consumption on the damper coil shape (v_x = 80[m/sec]).

3.2. The installation position of the damper coil

The optimal installation position of the damper coils in the passive and the semi-active damper system at low verbosity is discussed. In this study, one of the two rows of damper coils in figure.2 are used. The damper coil is installed at z_d which is the distance from center of the SC coil to that of the damper coil. The analysis range for z_d is -0.40 - 0.40 [m]. Also, the optimal shape of the damper coil at the low velocity is used.

Figure 6 shows dependence of energy consumption on the installation position of the damper coil. In the passive damper system, the optimal installation position of the damper coil is z_{du} = 0.28 and z_{dl} = -0.13[m] at v_x = 60[m/sec] and z_{du} = 0.27 and z_{dl} = -0.14[m] at v_x = 80[m/sec]. The optimal installation positions at v_x = 60[m/sec] are 0.01[m] higher than that at the normal operation velocity. The optimal installation positions of the damper coil at v_x = 80[m/sec] are as same as that at the normal operation velocity.

In the semi-active damper system, the optimal installation position at v_x = 60 and 80[m/sec] are z_{du} = 0.29 and z_{dl} = -0.12[m]. The optimal installation position at v_x = 60 and 80[m/sec] are 0.01[m] higher than that at the normal operation velocity.

The optimal installation positions in the semi-active damper system are up to 0.02[m] higher than that in the passive damper system. The optimal installation positions depend on the damper coil system. The optimal installation position is higher at low velocity than that at the normal operation velocity.

In case of semi-active damping system, the coil is only switched ON only it generates damping force. The difference of the shape and the installation position between the passive and the semi-active damper system is caused by the switching function of the semi-active damper system.

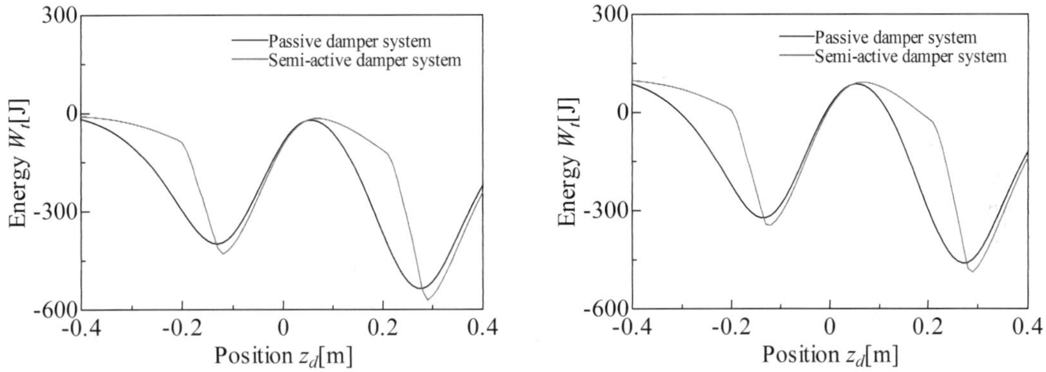

(a) v_x=60[m/sec] (b) v_x = 80[m/sec]

Figure 6. Dependence of energy consumption on the installation position of the damper coil.

3.3. Running simulation

The improvement of the damping factor at the low velocity range is studied. The optimal damper coil at low velocity is used. The bogie runs at the constant v_x = 60[m/sec]. The initial position of the bogie is the centre of the levitation coil (z = 0). When the support of the bogie is released at t = 0.2[sec], vertical and pitching oscillation occurs.

Figure 7 and Figure 8 show the running simulation in vertical and pitching oscillation in the passive damper system. The oscillation becomes small slowly. Using the damper coil optimized at v_x = 60[m/sec], the vertical oscillation is up to 1.6% smaller and the pitching oscillation is up to 4.0% smaller than that at the normal operation velocity. However, after the t = 12.0[sec], using the damper coil optimized at v_x = 60[m/sec], the amplitude of the oscillation is almost same as that at the normal operation velocity.

Figure 9 and Figure 10 show the running simulation in vertical oscillation and pitching oscillation in the semi-active damper system. The vertical and pitching oscillations converge more quickly in the semi-active damper system than that in the passive damper system. Using the damper coil optimized at v_x = 60[m/sec], the vertical oscillation is up to 13.6% smaller and the pitching oscillation is up to 11.0% smaller than that at the normal operation velocity. However, after the t = 6.2[sec], using the damper coil optimized at v_x = 60[m/sec], the amplitude of the oscillation is almost same as that at the normal operation velocity. The optimized damper coils improve the damping factor at v_x = 60[m/sec]. Especially in the semi-active damper system, the damping factor improves compared to that in the passive damper system.

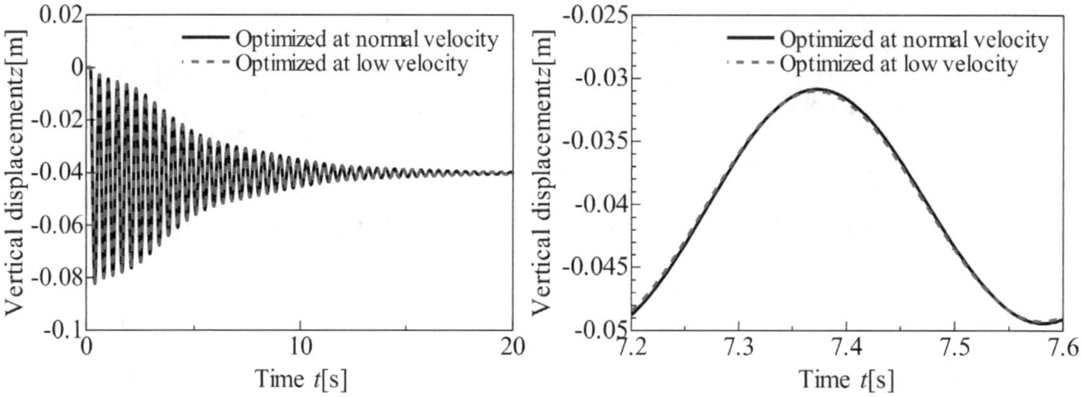

(a) Vertical oscillation (b) Enlarged (t = 7.2 - 7.6[sec])

Figure 7. Running simulation in vertical oscillation (passive).

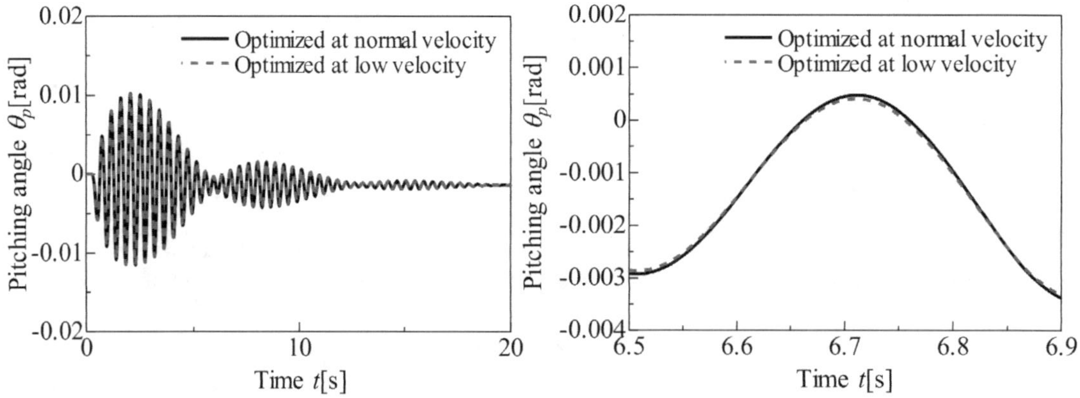

(a) Pitching oscillation (b) Enlarged (t = 6.5 - 6.9[sec])

Figure 8. Running simulation in pitching oscillation (passive) .

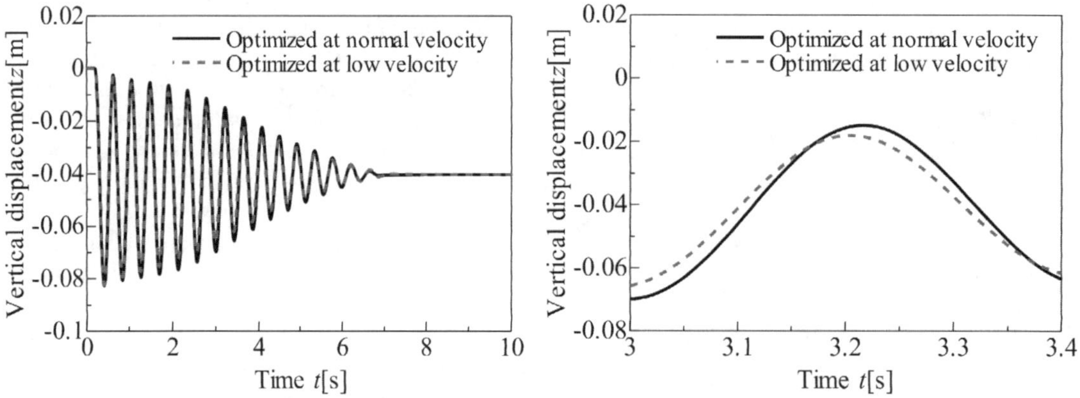

(a) Vertical oscillation (b) Enlarged (t = 3.0 - 3.4[sec])

Figure 9. Running simulation in vertical oscillation (semi-active).

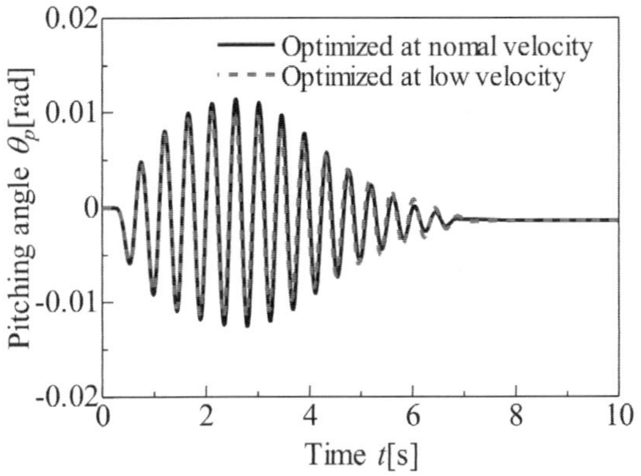

(a) Pitching oscillation

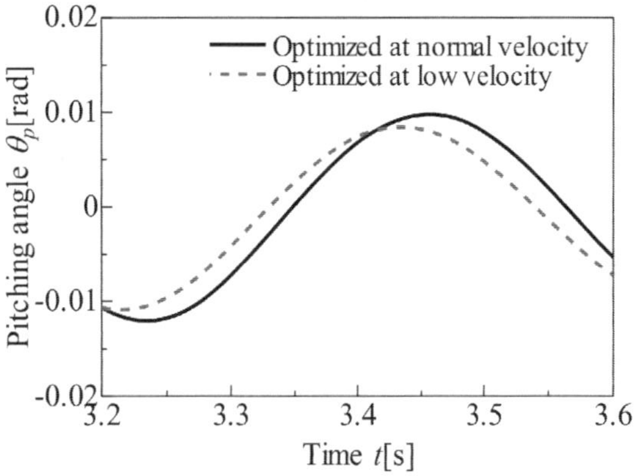

(b) Enlarged (t = 3.2 - 3.6[sec])

Figure 10. Running simulation in pitching oscillation (semi-active).

4. Conclusion

In this study, the shape and installation position of the damper coil at low velocity range is studied. The optimal height does not depend on type of the damper coil system and the velocity. The optimal width of the damper coil at the low velocity in the semi-active damper system is 8.9% smaller than that in the passive one. The optimal installation position is higher at the low velocity.

The improvement of damping factor by optimizing the damper coil is discussed. In the passive damper system, using the damper coil optimized, the vertical oscillation is up to 1.6% smaller and the pitching oscillation is up to 4.0% smaller. In the semi-active damper system,, the vertical oscillation is up to 13.6% smaller and the pitching oscillation is up to 11.0% smaller. Optimization of the damper coil at the low velocity improves the damping factor.

References

[1] Powell J.R and Dandy G.T, 1966 *High speed transport by magnetically suspended trains*, (ASME meeting) Paper 66-WA/RR-5, N.Y.

[2] Seki A, Kitano J and Miyamoto S, 1996 *Dynamics of the bogie of a maglev system with guideway irregularities*, (IEEE Trans on Magnetics) vol.32, No.5, pp5043-5045.

[3] Okubo T, Ohashi S and Ueda N, Jun. 2016, *Effective Control Method of the Active Damper System against the Multi Directional Vibration in the Superconducting Magnetically Levitated Bogie*, (IEEE Trans. on Appl. Supercond) Vol.26, No.4, Art. no. 3601804.

[4] Betsunoh R and Ohashi S 2021 *Improvement of the Damping Factor at Low Speed in Electrodynamic Suspension System for the JR Maglev , J. phys.:Conf. Ser.* 1975 012031.

[5] Yonezu T, Watanabe K, Suzuki E and Sasakawa T, 2017 *Study on Electromagnetic Force Characteristics Acting on Levitation/Guidance Coils of a Superconducting Maglev Vehicle System* (IEEE Trans. on Magn) Vol. 53, no. 11.

[6] Thornton R Nov. 2009 *Efficient and Affordable Maglev Opportunities in the United States* (*Proc. IEEE*) vol.97, No.11, pp. 1901-1921.

[7] Ohashi S and Ueshima T 2012 *Control Method of the Semi-Active Damper Coil System in the Superconducting Magnetically Levitated Bogie Against Vertical and Pitching Oscillation* (IEEE Trans on Magn) vol.48, No.11, pp4542-4544.

[8] Higahi K, Ohashi S, Ohsaki H and Masada E 1997 *Magnetic damping of the Electrodynamic Suspention Type Superconducting Levitation System* (Transaition of IEEJ) Vol-D-117 No.8, pp1015-1023, (in Japanese).

Evaluation of the effect of gravity on the rotational loss of superconducting magnetic bearings for space applications

Satsuki Okumura[1], Takemi Onoue[1], Doa Ahmad[1], Yutaka Terao[1], Hiroyuki Ohsaki[1], Yuki Sakurai[2], Tomotake Matsumura[2] and Nobuhiko Katayama[2]

[1]Graduate School of Frontier Sciences, The University of Tokyo, 5-1-5 Kashiwanoha, Kashiwa, Chiba 277-8561, Japan
[2]Kavli Institute for the Physics and Mathematics of the Universe, The University of Tokyo, 5-1-5 Kashiwanoha, Kashiwa, Chiba 277-8583, Japan

E-mail: satsuki-okumura@g.ecc.u-tokyo.ac.jp

Abstract. We study the relationship between the levitation gap change and losses in a superconducting magnetic bearing (SMB) system. We develop a polarization modulator bearing for a cosmic microwave background (CMB) polarization experiment. The heat generated by the polarization modulator needs to be kept very low. Therefore, it is essential to model the effect of the levitation gap change of a rotor due to gravity on the rotational loss. We have constructed a prototype small experimental platform of an axial-flux SMB with an array of high-temperature superconductors, YBCO, and a permanent magnet ring. In this system, the weight of the rotor can be set. We measured the effect of the change in the levitation gap of the rotor due to gravity on the loss. When the levitation force was 5.5 N and 16.0 N, the losses generated in the rotor were 9.1×10^{-3} W and 1.6×10^{-2} W, respectively. It was confirmed that the loss increases with the increase in the levitation force of the SMB.

1. Introduction

Since superconducting magnetic bearings (SMBs) [1,2] have no mechanical contact, their losses are lower than those of typical mechanical bearings. Therefore, SMBs have been planned to be applied to flywheels [3,4,5], cryogenic pumps, polarization modulators for astronomical observations, etc. One unique application is to use in space, such as a polarization modulator for LiteBIRD [6] to observe the cosmic microwave background (CMB) polarization. The rotation mechanism with a limited cooling power is strongly required to have very low heat generation. Therefore, we are developing a low-loss SMB.

Since there is no mechanical contact in the SMB, losses are expected to be extremely low. However, electromagnetic losses occur in the SMB due to the electromagnetic coupling between the stator and the rotor. The rotor side of the SMB has several permanent magnets, which generate a small inhomogeneous magnetic field, causing losses in the superconductors on the stator side. On the other hand, the stator has multiple YBCO bulks, which generate an inhomogeneous magnetic field on the surface of the rotor due to the superconducting current when the levitation force is generated. This generates hysteresis losses in magnetic materials and eddy current losses in conducting materials in the rotor. Therefore, the losses generated in a superconducting magnetic bearing consist of hysteresis losses and eddy current losses in the rotor, and superconducting losses in the stator.

The operating conditions of the SMB in space are different from those on the ground because of gravity. In the presence of gravity, the rotor is displaced from its cooling position in the magnetic field, which causes an electric current to flow in the superconductor, generating a levitation force. This superconducting current generates a non-uniform magnetic field in the rotational direction, which causes hysteresis and eddy current losses in the rotor. On the other hand, in the space environment, there is no change in the position of the rotor due to the absence of gravity, and thus no losses due to this mechanism are expected.

A prototype small experimental platform of axial flux type SMB was built for experiments to evaluate the effect of gravity on the losses. In this experimental system, the vertical position of the rotor can be set, and the levitation force and rotational loss are measured. In this paper, the measurement results are reported and the characteristics of the rotational loss against the levitation force of SMB obtained in liquid nitrogen tests are described. These experiments will be used in the future to verify the numerical analysis of superconducting magnetic bearings.

2. Experimental Setup

For the SMB characterization measurements, we constructed an experimental apparatus shown in Figure 1. The main part of this apparatus is cooled by liquid nitrogen. A superconducting ring has an outer diameter of 95 mm, an inner diameter of 55 mm, and a height of 20 mm. 10 YBCO bulk superconductors were arranged in a ring shape and fixed in an epoxy resin holder. Figure 1 (a) shows the trapped magnetic flux density distribution of the bulk superconductor when magnetized at 2 T. The magnetic flux density distribution is not uniform in the circumferential direction. The permanent magnet is a one-piece ring magnet with an outer diameter of 85 mm, an inner diameter of 65 mm, and a height of 10 mm. The magnetic field of a permanent magnet is uniform in the circumferential direction. We prepared stainless steel (SUS304) weights ranging from 0.0 N to 10.5 N to vary the weight of the rotor. This can change the total weight of the rotor from 5.5 N to 16.0 N.

The load cell LMB-A-10N of KYOWA ELECTRONIC INSTRUMENTS CO., LTD. is installed between stage1 and stage2 for the measurement of the load applied to stage1. In the levitation force measurement experiment, the prototype small experimental platform of the axial-flux SMB shown in Figure 1 is placed in a Styrofoam container. The initial gap between the stator and the rotor is fixed at 5 mm by the z-axis stage. Liquid nitrogen is then poured into the Styrofoam container to cool the SMB system. We measured the levitation force by varying the gap with the z-axis stage.

In the spin-down test, the part surrounded by the blue frame in Figure 1 (b) and the shaft are removed. An optical encoder with an infrared LED and a silicon photodiode placed face to face is installed in stage4 to measure the rotation speed. An encoder disk with 90 slots is installed on top of the rotor magnet ring. The experimental platform is placed in the Styrofoam container for the spin-down test in the same way as for the levitation force measurement. The initial gap is set to 5 mm by the jig, and the center is aligned. In the spin-down test, the rotation of the rotor is applied by hand. We measured the time variation of the rotation speed for each of the different rotor weights.

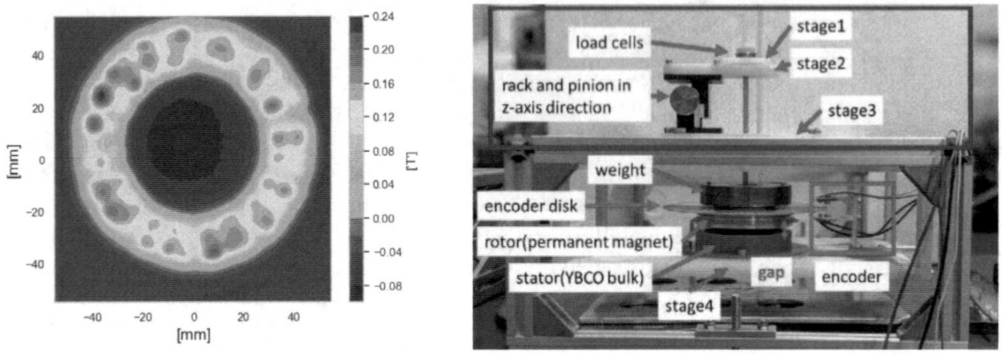

(a) (b)

Figure 1. (a) Trapped field distribution in the ring-shaped bulk superconductor. (b) Prototype small experimental platform for the axial flux type SMB.

3. Basic SMB characterization

3.1. Levitation force measurement

To investigate the static levitation force characteristics of the SMB used in this experiment, we measured the vertical force acting on the mechanical support, stage1, with the load cells when the vertical position of the rotor was changed by 0.2 mm in the direction of the arrow shown in Figure 2 (a) using a z-axis stage. The load cells can measure the load applied to stage1. The weight of the rotor is 15.8 N. When the rotor is cooled, it is supported by stage1 and the SMB. When the z-axis stage is lowered, the load on stage1 becomes smaller, and the levitation force of the SMB becomes larger. Therefore, the levitation force of the SMB can be obtained by subtracting the measured value of the load cell acting on stage1 from the measured value of the load cell before cooling. In this experiment, the measurement was carried out for 20 seconds, starting after 5 seconds to keep the levitation force almost constant because of the transient phenomenon of gap change in each condition. We performed these steps four times. Figure 2 (b) shows the results of the relationship between the gap and the levitation force. The horizontal axis shows the gap [mm], and the vertical axis shows the levitation force [N]. The levitation force differs in the process of increasing and decreasing the gap. This result indicates the hysteresis characteristics of the ring-shaped bulk superconductor used in this experiment.

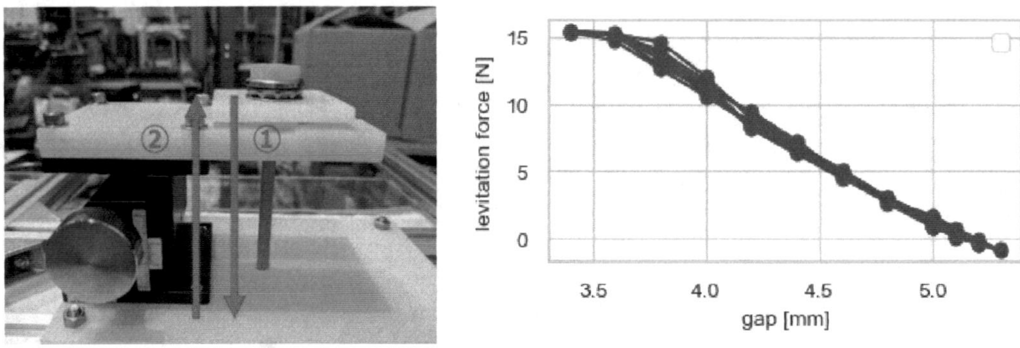

(a) (b)

Figure 2. (a) Moving direction of z-axis stage. (b) Quasi-static levitation force characteristics of SMB.

3.2. Levitation Force-Loss Characteristics of SMB

We performed a spin-down test to measure the rotational loss of an SMB. The rotor is accelerated by hand at an arbitrary rotor weight. The rotor is then decelerated by magnetic damping and air resistance. The rotational loss $P_{loss}(t)$ can be described as

$$P_{loss}(t) = -I \times \frac{d\omega(t)}{dt} \times \omega(t)$$

where I is the moment of inertia of the rotor and $\omega(t)$ is the rotational angular velocity.

Spin-down tests were conducted by varying the weight of the rotor from 5.5 N to 16 N. Figure 3 (a) shows the rotational speed during the spin-down test when the weight of the rotor is 11.8 N. We performed the spin-down test several times. The loss characteristics at each gap, which is determined by the rotor weight, were obtained from Figure 3 (a) and shown in Figure 3 (b). The rotor is decelerated by magnetic and air friction. The spin-down tests under the same conditions except for the rotor weight were conducted. The weights of the rotors were 5.5 N, 7.6 N, 9.7 N, and 16.0 N. The similar rotational speed decay was observed for all rotor weights. The losses increased with increasing rotational speed and weight.

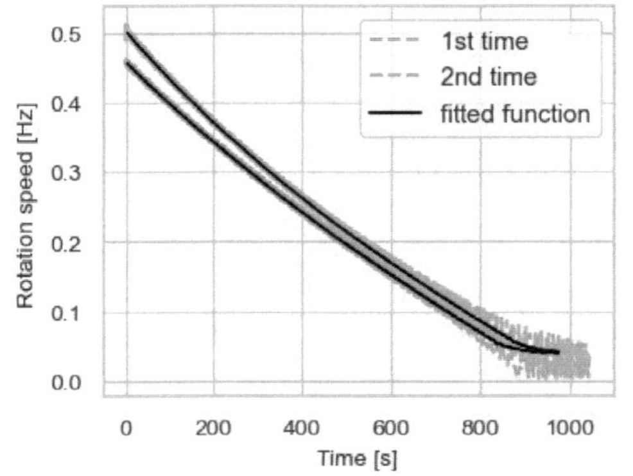

(a)

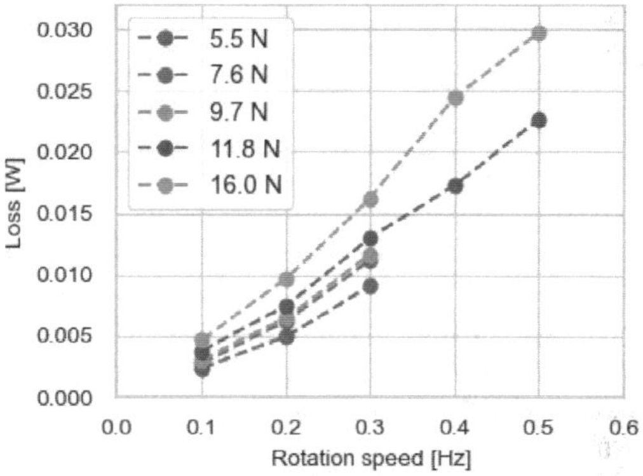

(b)

Figure 3. (a) Spin-down test data of the SMB when the rotor weight is 11.8N. The dashed lines are the measured data, and the solid black lines are the fitted function. (b) Loss of SMB as a function of rotational speed. Red, blue, green, purple, and orange are the results obtained with rotor weights of 5.5 N, 7.6 N, 9.7 N, 11.8 N, and 16.0 N, respectively.

Figure 4 shows the characteristics of the loss as a function of the levitation force of the SMB at 0.3 Hz. The loss increases with the increase in the levitation force. This is because the more significant the force generated by the SMB, the larger the superconducting current flowing through the superconductor. The superconducting current generates a non-uniform magnetic field distribution along the rotational direction. This magnetic field distribution causes larger eddy current losses on the rotor side.

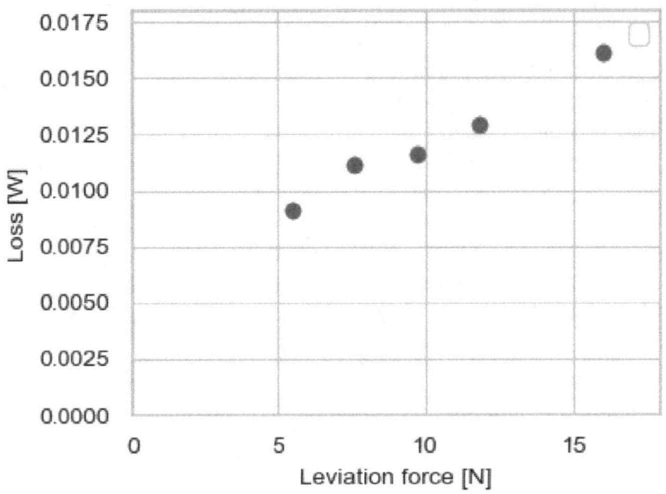

Figure 4. Loss characteristics as a function of the levitation force of the SMB at 0.3 Hz.

4. Conclusion

In order to advance the numerical analysis of superconducting bearings for LiteBIRD in detail in the future, we built a prototype small experimental platform for axial flux type SMB that can verify the superconducting characteristics in detail. In the LiteBIRD SMB the extremely low loss in the rotor is required. The rotor loss is mainly caused by the inhomogeneous magnetic field generated by the superconducting currents. We completed the experimental preparation, and we obtained the first experimental data, especially in verifying the loss generated on the rotor side. When the levitation force was 5.5 N and 16.0 N, the losses generated in the rotor were 9.1×10^{-3} W and 1.6×10^{-2} W, respectively. The larger the levitation force of the SMB is, the higher the loss is. In the future, we will study the characteristics of superconducting magnetic bearings in conjunction with numerical analysis.

References

[1] John R H 2000 *Supercond. Sci. Technol.* **13** R1
[2] Kazuyuki D, Issei M, Takumi I and Makoto K 2005 *Physica C* **426-431** 826-833
[3] Kazuyuki D, Akira M, Tetsuya U, Kenzo M, Hiromasa H, Ryoichi T and Hiroo K 2001 *Physica C* **357-360** 882-885
[4] Kazuyuki D, Kenzo M, Ryoichi T, Hironori K and Hiromasa H 2002 *Physica C* **378-381** 858-863
[5] Issei M, Kazuyuki D, Takumi I and Mitsuru U 2005 *IEEE Transaction Applied Superconductivity* **15** 2257-2260
[6] Tomotake M *et al.* 2014 *J. Low Temp. Phys.* **176** 733–740

AUTHOR INDEX